Bitterness
Perception, Chemistry and Food Processing

IFT Press

The IFT Press series reflects the mission of the Institute of Food Technologists – to advance the science of food contributing to healthier people everywhere. Developed in partnership with Wiley Blackwell, IFT Press books serve as leading-edge handbooks for industrial application and reference and as essential texts for academic programs. Crafted through rigorous peer review and meticulous research, IFT Press publications represent the latest, most significant resources available to food scientists and related agriculture professionals worldwide.

Founded in 1939, the Institute of Food Technologists is a nonprofit scientific society with 22,000 individual members working in food science, food technology, and related professions in industry, academia, and government. IFT serves as a conduit for multidisciplinary science thought leadership, championing the use of sound science across the food value chain through knowledge sharing, education, and advocacy.

WILEY Blackwell

Bitterness

Perception, Chemistry and Food Processing

First Edition

Edited by

Michel Aliani & Michael N. A. Eskin

Manitoba, Canada

IFT Press

WILEY Blackwell

Registered Offices
John Wiley & Sons, Inc., 111 River Street, Hoboken, NJ 07030, USA

Editorial Office
9600 Garsington Road, Oxford, OX4 2DQ, UK

For details of our global editorial offices, customer services, and more information about Wiley products visit us at www.wiley.com.

Wiley also publishes its books in a variety of electronic formats and by print-on-demand. Some content that appears in standard print versions of this book may not be available in other formats.

Library of Congress Cataloging-in-Publication Data

Names: Aliani, Michel, editor. | Eskin, N. A. M. (Neason Akivah Michael), editor.
Title: Bitterness : perception, chemistry and food processing / edited by Michel Aliani, Manitoba, Canada; Michael N. A. Eskin, Manitoba, Canada.
Other titles: Bitterness (John Wiley & Sons)
Description: First edition. | Hoboken : John Wiley & Sons, Inc., 2017. | Series: Institute of food technologists series | Includes bibliographical references and index.
Identifiers: LCCN 2016049478 (print) | LCCN 2016049871 (ebook) | ISBN 9781118590294 (hardback) | ISBN 9781118590317 (pdf) | ISBN 9781118590232 (epub)
Subjects: LCSH: Bitterness (Taste) | BISAC: TECHNOLOGY & ENGINEERING / Food Science.
Classification: LCC TX546 .B58 2017 (print) | LCC TX546 (ebook) | DDC 664/.07–dc23
LC record available at https://lccn.loc.gov/2016049478

Cover Design: Wiley
Cover Images: Background Image: fcafotodigital/Gettyimages; Inset Images (from top to bottom): luckyraccoon/shutterstock; ansonsaw/Gettyimages; Brent Hofacker/shutterstock

Typeset in 10/13pt TimesLTStd by SPi Global, Chennai, India
Printed and bound in Malaysia by Vivar Printing Sdn Bhd

10 9 8 7 6 5 4 3 2 1

Titles in the series

- *Natural Food Flavors and Colorants*
  by Mathew Attokaran, PhD
- *Food Carotenoids: Chemistry, Biology and Technology*
  by Delia B. Rodriguez-Amaya
- *Biofilms in the Food Environment, 2nd Edition*
  by Anthony L. Pometto III, Ali Demirci
- *Spray Drying Techniques for Food Ingredient Encapsulation*
  by C. Anandharamakrishnan, Padma Ishwarya S.
- *Trait-Modified Oils in Foods*
  by Frank T. Orthoefer (Editor), Gary R. List (Editor)
- *Membrane Processing for Dairy Ingredient Separation*
  by Kang Hu (Editor), James Dickson (Editor)
- *Nanotechnology and Functional Foods: Effective Delivery of Bioactive Ingredients*
  by Cristina Sabliov (Editor), Hongda Chen (Editor), Rickey Yada (Editor)
- *Anti-Ageing Nutrients: Evidence-Based Prevention of Age-Associated Diseases*
  by Deliminda Neves (Editor)
- *Food Industry Design, Technology and Innovation*
  by Helmut Traitler, Birgit Coleman, Karen Hofmann
- *Food Oligosaccharides: Production, Analysis and Bioactivity*
  by F. Javier Moreno (Editor), María Luz Sanz (Editor)
- *Food Texture Design and Optimization*
  by Yadunandan Lal Dar (Editor), Joseph M. Light (Editor)
- *Mathematical and Statistical Methods in Food Science and Technology*
  by Daniel Granato (Editor), Gaston Ares (Editor)
- *Bioactive Compounds from Marine Foods: Plant and Animal Sources*
  by Blanca Hernandez-Ledesma (Editor), Miguel Herrero (Editor)
- *Resistant Starch: Sources, Applications and Health Benefits*
  by Yong-Cheng Shi (Editor), Clodualdo C. Maningat (Editor)
- *Processing and Nutrition of Fats and Oils*
  by Ernesto M. Hernandez (Original Author), Afaf Kamal-Eldin (Original Author)
- *Practical Ethics for Food Professionals: Ethics in Research, Education and the Workplace*
  by J. Peter Clark (Editor), Christopher Ritson (Editor)
- *Advances in Dairy Ingredients*
  by Geoffrey W. Smithers (Editor), Mary Ann Augustin (Editor)
- *Improving Import Food Safety*
  by Wayne Ellefson (Editor), Lorna Zach (Editor), Darryl Sullivan (Editor)
- *Food Irradiation Research and Technology, 2nd Edition*
  by Xuetong Fan (Editor), Christopher H. Sommers (Editor)
- *Coffee: Emerging Health Effects and Disease Prevention*
  by Yi-Fang Chu (Editor)

WILEY Blackwell

Contents

List of Contributors

Dr. Ayyappan Appukutan Aachary
Department of Human Nutritional Sciences, University of Manitoba, Winnipeg, Canada

Dr. Michel Aliani
Department of Human Nutritional Sciences, University of Manitoba, Winnipeg, Manitoba, Canada

Dr. Rotimi Aluko
Department of Human Nutritional Sciences, U of Manitoba, Winnipeg, Canada

Dr. Daniel Baumgartner
Agroscope, Institute for Food science IFS, Schloss, Switzerland

Dr. Malik Behrens
Department of Molecular Genetics, German Institute of Human Nutrition, Potsdam, Rebbruecke, Nuthenal, Germany.

Raj Bhullar
Department of Oral Biology, Faculty of Dentistry, University of Manitoba, Winnipeg, Canada

Dr. Prashen Chelikani
Department of Oral Biology, Faculty of Dentistry, University of Manitoba, Winnipeg, Canada

Dr. Geraldine Dowling
Department of Pharmacology and Therapeutics, School of Medicine, Trinity Centre for Health Sciences, St. James's Hospital, James's Street, Dublin, Ireland

Dr. Michael N. A. Eskin
Department of Human Nutritional Sciences, University of Manitoba, Winnipeg, Canada

Dr. Erin Goldberg
Department of Human Nutritional Sciences, University of Manitoba, Winnipeg, Manitoba, Canada

Jennifer Grant
Department of Human Nutritional Sciences, University of Manitoba, Winnipeg, Canada

Dr. Ernst Hoehn
Swiss Federal Research Station

Dr. Wolfgang Meyerhof
Department of Molecular Genetics, German Institute of Human Nutrition, Potsdam, Rebbruecke, Nuthenal, Germany

Donna Ryland
Department of Human Nutritional Sciences, University of Manitoba, Winnipeg, Manitoba, Canada

Dr. Peng Shi
State Key Laboratory of Genetic Resources and Evolution, Kunming Institute of Zoology, Chinese Academy of Sciences, Kunming, China.

Nisha Singh
Department of Oral Biology, Faculty of Dentistry, University of Manitoba, Winnipeg, Canada

Jasbir Upadhyaya
Department of Oral Biology, Faculty of Dentistry, University of Manitoba, Winnipeg, Canada

Preface

Bitterness is one of the most interesting and least studied/understood of all the human tastes. It produces aversive reactions because it was originally associated with the plant source being poisonous. In fact, it was considered a defence mechanism for avoiding the ingestion of such harmful substances so that early human survival was based on the knowledge and ability to discriminate between edible plants particularly those with potentially harmful effects. With the advent of modern technology our understanding of bitterness is far more sophisticated and that we now know that not all bitter compounds are poisonous. In fact there are many foods in which bitterness is quite acceptable such as in some cheeses and beverages. In this book we have attempted to provide a comprehensive review of bitterness, from the novel genes in humans responsible for the expression of bitterness to methods used to remove or reduce bitterness in functional foods and nutraceuticals.

The book is organized into four sections. The first section covers the biology of bitterness perception with chapter 1 discussing the biochemistry of the 25 human bitter taste receptors of the TAS2R gene family. Chapter 2 examines the physiological aspects of bitterness while chapter 3 discusses human bitterness from an evolutionary perspective.

Section II covers the chemistry of bitterness with chapter 4 detailing those secondary plant metabolites responsible for the bitterness of selected fruits and vegetables. The compounds responsible for the bitterness of such beverages as tea, coffee, cocoa, wine and cider are reviewed in chapter 5, whereas 'food protein-derived bitter peptides' is the subject of chapter 6.

The analysis of bitterness, both sensory and chemical, is detailed in section III. Chapter 7 is a comprehensive review of sensory methods for assessing the bitterness of foods and beverages while chapter 8 is focused on the application of mass spectrometry for identifying bitter compounds. The final chapter in this section, chapter 9, discusses the ability of the electronic tongue to analyze bitterness and its correlation with sensory analysis.

The final section, section IV, covers the physical and chemical methods available for removing or masking bitterness in functional foods and nutraceuticals. The recent development of bitter blockers is also discussed as it provides a healthy alternative to adding sugar or salt for masking bitterness.

We hope this book will provide useful information to food scientists as well as those working in the food and flavor industries. We are grateful to colleagues from around the world for their important contributions to this book and acknowledge the excellent editorial assistance provided by the staff of Wiley.

Michel Aliani and Michael N. A. Eskin

SECTION I

THE BIOLOGY OF BITTERNESS PERCEPTION

1 Biochemistry of Human Bitter Taste Receptors

Jasbir Upadhyaya, Nisha Singh, Raj Bhullar, and Prashen Chelikani

1.1 INTRODUCTION

The gustatory system has been selected during evolution to detect nutritive and beneficial compounds as well as harmful substances. Humans, and probably other mammals, can taste many compounds but distinguish between five basic tastes which are sweet, bitter, sour, salt and umami. Sour and salt tastes are thought to be perceived via cation channels (Heck *et al.*, 1984; Kinnamon *et al.*, 1988; Ugawa *et al.*, 1998). In contrast, sensation of bitter, sweet and umami tastes is initiated by the interaction of taste molecules with G protein-coupled receptors (GPCRs) (Adler *et al.*, 2000; Gilbertson *et al.*, 2000; Sainz *et al.*, 2001). Bitter taste, among all tastes, is believed to have evolved as a central warning signal against the ingestion of potentially toxic substances. The molecular events in the perception of taste start at the apical surface of taste receptor cells (TRCs) found in taste buds in the mouth. Taste buds are found in taste papillae located on the tongue, the palate, and to a lesser extent the epiglottis, pharynx and larynx, and each taste bud is formed of 50-100 TRCs (Lalonde and Eglitis, 1961; Miller, 1986; Brouwer and Wiersma, 1978). The interaction of tastants with taste receptors, located in the membrane of TRCs, initiates signaling cascades which are transmitted to the brain through sensory afferents and perceived as taste (Chen *et al.*, 2011).

1.2 BITTER TASTE RECEPTORS: T2Rs

In humans, bitter taste is perceived by 25 members of the GPCR superfamily, referred to as T2Rs, which are 291 to 334 amino acids long (Adler *et al.*, 2000, Chandrashekar *et al.*, 2000, Matsunami *et al.*, 2000). These taste receptors, discovered a little more than a decade ago, encode for intronless genes which are referred to as TAS2Rs. The HUGO gene nomenclature of TAS2R is used wherever the gene is mentioned. Except for the TAS2R1 gene, which is localized on chromosome 5p, all other TAS2Rs are organized in the genome in clusters on human chromosomes 7q and 12p, and are genetically linked to loci that influence bitter perception (Conte *et al.*, 2002). Additionally, there are a large number of TAS2R

Bitterness: Perception, Chemistry and Food Processing, First Edition. Edited by Michel Aliani & Michael N. A. Eskin.

pseudogenes and more than 80 single nucleotide polymorphisms (SNPs) among individual TAS2R genes (Conte *et al.*, 2002; Kim *et al.*, 2005). The classification of T2Rs within the GPCR family is unclear, with some describing them as a separate family (Horn *et al.*, 2003), whereas other classification systems have grouped them with the frizzled receptors (Fredriksson *et al.*, 2003). The International Union of Basic and Clinical Pharmacology (IUPHAR) list Frizzled receptors as a separate GPCR family, Class F, and this class does not include T2Rs (Sharman *et al.*, 2013). T2Rs are relatively divergent, showing ~25–90% amino acid identity (Adler *et al.*, 2000; Matsunami *et al.*, 2000). This variability corresponds well with an ability to interact with chemically diverse ligands associated with bitter tastes. A single bitter compound is capable of activating multiple T2Rs and each T2R can be activated by multiple bitter compounds (Meyerhof *et al.*, 2010). Like all GPCRs, T2Rs contain seven transmembranes (TMs), three extracellular loops (ECLs) and three intracellular loops (ICLs), with a short extracellular N- and an intracellular C-terminus (Fig. 1.1). The other class of taste GPCRs, which codes for sweet and umami receptors (T1Rs), belongs to the class C GPCR family (Lagerstrom and Schioth, 2008). Sweet and umami tastes are mediated by three GPCRs that combine to form two heterodimeric receptors, T1R1/T1R3 for umami and T1R2/T1R3 for sweet-tasting

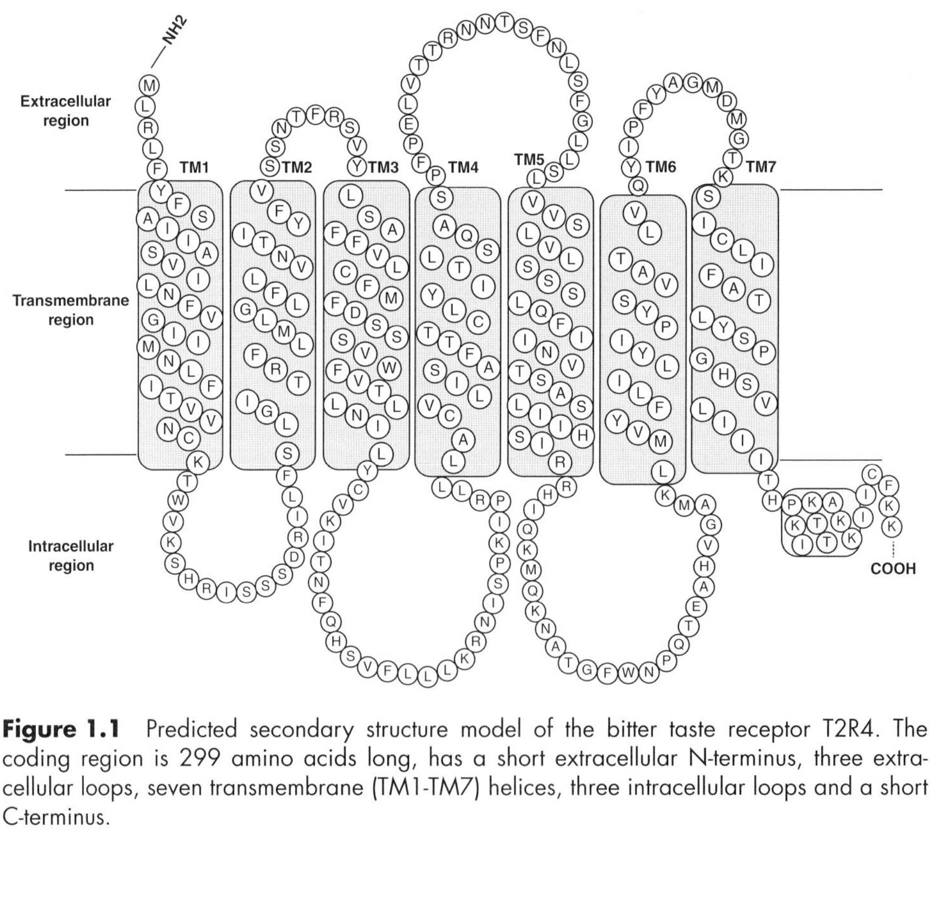

Figure 1.1 Predicted secondary structure model of the bitter taste receptor T2R4. The coding region is 299 amino acids long, has a short extracellular N-terminus, three extracellular loops, seven transmembrane (TM1-TM7) helices, three intracellular loops and a short C-terminus.

compounds (Li *et al.*, 2002; Nelson *et al.*, 2001, 2002; Zhao *et al.*, 2003). In contrast to the short N-terminus of T2Rs, T1Rs are characterized by a long N-terminus, also known as Venus flytrap, which forms the primary or orthosteric ligand binding site (Pin *et al.*, 2003). Differences in ligand specificity between species has been reported for the sweet and umami receptors (Xu *et al.*, 2004; Li *et al.*, 2002; Nelson *et al.*, 2002). Human T1R1/T1R3 specifically responds to L-Glu, whereas mouse T1R1/T1R3 responds more strongly to other L-amino acids than to L-Glu. In a recent study, the residues in the extracellular Venus flytrap domain of T1R1 which are crucial for amino acid recognition in the human- and mouse-type responses were identified (Toda *et al.*, 2013). In contrast to the low amino acid identity in the N- and C-termini and the ECLs, sequence conservation is more in the TMs and ICLs of T2Rs. The TMs and ECLs are the predicted regions of ligand binding in T2Rs and ICLs are the regions for G-protein interaction (Adler *et al.*, 2000).

1.3 T2R SIGNAL TRANSDUCTION

Long before the discovery of T2Rs, the involvement of taste-specific Gα protein, Gα-gustducin, in bitter receptor mediated transduction mechanism was demonstrated (Wong *et al.*, 1996). The generation of α-gustducin knock-out mice resulted in dramatic reduction of their bitter tasting abilities. Moreover, T2Rs were shown to functionally couple to transducin (He *et al.*, 2002) in vivo as well as to other Gi/Go proteins in vitro (Ozeck *et al.*, 2004). The mechanism involved in the perception of bitter taste and the second messengers or other downstream components of T2R signaling pathway were also known before the T2Rs were discovered in 2000 (Kurihara *et al.*, 1994; Spielman *et al.*, 1996; Chandrashekar *et al.*, 2000). A cation channel, transient receptor potential melastatin subtype 5 channel (TRPM5), was found coexpressed with other taste signaling molecules in taste tissue (Perez *et al.*, 2002).

The canonical T2R signal transduction pathway is described below. The binding of a bitter-tasting compound, also referred to as an agonist, on the extracellular surface of a T2R causes conformational changes in the receptor, and this in turn activates the heterotrimeric G-protein complex, α-gustducin, β1/3 and γ13 on the intracellular surface of the receptor. The βγ-subunits activate the enzyme phospholipase Cβ2 (PLC β2) which hydrolyzes inositol phospholipid (PIP_2) resulting in the production of 1,4,5-triphosphate (IP_3) and diacylglycerol (DAG). Generation of IP_3 activates IP_3 receptors on the membrane of endoplasmic reticulum (ER), thus opening the calcium release channels and causing transient increase in intracellular calcium. This opens the monovalent selective TRPM5 channels, leading to sodium influx, membrane depolarization and thus release of ATP as a neurotransmitter to activate the gustatory afferents (Finger *et al.*, 2005) (Fig. 1.2). Gα-gustducin activates phosphodiesterases (PDEs) which lead to a reduction in cAMP production (McLaughlin *et al.*, 1992; Spielman, 1998).

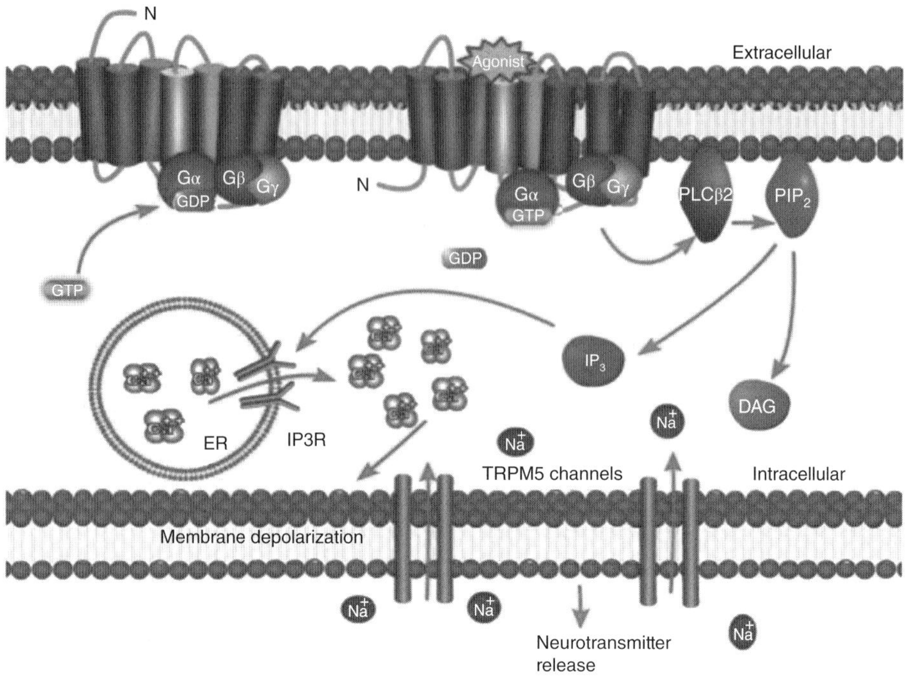

Figure 1.2 Bitter taste signaling pathway (IP_3 pathway) Abbreviations: PLCβ2, phospholipase C β2; PIP_2, phospatidyl-inositol-biphosphate; DAG, diacylglycerol; IP_3, inositol triphosphate; ER, endoplasmic reticulum; Ca^{2+}, calcium; Na^+, sodium.

1.4 BITTER TASTE PERCEPTION AND T2R POLYMORPHISMS

The sensitivity of humans to the perception of some bitter compounds varies greatly (Bartoshuk, 2000a, 2000b). This variable bitter taste perception is the best-known example of genetic variation in oral sensation. A vast number of structurally diverse compounds elicit bitter taste in humans and many bitter substances can be detected at concentrations roughly 1000-fold lower than substances that stimulate other basic tastes (Meyerhof, 2005). Studies on the genetics of taste perception for phenylthiocarbamide (PTC) began in the early 1930s with the accidental finding by A. L. Fox that crystals of PTC tasted very bitter to some people but not to others (Fox, 1932). Thus, 6-n-propyl-2-thiouracil (PROP) and PTC, which share thiocyanate (N—C=S) moiety, taste bitter to some people but are tasteless to others (Fox, 1932).

Sensitivity to PTC/PROP is an inherited trait, and PROP sensitivity was linked with lower acceptability of other bitter compounds and lower reported liking for some bitter foods. Based on the detection thresholds for PTC/PROP solutions, people were categorized into supertasters, tasters and non-tasters. Similarly, inbred mouse strains differ in their ability to detect certain bitter taste stimuli, such as sucrose octaacetate (SOA) and cycloheximide. Genetic studies in humans have demonstrated that the ability to detect PROP is determined by a locus on chromosome 5p15 (Reed *et al.*, 1999).

How humans respond to different bitter tasting compounds is an important question in the field of bitter taste research. Missense mutations were found in the sequences of T2R5 in mouse strains deficient to cycloheximide sensitivity (Chandrashekar *et al.*, 2000). These genetic variants, found in bitter-insensitive mouse strains, also were less responsive in cell-based assays compared with alleles from bitter-sensitive strains. This demonstrated that alleles of a taste receptor can change both behavioral and cellular responses to bitter compounds. A similar discovery was made in humans when naturally occurring alleles of the TAS2R38 gene, which is localized to chromosome 7q, were reported to be responsible for individual differences in the ability of humans to taste PTC and PROP (Mennella *et al.*, 2005). Three polymorphic variants in T2R38 (proline or alanine at position 49, alanine or valine at position 262, and valine or isoleucine at position 296) gave rise to five common haplotypes that accounted for 55-85% of the variance in PTC sensitivity (Bufe *et al.*, 2005). The taster haplotype or PROP-sensitive individuals possess one or two dominant alleles (***p***roline-***a***lanine-***v***aline; PAV/PAV), or PAV/AVI (***a***lanine-***v***aline-***i***soleucine), whereas insensitive individuals are recessive for the trait, AVI/AVI (Bufe *et al.*, 2005). The ability to taste PTC/PROP may protect against cigarette smoking and has also been linked to decreased alcohol consumption (Cannon *et al.*, 2005; Duffy *et al.*, 2004).

Until recently, TAS2R38 was considered the only bitter taste gene that exhibits prominent phenotypic variation in humans. But variation in bitter receptor sequence is not confined to the TAS2R38 locus. Human TAS2Rs have more genetic variation within and between populations than do most other genes (Kim *et al.*, 2005). One possible explanation is that genes adapt to local conditions especially to the bitter toxins in food. SNPs in some other TAS2R genes have recently been identified. For example, a missense mutation in the TAS2R16 gene, which encodes the β-glucopyranoside receptor or T2R16, reduces sensitivity of the receptor to bitter-taste stimuli which has been associated with risk for alcohol dependence (Bufe *et al.*, 2002; Wang *et al.*, 2007). Polymorphism in the TAS2R43 gene allele makes people very sensitive to bitterness of the natural plant compounds aloin and aristolochic acid (Pronin *et al.*, 2007). TAS2R43 and TAS2R44 gene alleles are also related to the bitterness perception of artificial sweetener, saccharin. Recently an SNP in the cluster of T2Rs on chromosome 12, which contributes to the variation in human bitterness perception of caffeine, was identified (Ledda *et al.*, 2014). Thus, it seems likely that the examination of multiple taste phenotypes might provide a more complete understanding of human eating behavior than a single taste phenotype.

1.5 LIGAND BINDING AND ACTIVATION MECHANISMS OF T2Rs

Bitter compounds are not only numerous but also structurally diverse. They include hydroxy fatty acids, peptides, amino acids, amines, N-heterocyclic compounds, ureas, thioureas, carbamides, esters, lactones, phenols, alkaloids, glycosides and many more. In contrast, only 25 T2Rs have been identified, raising the question as to how the vast array of bitter compounds can be detected by such a limited number of receptors. While many T2Rs remain poorly characterized, the ligand specificity of several T2Rs was explored in the past decade (Chandrashekar *et al.*, 2000; Bufe *et al.*, 2002, Kim *et al.*, 2003; Behrens *et al.*, 2004; Kuhn *et al.*, 2004; Pronin *et al.*, 2004; Brockhoff *et al.*, 2007; Sainz *et al.*, 2007; Dotson *et al.*, 2008; Maehashi *et al.*, 2008; Upadhyaya *et al.*, 2010; Meyerhof *et al.*, 2010). Whereas some receptors recognize only a single or few compounds, others respond to multiple compounds. The affinity of T2Rs for their respective bitter ligands is low, with EC_{50} values in the high micromolar to low millimolar range (Meyerhof *et al.*, 2010). Thus, bitter compounds activate various T2Rs in different concentration ranges, differences usually being in the range of 10- to 100-fold. However, knowledge of the structural determinants of T2Rs is crucial to provide insights into the molecular basis of bitter sensing and to design new taste modifiers. Molecular modeling and site-directed mutagenesis studies were pursued to characterize the ligand-binding pocket of some T2Rs. The 3D structure of T2R38, also referred to as the PTC receptor, was predicted using computational method *MembStruk* and homology modeling (Floriano *et al.*, 2006). *Hierdock* and *ScanBindSite* computational tools were then used to generate models of PTC bound to T2R38 to predict the binding site and binding energy. According to these models, PTC binds at a site distant from the variant amino acids P49A, A262V and V296I (Floriano *et al.*, 2006). It is also suggested that the inability of humans to taste PTC is due to a failure of G-protein activation rather than decreased binding affinity of the receptor for PTC. This study emphasizes the role of TM6 and TM7 in PTC receptor function. The introduction of bulkier side chains in the nontaster variant alters the packing of TMs 6 and 7, which might render the movement of TM6 more difficult (Biarnes *et al.*, 2010). A recent study predicted the 3D structure of T2R38 using BiHelix and SuperBiHelix Monte Carlo methods (Tan *et al.*, 2012). This study suggests that the residue 262 is involved in interhelical hydrogen bond network which stabilizes the receptor in tasters ($hTAS2R38_{PAV}$, $hTAS2R38_{AAI}$, and $hTAS2R38_{PVV}$), but not in the non-tasters ($hTAS2R38_{AVI}$) (Tan *et al.*, 2012).

In a study by Pronin *et al.*, chimeric receptors for T2R43 and T2R44 were generated in an effort to identify the residues involved in ligand recognition (Pronin *et al.*, 2004). T2R43 is activated by 6-nitrosaccharin and N-isopropyl-2-methyl-5-nitrobenzenesulfonamide (IMNB), a bitter derivative of saccharin. Whereas, T2R44 is activated by denatonium and 6-nitrosaccharin. The amino acid sequences of T2R43 and T2R44 are 89% identical and 15 of the 34 amino acid differences among them are concentrated in ECL1 and ECL2, while ECL3 is completely conserved. T2R43 and T2R44 chimeras were generated by swapping their ECLs–1 and –2. There are only four amino acid differences between T2R43 and T2R44 in ECL1. Functional studies revealed that ECL1 is very important for receptor activation, as replacing these residues of T2R43 with those of T2R44 is sufficient to render T2R43 insensitive to IMNB. On the other hand, replacing both ECL1 and ECL2 in T2R43 with T2R44 loops eliminated most of the activation by 6-nitrosaccharin. Recently, ligand docking simulations and functional analysis using point mutants of T2R16 were performed to identify binding sites of the receptor to β-glucopyranosides (Sakurai *et al.*, 2010). Seven amino acid residues in TMs 3, 5 and 6 were involved in ligand recognition. Amino acid residues Glu86, Trp94 and Gln177 were involved in salicin recognition, whereas His181 and hydrophobic residues, Phe93, Phe240 and Ile243 likely contributed to formation of the binding site. With the generation of chimeric and mutant receptors, followed by functional analysis, the amino acid residues critical for the activation of T2R46, T2R43 and T2R31 were identified (Brockhoff *et al.*, 2010). The construction of receptor chimeras demonstrated that agonist selectivity was predominantly determined by TM7 region of the receptors. Exchange of two residues within TM7 between T2R46, activated by strychnine, and T2R31, activated by aristolochic acid, was sufficient to invert the agonist selectivity.

Fermentation of protein-rich foods results in the formation of bitter peptides, which are responsible for the bitter taste of fermented food. Bitter casein digests were able to activate T2R1, T2R4, T2R14 and T2R16 in a heterologous expression system (Maehashi *et al.*, 2008). Two bitter dipeptides, Gly-Phe (glycine-phenylalanine) and Gly-Leu (glycine-leucine), activated T2R1 more strongly, whereas they evoked no or weak responses in other receptors. The ability of bitter di- and tri-peptides to activate T2R1 was tested further (Upadhyaya *et al.*, 2010). Results revealed that bitter tri-peptides also activated T2R1 and were more potent than the tested di-peptides. Among all the tested peptides, Phe-Phe-Phe (phenylalanine-phenylalanine-phenylalanine) activated T2R1-expressing cells the most, at concentrations of 0.125–1 mM that humans also perceive as bitter, with an EC_{50} value of 370 μM. Phe-Phe-Phe consists of hydrophobic amino acids and the bitter taste of a peptide is more apparent when the hydrophobic amino acid is located at the C-terminus. For the tri-peptides, the middle amino acid residue is considered more important than both the C- and N-terminal amino acids (Wu and Aluko, 2007). In addition, some peptides with ACE (angiotensin-converting enzyme)-inhibitory activity were also able to activate T2R1. Homology modeling and docking studies showed that amino acid residues from TMs 1-3, TM 7 and from ECL1 and ECL2 contributed in forming the ligand binding pocket of T2R1 for the peptide ligands (Upadhyaya *et al.*, 2010). In another study of T2R1, molecular modeling and site-directed mutagenesis studies revealed that two asparagines, Asn66 and the highly conserved Asn24, are important for dextromethorphan (DXM)-induced receptor signaling (Singh *et al.*, 2011). Asn24 plays a crucial role in receptor activation by mediating a hydrogen-bond network connecting TM1-TM2-TM7, whereas Asn66 is essential for bonding to DXM. There is a unique signature sequence of T2Rs, the LXXSL motif. It plays a predominantly structural role in stabilizing the helical conformation of TM5 at the cytoplasmic end and a functional role by influencing the conformation of ICL3. Replacement of the conserved residues in this motif with bulky β-branched amino acids results in protein misfolding and/or non-functional receptor (Singh *et al.*, 2011).

Recently, the role of ICL3 in quinine-mediated activation of bitter taste receptor T2R4 was demonstrated using alanine scan mutagenesis (Pydi *et al.*, 2013). ICL3 of T2R4 consists of 23 amino acid residues which were mutated to alanine. Only 14 of the 23 mutants displayed quinine-induced signaling in a concentration-dependent manner. Three mutants, Q216A, T230A and V234A, showed an increased response to quinine. Six mutants, R213A, Q219A, K220A, Q229A, E231A and H233A, showed no detectable or statistically significant increase in intracellular calcium mobilization, suggesting that they may have an important role in receptor activation. Whereas mutants I215A, F225A and P228A displayed altered receptor activation and/or defective ligand binding. Some mutants showed statistically significant basal signaling or constitutive activity. H214A, which is present in 24 of the 25 human T2Rs, showed the highest constitutive activity (i.e., in the absence of agonist). A recent study identified a conserved KLK/R motif in the C-terminus of T2Rs. This KLK motif was suggested to perform a critical functional role involving trafficking and activation in T2R4 (Upadhyaya et al., 2015).

1.6 NUTRIGENOMICS OF TASTE

The PROP phenotype serves as a general marker for bitterness perception which influences general food preferences and dietary behavior with subsequent links to body weight and chronic disease risk. Strong bitter taste is closely associated with the presence of toxins and is aversive. However, moderate bitter taste is appealing and expected in a variety of foods including beer, wine, chocolates and many cheeses. Fischer and colleagues noted that PTC tasters tended to manifest a thin and angular body type, whereas non-tasters tended to have generous body proportions (Fischer *et al.*, 1966). Studies in overweight middle-aged women have provided convincing evidence linking PROP status with body weight (Goldstein *et al.*, 2005). Goldstein *et al.* showed that non-taster women were heavier than supertaster women by ~6 BMI (body mass index) units. Anatomical evidence demonstrates that individuals who differ in taste sensitivity to PTC/PROP also differ in the density of fungiform taste papillae on the anterior surface of tongue (Bartoshuk *et al.*, 1994; Essick *et al.*, 2003; Tepper and Nurse, 1997). Non-tasters have the lowest density of fungiform papillae, whereas supertasters have the highest density.

Isothiocyanates, the breakdown products of glucosinolates that are widely distributed in plants, interfere with the uptake of iodine by the thyroid gland, leading to goiter, and cretinism in its extreme form. Although iodine deficiency is the primary cause of this disease, goitrogens in the food supply can play a contributing role particularly when dietary iodine is low. It was shown that a large percentage of athyroidic cretins in a clinical population in the United States were PTC non-tasters (Shepard, 1961). Investigation of the role of PROP status in children's selection and consumption of vegetables showed that non-taster children consumed more bitter vegetables overall than taster children (Bell and Tepper, 2006). PROP status has also been linked to sweet taste preference in children. Taster preschool children showed greater preferences for sweets than non-taster children (Keller and Tepper, 2004). The perception of oral irritation from capsaicin (chili pepper), cinnamaldehyde (from cinnamon), and carbonation is influenced by PROP sensitivity (Karrer and Bartoshuk, 1991; Prescott and Swain-Campbell, 2000; Prescott *et al.*, 2004). Individual differences in fat perception have been linked to PROP taster status and taste bud density. A study in college students revealed that medium and supertasters reliably discriminated a high-fat from a low-fat dressing, whereas non-tasters could not distinguish the two samples (Tepper and Nurse, 1997). Study by Keller and coworkers in preschool children demonstrated that this phenotype might have greater influence on preferences for fats in females than males (Keller *et al.*, 2002). Discretionary fat intake did not differ between taster and non-taster boys.

Few studies have examined associations between PROP status and disease risk, though the data addressing this issue is scarce. No associations were reported between T2R38 polymorphisms and cardiovascular risk in the elderly women or between PROP status and lipid profiles in the breast cancer patients (Timpson *et al.*, 2005; Drewnowski *et al.*, 2007). However, a modest association between greater sensitivity to PROP and a higher number of colon polyps was found in older men undergoing routine screening for colon pathology (Basson *et al.*, 2005).

Dental caries is the most common chronic disease of childhood that is neither self-limiting nor amenable to short-term pharmacological management (Edelstein and Douglass, 1995). Effective dentistry requires early identification of children at higher risk for caries so they may receive early and intense preventive intervention. The individual differences in PROP sensitivity have been linked to dental caries and can be used as an important tool to determine the taster status in relation to caries experience in children (Rupesh and Nayak, 2006; Verma *et al.*, 2006; Pidamale *et al.*, 2012; Hedge and Sharma, 2008). A comprehensive review of the role of diet and dental caries reaffirmed that sucrose is the most important dietary item associated with dental caries (Habibian *et al.*, 2001). Non-taster children may have higher concentration and frequencies of sugar intake compared to children who are medium or supertasters and are therefore more susceptible to dental caries (Anliker *et al.*, 1991). Whereas supertasters and medium tasters are more likely to avoid sweet food, thus making them less prone to dental decay. *Streptococcus mutans* levels were also shown to increase from tasters to non-tasters, thus placing them at higher risk of developing caries (Verma *et al.*, 2006).

1.7 BITTER TASTE BLOCKERS

The sense of taste has a significant impact on food selection, nutrition and health. It is, therefore, highly desirable to modulate bitter taste perception and bitter taste receptors so that beneficial food and medicines may be rendered more palatable. In addition to having an important role in food and nutraceutical industries, bitter taste blockers could be beneficial as chemical probes to examine the role of T2R function in gustatory and non-gustatory tissues. The T2R antagonist, GIV3727, was able to inhibit the activation of T2R44 by saccharin and acesulfame-K (Slack *et al.*, 2010). This compound also inhibited five additional T2Rs, including the closely related T2R43. It appears the –COOH moiety is essential for antagonist activity of GIV3727 since replacement of this group with an ester or corresponding alcohol abolished its activity. Two residues in TM7 are important for antagonist activity in T2R43/T2R44. Shortly after this study, probenecid, an approved inhibitor of Multidrug Resistance Protein 1 (MRP1) transporter, was shown to inhibit T2R16, T2R38 and T2R43 in a non-competetive (allosteric) mechanism (Greene *et al.*, 2011). And, two natural sesquiterpene lactones from edible plants, 3β-Hydroxydihydrocostunolide (3HDC) and 3β-Hydroxypelenolide (3HP), were identified which blocked the responses of T2R46 receptor (Brockhoff *et al.*, 2011). Besides T2R46, 3HDC also inhibited T2R30 and T2R40, and 3HP inhibited T2R30, T2R43 and T2R44. Recent studies characterized few novel bitter blockers, γ-aminobutyric acid (GABA), abscisic acid and Na,Na-bis(carboxymethyl)-L-lysine (BCML). These acted as competitive inhibitors of quinine-activated human T2R4, sharing the same orthosteric site as agonist quinine (Pydi et al., 2014, Pydi et al., 2015). Though there is a vast number of bitter agonists known for T2Rs, the knowledge of bitter taste blockers or T2R antagonists and inverse agonists is limited. Hence there is an urgent need to discover more natural or synthetic blockers for T2Rs to increase the consumption of healthy bitter foods and for drug compliance.

1.8 EXPRESSION OF T2Rs IN EXTRAORAL TISSUES

With the molecular identification of taste GPCRs, it has become clear that taste signaling is not limited to taste buds, but occurs in many extraoral tissues and has additional functions apart from taste. Shortly after the discovery of T2Rs in taste tissue, their expression was demonstrated in the gastrointestinal tract (GIT) and enteroendocrine STC-1 cells of rodents and humans (Wu *et al.*, 2002; Rozengurt, 2006), where they are involved in the chemosensation of nutrients. Gα-gustducin and Gα-transducin were also expressed in these tissues, suggesting that a taste-sensing mechanism may also exist in the GIT. Addition of bitter compounds like denatonium, PTC, PROP, caffeine and cycloheximide to STC-1 cell cultures promoted rapid $[Ca^{2+}]_i$ responses (Wu *et al.*, 2002; Masuho *et al.*,

2005). In addition, activation of T2Rs stimulated the secretion of hunger hormone ghrelin (Janssen *et al.*, 2011) via the gustatory G-protein, α-gustducin.

In the airway epithelium, expression of T2Rs was revealed in chemosensory receptor cells of the nasal epithelium and in ciliated epithelial cells (Shah *et al.*, 2009; Tizzano *et al.*, 2011; Masuho *et al.*, 2005). Application of bitter substances to the nasal epithelium activated the trigeminal nerve and elicited protective reflexes like apnea to prevent inhalation of bacteria further into the respiratory system or sneezing and coughing to expel bacteria from the nasal cavity. Exposure of T2Rs in motile cilia of human airway epithelial cells with bitter compounds stimulated ciliary beat frequency (Shah *et al.*, 2009), thus initiating a defensive mechanical mechanism to eliminate the offending compound. In the human airway smooth muscle (ASM), T2Rs lead to ASM relaxation and bronchodilation (Deshpande *et al.*, 2010). Bitter tastants like denatonium, saccharin and chloroquine caused relaxation of mouse isolated ASM preparations, and dilation of airways that was three-fold greater than the presently used β-agonists. This relaxation by T2Rs was due to increased $[Ca^{2+}]_i$ that was suggested to activate large conductance potassium channels (BK_{Ca}) and result in hyperpolarization of the cell membrane. Additional studies showed that bronchodilatory effects of T2R agonists were not impeded by β_2-AR desensitization (An *et al.*, 2012). These findings have reinforced the role of T2Rs as potential novel targets in asthma pharmacotherapy. Expression of the 25 human TAS2Rs was revealed in the pulmonary artery smooth muscle cells where they are functional and activated by bitter compounds (Upadhyaya et al., 2014). This study suggests that T2Rs in the vasculature might be involved in regulating the vascular tone (Upadhyaya et al., 2014).

Regulation of the mucosal innate defense of human and mouse upper respiratory epithelium by activation of T2R38 was recently demonstrated (Lee *et al.*, 2012, 2014). Gram-negative respiratory pathogens like *Pseudomonas aeruginosa* produce acyl-homoserine lactones (AHLs) as signals for their population density (quorum sensing). AHLs are chemically related to bitter sesquiterpene lactones and activate T2R38 in upper respiratory epithelium. Receptor activation causes calcium and nitric oxide (NO) signaling resulting in stimulation of mucociliary clearance, the major physical respiratory defense against inhaled pathogens. Genetic variation in T2R38 has also been linked to individual differences in susceptibility to respiratory infection.

1.9 CONCLUSION

With the deorphanization of T2Rs, studies of the mechanisms of their interaction with bitter agonists have started revealing how these receptors are able to sense such a vast array of bitter compounds. Knowledge of their ligand bound structure would further help in the identification or design of taste modulators like bitter blockers.

T2R blockers could have widespread utility in antioxidant and/or nutrient-fortified food and beverages, and in pharmaceutical and nutraceutical industry.

Elucidation of the biochemistry of bitter taste signal transduction plays an important role in understanding how humans perceive bitter taste. The next step includes deciphering how the taste signal is terminated. A study by Robinette *et al.* has demonstrated a 30% desensitization of T2R function with quinine pretreatment and subsequent exposure in airway smooth muscle (Robinett *et al.*, 2011). Another study, using molecular and pharmacological techniques, showed that T2R4 does not get internalized upon agonist exposure (Upadhyaya et al., 2016). Instead, treatment with bitter compound quinine caused a two-fold increase in surface expression of T2R4 which was Brefeldin A-sensitive. Quinine pretreatment led to a reduction in subsequent calcium responses to $35 \pm 5\%$ compared to the control untreated cells. This study thus, discovered a novel pharmacochaperone role of quinine and provides insights into the possible mechanism of T2R desensitization (Upadhyaya et al., 2016). However, data of T2R desensitization is very scarce, and the potential molecular mechanisms involved in desensitization like receptor internalization, phosphorylation by the respective kinases, β-arrestin binding, leading to uncoupling of receptor-G protein complex, remain poorly characterized. Before the introduction of T2Rs as novel therapeutic targets, it is very crucial that their desensitization mechanisms be probed in detail. T2Rs have a low affinity for their respective ligands. In recent studies, T2R agonists were used at a concentration 50-100 times higher than β-agonists (An *et al.*, 2012; Pulkkinen *et al.*, 2012). Thus, elucidation of the signaling mechanisms utilized downstream of T2Rs, may allow the synthesis of more specific and potent bitter compounds and/or blockers.

ACKNOWLEDGEMENT

This work was supported by a Discovery grant from the Natural Sciences and Engineering Research Council of Canada (RGPIN 356285) and the MMSF Allen Rouse Career Award to PC.

REFERENCES

Adler, E., Hoon, M. A., Mueller, K. L., Chandrashekar, J., Ryba, N. J. & Zuker, C. S. (2000). A novel family of mammalian taste receptors. *Cell* **100,** 693–702.

An, S. S., Wang, W. C., Koziol-White, C. J., Ahn, K., Lee, D. Y., Kurten, R. C., Panettieri, R. A., JR . & Liggett, S. B. (2012). TAS2R activation promotes airway smooth muscle relaxation despite beta(2)-adrenergic receptor tachyphylaxis. *Am J Physiol Lung Cell Mol Physiol* **303,** L304–L311.

Anliker, J. A., Bartoshuk, L., Ferris, A. M. & Hooks, L. D. (1991). Children's food preferences and genetic sensitivity to the bitter taste of 6-n-propylthiouracil (PROP). *Am J Clin Nutr* **54,** 316–320.

Bartoshuk, L. M. (2000a). Comparing sensory experiences across individuals: recent psychophysical advances illuminate genetic variation in taste perception. *Chem Senses* **25,** 447–460.

Bartoshuk, L. M. (2000b). Psychophysical advances aid the study of genetic variation in taste. *Appetite* **34,** 105.

Bartoshuk, L. M., Duffy, V. B. & Miller, I. J. (1994). PTC/PROP tasting: anatomy, psychophysics, and sex effects. *Physiol Behav* **56,** 1165–1171.

Basson, M. D., Bartoshuk, L. M., Dichello, S. Z., Panzini, L., Weiffenbach, J. M. & Duffy, V. B. (2005). Association between 6-n-propylthiouracil (PROP) bitterness and colonic neoplasms. *Dig Dis Sci* **50,** 483–489.

Behrens, M., Brockhoff, A., Kuhn, C., Bufe, B., Winnig, M. & Meyerhof, W. (2004). The human taste receptor hTAS2R14 responds to a variety of different bitter compounds. *Biochem Biophys Res Commun* **319,** 479–485.

Bell, K. I. & Tepper, B. J. (2006). Short-term vegetable intake by young children classified by 6-n-propylthoiuracil bitter-taste phenotype. *Am J Clin Nutr* **84,** 245–251.

Biarnes, X., Marchiori, A., Giorgetti, A., Lanzara, C., Gasparini, P., Carloni, P., Born, S., Brockhoff, A., Behrens, M. & Meyerhof, W. (2010). Insights into the binding of Phenyltiocarbamide (PTC) agonist to its target human TAS2R38 bitter receptor. *PLoS One* **5,** e12394.

Brockhoff, A., Behrens, M., Massarotti, A., Appendino, G. & Meyerhof, W. (2007). Broad tuning of the human bitter taste receptor hTAS2R46 to various sesquiterpene lactones, clerodane and labdane diterpenoids, strychnine, and denatonium. *J Agric Food Chem* **55,** 6236–6243.

Brockhoff, A., Behrens, M., Niv, M. Y. & Meyerhof, W. (2010). Structural requirements of bitter taste receptor activation. *Proc Natl Acad Sci U S A* **107,** 11110–11115.

Brockhoff, A., Behrens, M., Roudnitzky, N., Appendino, G., Avonto, C. & Meyerhof, W. (2011). Receptor agonism and antagonism of dietary bitter compounds. *J Neurosci* **31,** 14775–14782.

Brouwer, J. N. & Wiersma, A. (1978). Location of taste buds in intact taste papillae by a selective staining method. *Histochemistry* **58,** 145–151.

Bufe, B., Breslin, P. A., Kuhn, C., Reed, D. R., Tharp, C. D., Slack, J. P., Kim, U. K., Drayna, D. & Meyerhof, W. (2005). The molecular basis of individual differences in phenylthiocarbamide and propylthiouracil bitterness perception. *Curr Biol* **15,** 322–327.

Bufe, B., Hofmann, T., Krautwurst, D., Raguse, J. D. & Meyerhof, W. (2002). The human TAS2R16 receptor mediates bitter taste in response to beta-glucopyranosides. *Nat Genet* **32,** 397–401.

Cannon, D. S., Baker, T. B., Piper, M. E., Scholand, M. B., Lawrence, D. L., Drayna, D. T., Mcmahon, W. M., Villegas, G. M., Caton, T. C., Coon, H. & Leppert, M. F. (2005). Associations between phenylthiocarbamide gene polymorphisms and cigarette smoking. *Nicotine Tob Res* **7,** 853–858.

Chandrashekar, J., Mueller, K. L., Hoon, M. A., Adler, E., Feng, L., Guo, W., Zuker, C. S. & Ryba, N. J. (2000). T2Rs function as bitter taste receptors. *Cell* **100,** 703–711.

Chen, X., Gabitto, M., Peng, Y., Ryba, N. J. & Zuker, C. S. (2011). A gustotopic map of taste qualities in the mammalian brain. *Science* **333,** 1262–1266.

Conte, C., Ebeling, M., Marcuz, A., Nef, P. & Andres-Barquin, P. J. (2002). Identification and characterization of human taste receptor genes belonging to the TAS2R family. *Cytogenet Genome Res* **98,** 45–53.

Deshpande, D. A., Wang, W. C., Mcilmoyle, E. L., Robinett, K. S., Schillinger, R. M., An, S. S., Sham, J. S. & Liggett, S. B. (2010). Bitter taste receptors on airway smooth muscle bronchodilate by localized calcium signaling and reverse obstruction. *Nat Med* **16,** 1299–1304.

Dotson, C. D., Zhang, L., Xu, H., Shin, Y. K., Vigues, S., Ott, S. H., Elson, A. E., Choi, H. J., Shaw, H., Egan, J. M., Mitchell, B. D., Li, X., Steinle, N. I. & Munger, S. D. (2008). Bitter taste receptors influence glucose homeostasis. *PLoS One* **3,** e3974.

Drewnowski, A., Henderson, S. A. & Cockroft, J. E. (2007). Genetic sensitivity to 6-n-propylthiouracil has no influence on dietary patterns, body mass indexes, or plasma lipid profiles of women. *J Am Diet Assoc* **107,** 1340–1348.

Duffy, V. B., Davidson, A. C., Kidd, J. R., Kidd, K. K., Speed, W. C., Pakstis, A. J., Reed, D. R., Snyder, D. J. & Bartoshuk, L. M. (2004). Bitter receptor gene (TAS2R38), 6-n-propylthiouracil (PROP) bitterness and alcohol intake. *Alcohol Clin Exp Res* **28,** 1629–1637.

Edelstein, B. L. & Douglass, C. W. (1995). Dispelling the myth that 50 percent of U.S. schoolchildren have never had a cavity. *Public Health Rep* **110,** 522–530; discussion 521, 531–533.

Essick, G. K., Chopra, A., Guest, S. & Mcglone, F. (2003). Lingual tactile acuity, taste perception, and the density and diameter of fungiform papillae in female subjects. *Physiol Behav* **80,** 289–302.

Finger, T. E., Danilova, V., Barrows, J., Bartel, D. L., Vigers, A. J., Stone, L., Hellekant, G. & Kinnamon, S. C. (2005). ATP signaling is crucial for communication from taste buds to gustatory nerves. *Science* **310,** 1495–1499.

Fischer, R., Griffin, F. & Rockey, M. A. (1966). Gustatory chemoreception in man: multidisciplinary aspects and perspectives. *Perspect Biol Med* **9,** 549–577.

Floriano, W. B., Hall, S., Vaidehi, N., Kim, U., Drayna, D. & Goddard, W. A., 3rd (2006). Modeling the human PTC bitter-taste receptor interactions with bitter tastants. *J Mol Model* **12,** 931–941.

Fox, A. L. (1932). The Relationship between Chemical Constitution and Taste. *Proc Natl Acad Sci U S A* **18,** 115–120.

Fredriksson, R., Lagerstrom, M. C., Lundin, L. G. & Schioth, H. B. (2003). The G-protein-coupled receptors in the human genome form five main families. Phylogenetic analysis, paralogon groups, and fingerprints. *Mol Pharmacol* **63,** 1256–1272.

Gilbertson, T. A., Damak, S. & Margolskee, R. F. (2000). The molecular physiology of taste transduction. *Curr Opin Neurobiol* **10,** 519–527.

Goldstein, G. L., Daun, H. & Tepper, B. J. (2005). Adiposity in middle-aged women is associated with genetic taste blindness to 6-n-propylthiouracil. *Obes Res* **13,** 1017–1023.

Greene, T. A., Alarcon, S., Thomas, A., Berdougo, E., Doranz, B. J., Breslin, P. A. & Rucker, J. B. (2011). Probenecid inhibits the human bitter taste receptor TAS2R16 and suppresses bitter perception of salicin. *PLoS One* **6,** e20123.

Habibian, M., Roberts, G., Lawson, M., Stevenson, R. & Harris, S. (2001). Dietary habits and dental health over the first 18 months of life. *Community Dent Oral Epidemiol* **29,** 239–246.

He, W., Danilova, V., Zou, S., Hellekant, G., Max, M., Margolskee, R. F. & Damak, S. (2002). Partial rescue of taste responses of alpha-gustducin null mice by transgenic expression of alpha-transducin. *Chem Senses* **27,** 719–727.

Heck, G. L., Mierson, S. & Desimone, J. A. (1984). Salt taste transduction occurs through an amiloride-sensitive sodium transport pathway. *Science* **223,** 403–405.

Hedge, A. M. & Sharma, A. (2008). Genetic sensitivity to 6-n-propylthiouracil (PROP) as a screening tool for obesity and dental caries in children. *J Clin Pediatr Dent* **33,** 107–111.

Horn, F., Bettler, E., Oliveira, L., Campagne, F., Cohen, F. E. & Vriend, G. (2003). GPCRDB information system for G protein-coupled receptors. *Nucleic Acids Res* **31,** 294–297.

Janssen, S., Laermans, J., Verhulst, P. J., Thijs, T., Tack, J. & Depoortere, I. (2011). Bitter taste receptors and alpha-gustducin regulate the secretion of ghrelin with functional effects on food intake and gastric emptying. *Proc Natl Acad Sci U S A* **108,** 2094–2099.

Karrer, T. & Bartoshuk, L. (1991). Capsaicin desensitization and recovery on the human tongue. *Physiol Behav* **49,** 757–764.

Keller, K. L., Steinmann, L., Nurse, R. J. & Tepper, B. J. (2002). Genetic taste sensitivity to 6-n-propylthiouracil influences food preference and reported intake in preschool children. *Appetite* **38,** 3–12.

Keller, K. L. & Tepper, B. J. (2004). Inherited taste sensitivity to 6-n-propylthiouracil in diet and body weight in children. *Obes Res* **12,** 904–912.

Kim, U., Wooding, S., Ricci, D., Jorde, L. B. & Drayna, D. (2005). Worldwide haplotype diversity and coding sequence variation at human bitter taste receptor loci. *Hum Mutat* **26,** 199–204.

Kim, U. K., Jorgenson, E., Coon, H., Leppert, M., Risch, N. & Drayna, D. (2003). Positional cloning of the human quantitative trait locus underlying taste sensitivity to phenylthiocarbamide. *Science* **299,** 1221–1225.

Kinnamon, S. C., Dionne, V. E. & Beam, K. G. (1988). Apical localization of K+ channels in taste cells provides the basis for sour taste transduction. *Proc Natl Acad Sci U S A* **85,** 7023–7027.

Kuhn, C., Bufe, B., Winnig, M., Hofmann, T., Frank, O., Behrens, M., Lewtschenko, T., Slack, J. P., Ward, C. D. & Meyerhof, W. (2004). Bitter taste receptors for saccharin and acesulfame K. *J Neurosci* **24,** 10260–10265.

Kurihara, K., Katsuragi, Y., Matsuoka, I., Kashiwayanagi, M., Kumazawa, T. & Shoji, T. (1994). Receptor mechanisms of bitter substances. *Physiol Behav* **56,** 1125–1132.

Lagerstrom, M. C. & Schioth, H. B. (2008). Structural diversity of G protein-coupled receptors and significance for drug discovery. *Nat Rev Drug Discov* **7,** 339–357.

Lalonde, E. R. & Eglitis, J. A. (1961). Number and distribution of taste buds on the epiglottis, pharynx, larynx, soft palate and uvula in a human newborn. *Anat Rec* **140,** 91–95.

Ledda, M., Kutalik, Z., Souza Destito, M. C., Souza, M. M., Cirillo, C. A., Zamboni, A., Martin, N., Morya, E., Sameshima, K., Beckmann, J. S., Le Coutre, J., Bergmann, S. & Genick, U. K. (2014). GWAS of human bitter taste perception identifies new loci and reveals additional complexity of bitter taste genetics. *Hum Mol Genet* **23**(1), 259–267.

Lee, R. J., Chen, B., Redding, K. M., Margolskee, R. F. & Cohen, N. A. (2014). Mouse nasal epithelial innate immune responses to Pseudomonas aeruginosa quorum-sensing molecules require taste signaling components. *Innate Immun* **20**(6), 606–617.

Lee, R. J., Xiong, G., Kofonow, J. M., Chen, B., Lysenko, A., Jiang, P., Abraham, V., Doghramji, L., Adappa, N. D., Palmer, J. N., Kennedy, D. W., Beauchamp, G. K., Doulias, P. T., Ischiropoulos, H., Kreindler, J. L., Reed, D. R. & Cohen, N. A. (2012). T2R38 taste receptor polymorphisms underlie susceptibility to upper respiratory infection. *J Clin Invest* **122,** 4145–4159.

Li, X., Staszewski, L., Xu, H., Durick, K., Zoller, M. & Adler, E. (2002). Human receptors for sweet and umami taste. *Proc Natl Acad Sci U S A* **99,** 4692–4696.

Maehashi, K., Matano, M., Wang, H., Vo, L. A., Yamamoto, Y. & Huang, L. (2008). Bitter peptides activate hTAS2Rs, the human bitter receptors. *Biochem Biophys Res Commun* **365,** 851–855.

Masuho, I., Tateyama, M. & Saitoh, O. (2005). Characterization of bitter taste responses of intestinal STC-1 cells. *Chem Senses* **30,** 281–290.

Matsunami, H., Montmayeur, J. P. & Buck, L. B. (2000). A family of candidate taste receptors in human and mouse. *Nature* **404,** 601–604.

Mclaughlin, S. K., Mckinnon, P. J. & Margolskee, R. F. (1992). Gustducin is a taste-cell-specific G protein closely related to the transducins. *Nature* **357,** 563–569.

Mennella, J. A., Pepino, M. Y. & Reed, D. R. (2005). Genetic and environmental determinants of bitter perception and sweet preferences. *Pediatrics* **115,** e216–e222.

Meyerhof, W. (2005). Elucidation of mammalian bitter taste. *Rev Physiol Biochem Pharmacol* **154,** 37–72.

Meyerhof, W., Batram, C., Kuhn, C., Brockhoff, A., Chudoba, E., Bufe, B., Appendino, G. & Behrens, M. (2010). The molecular receptive ranges of human TAS2R bitter taste receptors. *Chem Senses* **35,** 157–170.

Miller, I. J., Jr. (1986). Variation in human fungiform taste bud densities among regions and subjects. *Anat Rec* **216,** 474–482.

Nelson, G., Chandrashekar, J., Hoon, M. A., Feng, L., Zhao, G., Ryba, N. J. & Zuker, C. S. (2002). An amino-acid taste receptor. *Nature* **416,** 199–202.

Nelson, G., Hoon, M. A., Chandrashekar, J., Zhang, Y., Ryba, N. J. & Zuker, C. S. (2001). Mammalian sweet taste receptors. *Cell* **106,** 381–390.

Ozeck, M., Brust, P., Xu, H. & Servant, G. (2004). Receptors for bitter, sweet and umami taste couple to inhibitory G protein signaling pathways. *Eur J Pharmacol* **489,** 139–149.

Perez, C. A., Huang, L., Rong, M., Kozak, J. A., Preuss, A. K., Zhang, H., Max, M. & Margolskee, R. F. (2002). A transient receptor potential channel expressed in taste receptor cells. *Nat Neurosci* **5,** 1169–1176.

Pidamale, R., Sowmya, B., Thomas, A. & Jose, T. (2012). Genetic sensitivity to bitter taste of 6-n Propylthiouracil: A useful diagnostic aid to detect early childhood caries in pre-school children. *Indian J Hum Genet* **18,** 101–105.

Pin, J. P., Galvez, T. & Prezeau, L. (2003). Evolution, structure, and activation mechanism of family 3/C G-protein-coupled receptors. *Pharmacol Ther* **98,** 325–354.

Prescott, J., Soo, J., Campbell, H. & Roberts, C. (2004). Responses of PROP taster groups to variations in sensory qualities within foods and beverages. *Physiol Behav* **82,** 459–469.

Prescott, J. & Swain-Campbell, N. (2000). Responses to repeated oral irritation by capsaicin, cinnamaldehyde and ethanol in PROP tasters and non-tasters. *Chem Senses* **25,** 239–246.

Pronin, A. N., Tang, H., Connor, J. & Keung, W. (2004). Identification of ligands for two human bitter T2R receptors. *Chem Senses* **29,** 583–593.

Pronin, A. N., Xu, H., Tang, H., Zhang, L., Li, Q. & Li, X. (2007). Specific alleles of bitter receptor genes influence human sensitivity to the bitterness of aloin and saccharin. *Curr Biol* **17,** 1403–1408.

Pulkkinen, V., Manson, M. L., Safholm, J., Adner, M. & Dahlen, S. E. (2012). The bitter taste receptor (TAS2R) agonists denatonium and chloroquine display distinct patterns of relaxation of the guinea pig trachea. *Am J Physiol Lung Cell Mol Physiol* **303,** L956–L966.

Pydi, S. P., Singh, N., Upadhyaya, J., Bhullar, R. P. & Chelikani, P. (2013). The third intracellular loop plays a critical role in bitter taste receptor activation. *Biochim Biophys Acta* **1838,** 231–236.

Pydi, S. P., Sobothiewicz, T., Billakanti, R., Bhullar, R. P., Loewen, M. C. & Chelikani, P. (2014). Amino acid derivatives as bitter taste receptor (T2R) blockers. *J Biol Chem* **289**, 25054–66.

Pydi, S. P., Jaggupilli, A., Nelson, K. M., Abrams, S. R., Bhullar, R. P., Loewen, M. C. & Chelikani, P. (2015). Abscisic Acid Acts as a Blocker of the Bitter Taste G Protein-Coupled Receptor T2R4. *Biochemistry* **54**, 2622–31.

Reed, D. R., Nanthakumar, E., North, M., Bell, C., Bartoshuk, L. M. & Price, R. A. (1999). Localization of a gene for bitter-taste perception to human chromosome 5p15. *Am J Hum Genet* **64,** 1478–1480.

Robinett, K. S., Deshpande, D. A., Malone, M. M. & Liggett, S. B. (2011). Agonist-promoted homologous desensitization of human airway smooth muscle bitter taste receptors. *Am J Respir Cell Mol Biol* **45,** 1069–1074.

Rozengurt, E. (2006). Taste receptors in the gastrointestinal tract. I. Bitter taste receptors and alpha-gustducin in the mammalian gut. *Am J Physiol Gastrointest Liver Physiol* **291,** G171–G177.

Rupesh, S. & Nayak, U. A. (2006). Genetic sensitivity to the bitter taste of 6-n propylthiouracil: a new risk determinant for dental caries in children. *J Indian Soc Pedod Prev Dent* **24,** 63–68.

Sainz, E., Cavenagh, M. M., Gutierrez, J., Battey, J. F., Northup, J. K. & Sullivan, S. L. (2007). Functional characterization of human bitter taste receptors. *Biochem J* **403,** 537–543.

Sainz, E., Korley, J. N., Battey, J. F. & Sullivan, S. L. (2001). Identification of a novel member of the T1R family of putative taste receptors. *J Neurochem* **77,** 896–903.

Sakurai, T., Misaka, T., Ishiguro, M., Masuda, K., Sugawara, T., Ito, K., Kobayashi, T., Matsuo, S., Ishimaru, Y., Asakura, T. & Abe, K. (2010). Characterization of the beta-D-glucopyranoside binding site of the human bitter taste receptor hTAS2R16. *J Biol Chem* **285,** 28373–28378.

Shah, A. S., Ben-Shahar, Y., Moninger, T. O., Kline, J. N. & Welsh, M. J. (2009). Motile cilia of human airway epithelia are chemosensory. *Science* **325,** 1131–1134.

Sharman, J. L., Benson, H. E., Pawson, A. J., Lukito, V., Mpamhanga, C. P., Bombail, V., Davenport, A. P., Peters, J. A., Spedding, M., Harmar, A. J. & Nc, I. (2013). IUPHAR-DB: updated database content and new features. *Nucleic Acids Res* **41,** D1083–D1088.

Shepard, T. H. (1961). Phenylthiocarbamide Non-Tasting among Congenital Athyrotic Cretins: Further Studies in an Attempt to Explain the Increased Incidence. *J Clin Invest* **40,** 1751–1757.

Singh, N., Pydi, S. P., Upadhyaya, J. & Chelikani, P. (2011). Structural basis of activation of bitter taste receptor T2R1 and comparison with Class A G-protein-coupled receptors (GPCRs). *J Biol Chem* **286,** 36032–36041.

Slack, J. P., Brockhoff, A., Batram, C., Menzel, S., Sonnabend, C., Born, S., Galindo, M. M., Kohl, S., Thalmann, S., Ostopovici-Halip, L., Simons, C. T., Ungureanu, I., Duineveld, K., Bologa, C. G., Behrens, M., Furrer, S., Oprea, T. I. & Meyerhof, W. (2010). Modulation of bitter taste perception by a small molecule hTAS2R antagonist. *Curr Biol* **20,** 1104–1109.

Spielman, A. I. (1998). Gustducin and its role in taste. *J Dent Res* **77,** 539–544.

Spielman, A. I., Nagai, H., Sunavala, G., Dasso, M., Breer, H., Boekhoff, I., Huque, T., Whitney, G. & Brand, J. G. (1996). Rapid kinetics of second messenger production in bitter taste. *Am J Physiol* **270,** C926–C931.

Tan, J., Abrol, R., Trzaskowski, B. & Goddard, W. A., 3rd (2012). 3D Structure Prediction of TAS2R38 Bitter Receptors Bound to Agonists Phenylthiocarbamide (PTC) and 6-n-Propylthiouracil (PROP). *J Chem Inf Model* **52,** 1875–1885.

Tepper, B. J. & Nurse, R. J. (1997). Fat perception is related to PROP taster status. *Physiol Behav* **61,** 949–954.

Timpson, N. J., Christensen, M., Lawlor, D. A., Gaunt, T. R., Day, I. N., Ebrahim, S. & Davey Smith, G. (2005). TAS2R38 (phenylthiocarbamide) haplotypes, coronary heart disease traits, and eating behavior in the British Women's Heart and Health Study. *Am J Clin Nutr* **81,** 1005–1011.

Tizzano, M., Cristofoletti, M., Sbarbati, A. & Finger, T. E. (2011). Expression of taste receptors in solitary chemosensory cells of rodent airways. *BMC Pulm Med* **11,** 3.

Toda, Y., Nakagita, T., Hayakawa, T., Okada, S., Narukawa, M., Imai, H., Ishimaru, Y. & Misaka, T. (2013). Two distinct determinants of ligand specificity in T1R1/T1R3 (the umami taste receptor). *J Biol Chem* **288**(52), 36863–36877.

Ugawa, S., Minami, Y., Guo, W., Saishin, Y., Takatsuji, K., Yamamoto, T., Tohyama, M. & Shimada, S. (1998). Receptor that leaves a sour taste in the mouth. *Nature* **395,** 555–556.

Upadhyaya, J., Pydi, S. P., Singh, N., Aluko, R. E. & Chelikani, P. (2010). Bitter taste receptor T2R1 is activated by dipeptides and tripeptides. *Biochem Biophys Res Commun* **398,** 331–335.

Upadhyaya, J. D., Singh, N., Sikarawar, A. S., Chakraborty, R., Pydi, S. P, Bhullar, R. P., Dakshinamurti, S. & Chelikani, P. (2014). Dextromethorphan mediated bitter taste receptor activation in the pulmonary circuit causes vasoconstriction. *PLoS One* **9,** e110373.

Upadhyaya, J., Singh, N., Bhullar, R. P. & Chelikani, P. (2015). The structure-function role of C-terminus in human bitter taste receptor T2R4 signaling. *Biochim Biophys Acta* **1848,** 1502–1508.

Upadhyaya, J. D., Chakraborty, R., Shaik, F. A, Jaggupilli, A., Bhullar, R. P. & Chelikani, P. (2016). The Pharmacochaperone Activity of Quinine on Bitter Taste Receptors. *PLoS One* **11,** e0156347.

Verma, P., Shetty, V. & Hegde, A. M. (2006). Propylthiouracil (PROP)--a tool to determine taster status in relation to caries experience, streptococcus mutans levels and dietary preferences in children. *J Clin Pediatr Dent* **31,** 113–117.

Wang, J. C., Hinrichs, A. L., Bertelsen, S., Stock, H., Budde, J. P., Dick, D. M., Bucholz, K. K., Rice, J., Saccone, N., Edenberg, H. J., Hesselbrock, V., Kuperman, S., Schuckit, M. A., Bierut, L. J. & Goate, A. M. (2007). Functional variants in TAS2R38 and TAS2R16 influence alcohol consumption in high-risk families of African-American origin. *Alcohol Clin Exp Res* **31,** 209–215.

Wong, G. T., Gannon, K. S. & Margolskee, R. F. (1996). Transduction of bitter and sweet taste by gustducin. *Nature* **381,** 796–800.

Wu, J. & Aluko, R. E. (2007). Quantitative structure-activity relationship study of bitter di- and tri-peptides including relationship with angiotensin I-converting enzyme inhibitory activity. *J Pept Sci* **13,** 63–69.

Wu, S. V., Rozengurt, N., Yang, M., Young, S. H., Sinnett-Smith, J. & Rozengurt, E. (2002). Expression of bitter taste receptors of the T2R family in the gastrointestinal tract and enteroendocrine STC-1 cells. *Proc Natl Acad Sci U S A* **99,** 2392–2397.

Xu, H., Staszewski, L., Tang, H., Adler, E., Zoller, M. & Li, X. (2004). Different functional roles of T1R subunits in the heteromeric taste receptors. *Proc Natl Acad Sci U S A* **101,** 14258–14263.

Zhao, G. Q., Zhang, Y., Hoon, M. A., Chandrashekar, J., Erlenbach, I., Ryba, N. J. & Zuker, C. S. (2003). The receptors for mammalian sweet and umami taste. *Cell* **115,** 255–266.

2 Physiological Aspects of Bitterness

Maik Behrens and Wolfgang Meyerhof

2.1 INTRODUCTION

The sense of taste guides the ingestive behavior of animals towards the consumption of energy-rich food free of potentially harmful substances. The caloric content of food is detected by sweet and umami sensors, the saltiness indicates the presence of important electrolytes, whereas sour and bitter detection facilitates cautionary responses to avoid the involuntary uptake of spoiled, unripe or toxic food items. Among these five basic taste qualities bitter taste is most complex both in terms of the number of bitter substances and the number of bitter taste receptors.

As the bitterness perception in humans, the genetics of bitter taste receptors, and the chemistry of bitterness are discussed in other chapters of this book, the present article will start with a description of the anatomical structures underlying bitter taste perception. We then will briefly elaborate on the signal transduction cascade found in bitter taste receptor cells. A larger section is devoted to the gustatory expression pattern of bitter taste receptor genes in the oral cavity and the question whether mammals might be able to discriminate between different bitter stimuli. The last section then summarizes the literature on the detection of bitter taste receptor gene expression in non-gustatory systems and the potential functional implications that these extra-oral TAS2Rs may have.

2.2 ANATOMY

The detection of taste stimuli in mammals is facilitated by sensory cells of epithelial origin located in the oral cavity. These cells reside in specialized morphological structures on the tongue surface, the nasoincisor ducts, the soft palate, epiglottis, larynx, and pharynx. Three types of taste papillae on the tongue surface housing

Bitterness: Perception, Chemistry and Food Processing, First Edition. Edited by Michel Aliani & Michael N. A. Eskin.

these sensory cells can be distinguished. The morphologically simplest form of gustatory papillae, the fungiform papillae, are distributed over the apical two-thirds of the tongue surface. The foliate and vallate papillae are more complex structures residing on the posterior tongue symmetrically on both sides and the center of the tongue, respectively. In contrast to the fungiform papillae the latter types of gustatory papillae are directly connected to the flow of saliva secreted from minor salivary glands into the trench-like structures of these papillae (Miller, 1995).

Within these gustatory structures taste cells are combined to groups of ~100 cells, the taste buds, which represent the functional units for taste detection in the oral cavity (Miller, 1995). Within the taste bud distinct cell types have been described based on their ultrastructural and cytological features. The type I cells, which are also known as "dark cells", are elongated cells with an electron-dense cytoplasm. It is believed that type I cells act as supporting cells fulfilling glia-like functions in the taste bud. Type II cells or "light cells" also exhibit an elongated shape, however, their cytoplasm appears electron-lucent. The type II cells act as taste receptor cells. The type III cells share structural similarities with type II cells, however, their electron-microscopic staining properties place them between "light" and "dark" cells (hence they are also known as "intermediate" cells). In contrast to type I and type II cells, type III cells, also known as presynaptic cells, form synapses with afferent nerve fibers (DeFazio *et al.*, 2006). In addition to these cell types basal cells (also known as type IV cells) of spherical shape are located at the basis of the taste bud. Basal cells are thought to represent stem cells, which have the capacity to replace the short-lived cell populations in the taste bud (Beidler and Smallman, 1965). While the classification of taste cells into the above mentioned cell types is still used to date, molecular markers or physiological properties have largely replaced ultrastructural and cytological features for categorization purposes. Briefly, type I cells are frequently classified by the expression of marker proteins associated with neurotransmitter clearance such as GLAST (Glutamate-Aspartate-Transporter) or NTPDase2 (an ecto-ATPase), type II cells by the expression of taste signaling molecules like taste 1 and taste 2 receptors, heterotrimeric G protein constituents such as α-gustducin, phospholipase Cβ2, and TRPM5 and type III cells by enzymes required for neurotransmitter synthesis (e.g., the glutamate decarboxylase GAD67 or aromatic L-amino acid decarboxylase AADC) and components of synaptic transmission (SNAP25) (for a recent review see (Chaudhari and Roper, 2010)).

Each taste bud possesses a single apical pore region, which is oriented towards the oral cavity and contains the microvilli of the taste receptor cells. This pore region represents the contact point between the tastants present in the food pulp and the taste receptor molecules located on the microvilli. As taste bud cells do not represent neurons, but secondary sensory cells of epithelial origin, the taste information is transmitted to the brain via afferent nerve fibers entering the taste bud from the basal side.

Branches of three cranial nerves contact different gustatory areas within the oral cavity to collect and transmit taste information. The fungiform papillae of the apical tongue and the soft palate receive afferent input from the VII. cranial nerve via the chorda tympani and the greater superficial petrosal branch, respectively. The foliate and vallate papillae of the posterior tongue are connected with afferents of the glossopharyngeal nerve (IX. cranial nerve), whereas the X. cranial nerve innervates epiglottal and laryngeal taste buds. The taste information from the various gustatory areas is first transmitted into the nucleus tractus solitarius (NTS) of the brain stem. Within the NTS taste information is connected with the generation of basic somatic reflexes such as salivation, swallowing or gag reflexes. From the NTS taste information is transmitted to the parvicellular portion of the ventroposteromedial nucleus of thalamus (gustatory thalamus). Next, the gustatory information is received by the insular/opercular cortex (primary gustatory cortex) and further, the orbitofrontal cortex (secondary gustatory cortex).

2.3 TASTE SIGNAL TRANSDUCTION

The bitter taste receptor cells harbor in addition to the bitter taste receptors the components required for subsequent signal amplification (Fig. 2.1). These include the subunits of heteromeric G proteins. The first identified and most prominent molecule involved in taste signal transduction is the G protein subunit α-gustducin (McLaughlin *et al.*, 1992). Although its critical involvement in the transduction of G protein-coupled taste receptor mediated signals, including bitter signals, has been shown already many years ago by experiments with α-gustducin knock-out mice (Wong *et al.*, 1996), other Gα-subunits such as α-transducin (Ruiz-Avila *et al.*, 1995), Gαi- (Asano-Miyoshi *et al.*, 2000; Kusakabe *et al.*, 2000), Gαs- (Kusakabe *et al.*, 2000) and Gαq-proteins (Kusakabe *et al.*, 1998; Tizzano *et al.*, 2008) have been identified in taste tissues and participate in bitter signaling. The other dominant components of the heterotrimeric complex are Gβ3 and Gγ13 (Huang *et al.*, 1999; Rossler *et al.*, 2000). Activation of the heterotrimeric G protein complex by tastants binding to a bitter taste receptor facilitates the exchange of α-gustducin-bound GDP for GTP and the subsequent dissociation into α-gustducin-GTP and Gβ3γ13. Gβ3γ13 in turn activates phospholipase Cβ2 (Zhang *et al.*, 2003) resulting in the generation of inositol-1,4,5-trisphosphate (Hwang *et al.*, 1990). This second messenger molecule leads to the activation of the IP_3-receptor type 3 (Clapp *et al.*, 2001; Hisatsune *et al.*, 2007; Miyoshi *et al.*, 2001) residing in the membrane of the endoplasmatic reticulum, which in turn causes a rise in intracellular calcium ion levels (Akabas *et al.*, 1988). The calcium ions bind to the transient receptor potential channel M5 (Perez *et al.*, 2002; Zhang *et al.*, 2003) facilitating the influx of cations depolarizing the cells. The cellular depolarization finally triggers the release of the neurotransmitter ATP

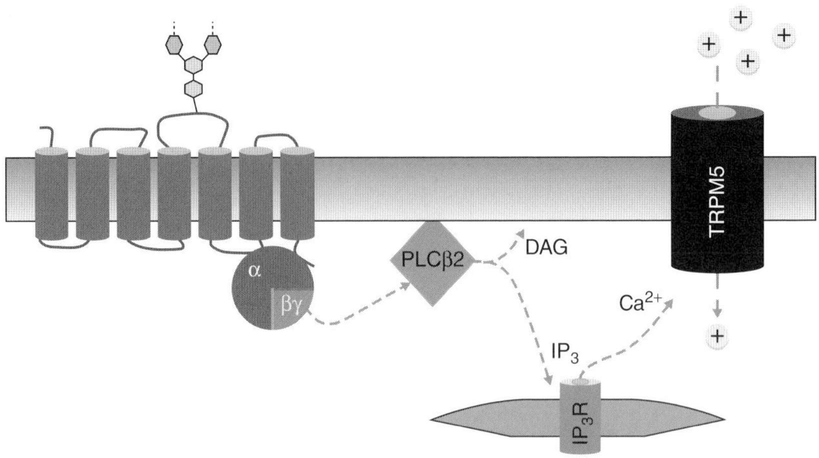

Figure 2.1 *Signal transduction in bitter taste receptor cells.* The bitter taste receptor (here shown as a seven transmembrane domain glycoprotein) transmits its activation via heterotrimeric G proteins (consisting of α-, β-, and γ-subunits). After the dissociation of the G protein complex, the βγ-subunits induce the turnover of phosphatidylinositol-4,5-bisphosphate leading to the generation of diacylglycerol (DAG) and inositol-1,4,5-trisphosphate (IP_3) by the activity of phospholipase Cβ2. The IP_3 triggers the release of calcium ions (Ca^{2+}) from intracellular stores via activation of the type III IP_3-receptor (IP_3R). Increasing intracellular calcium ion levels in turn result in the opening of the cation channel TRPM5 leading to depolarization and subsequent release of the neurotransmitter ATP (not shown).

through the ion channel CALHM1 (Taruno *et al.*, 2013), or connexin/pannexin type hemichannels (Huang *et al.*, 2007; Romanov *et al.*, 2007).

2.4 GUSTATORY BITTER TASTE RECEPTOR GENE EXPRESSION

Numerous studies have investigated the expression of the various taste receptor genes in gustatory tissues. Over the years, spearheaded by the laboratory of Charles Zuker, it became clear that for each of the five basic taste qualities specific populations of sensory cells exist that are devoted to detect stimuli of only one taste quality. The population of sweet receptor cells expresses the TAS1R2 and TAS1R3 genes coding for the two subunits of the sweet taste receptor. The umami-sensing cells also express the TAS1R3 gene which is coding for the common subunit of both, sweet and umami receptors, but specifically express the TAS1R1 gene. The population of bitter taste receptor cells was shown to exclusively express TAS2R genes. All of the above mentioned taste receptor genes code for G protein-coupled receptors (GPCRs) which share common intracellular effector molecules occurring in type II taste receptor cells. Hence, type II cells represent the sweet, umami, and bitter taste receptor cells within taste buds.

Recently, it was demonstrated that a subset of cells expressing the α-subunit of the epithelial sodium channel respond selectively to low concentrations of sodium chloride and hence represent the salt taste specific population (Chandrashekar *et al.*, 2010). Although the determination of the exact identity of these cells requires further clarification, it was clearly demonstrated that the population of salt-sensitive cells is distinct from taste bud cells devoted to the other taste qualities.

Also sour taste cells represent a unique cell population within the taste bud characterized by the expression of polycystic-kidney-disease-like ion channel 2 L 1 (PKD2L1), a molecule that has been suspected to be an integral part of the still unidentified sour taste receptor (Huang *et al.*, 2006; Ishimaru *et al.*, 2006; LopezJimenez *et al.*, 2006). It has been demonstrated that type III cells represent the sour responsive cells (Huang *et al.*, 2008, 2009).

Among the taste receptor cells the bitter sensitive cells are special because they express numerous taste receptor molecules. Already at the time-point of the discovery of bitter taste receptor genes, expression studies were performed, resulting in controversial interpretations of the results. Whereas one study concluded that each mouse bitter taste receptor cell expresses nearly the full complement of *Tas2r* genes and hence, a rather uniform population of bitter taste receptor cells exist (Adler *et al.*, 2000), another study found a lesser degree of overlap among Tas2rs suggesting a heterogeneous bitter taste receptor cell population (Matsunami *et al.*, 2000). These controversial observations have very important implications for bitter taste physiology. A uniform population of bitter taste receptor cells that co-expresses all bitter taste receptor genes would preclude the possibility to discriminate among different bitter compounds because the peripheral sensor cells would act as universal bitter sensors being sensitive to all bitter stimuli. In marked contrast, a heterogeneous population of bitter taste receptor cells characterized by cells that express distinct subsets of bitter taste receptor genes would constitute a prerequisite for bitter compound discrimination. Not surprisingly, this important point has been addressed by a number of previous studies. An important contribution to solve this issue has been a functional study from the Roper laboratory (Caicedo and Roper, 2001). Performing calcium-imaging analyses with five prototypical bitter stimuli on lingual slices the authors observed that the majority of bitter responsive cells in rat foliate papillae responded only to one or two of five prototypical bitter stimuli. Whereas few rat bitter taste receptor cells were activated by three and even four of the bitter stimuli, not a single cell responded to all five bitter compounds clearly demonstrating the heterogeneous response properties for this cell population. A similar study performed on individual mouse fungiform taste cells observed also different response patterns among bitter responsive cells, however, the diversity was less pronounced (Yoshida *et al.*, 2009). On the contrary, an elegant series of experiments using transgenic expression of the essential signaling component PLCβ2 under the control three different bitter taste receptor gene promoters resulted in each case in the rescue of the bitter tasting abilities in bitter taste-blind PLCβ2-knock-out mice (Mueller *et al.*, 2005). The fact that the three bitter taste receptor gene promoters were able to drive the expression of PLCβ2 in a sufficient number of bitter taste receptor cells to restore the responsiveness for several prototypical bitter compounds confirmed a significant overlap in Tas2r gene expression. Obviously, the population of bitter taste receptor cells in rodents is, on the one hand sufficiently heterogeneous to allow individual response characteristics of single cells, on the other hand sufficiently homogeneous to have prevented apparent bitter tasting deficits in the mouse models generated by Mueller and colleagues. In fact, another study performed in human gustatory tissue would be compatible with both of the above mentioned experimental outcomes (Behrens *et al.*, 2007). The study monitored the expression pattern of all 25 functional human TAS2R genes in circumvallate papillae by in situ hybridization experiments. It was shown that the 25 *TAS2R* genes are expressed with different frequencies and at apparent different expression levels in bitter taste receptor cells pointing to a heterogenous cell population in human. By double-labeling in situ hybridization the authors demonstrated directly that selected pairs of TAS2Rs can occur in the same cells as well as in separate cells. This observation would be in full agreement with a functionally heterogeneous bitter taste receptor cell population as reported by Caicedo and Roper (Caicedo and Roper, 2001). In this study (Behrens *et al.*, 2007) the authors estimated that on average 4 to 11 *TAS2R* genes are co-expressed in any given bitter taste receptor cells, which would not exclude the experimental result as reported by Mueller *et al.* (Mueller *et al.*, 2005) if coordinate gene regulation is assumed, even though not a single taste receptor cell in the oral cavity may express an identical repertoire of *TAS2R* genes. Because specific antisera for the detection of bitter receptor protein was lacking in the early years of bitter research, most data concerning the expression of bitter taste receptor genes were obtained by analyzing mRNA. This, however, does not allow investigating the subcellular localization of receptor protein in their natural environment, the bitter taste receptor cell. The recent availability of a specific commercial antiserum raised against human TAS2R38 allowed the analysis of this receptor in human circumvallate papillae (Behrens *et al.*, 2012). Somewhat surprisingly, this receptor showed a rather equal distribution along the cellular membrane compartment suggesting no particular enrichment at the apical side of bitter taste receptor cells (Fig. 2.2).

While a heterogeneous population of bitter taste receptor cells is an indispensable prerequisite for a potential discrimination between different bitter stimuli, other criteria have to be fulfilled as well. Firstly, the activation pattern of the bitter taste receptor cells evoked by an individual bitter stimulus has to reach the central nervous system without converging already peripherally. Secondly, the response properties of bitter taste receptor cells should remain somewhat stable despite the rapid turnover rate of type II cells to allow the "specialization" of the corresponding afferent nerve fibers. While at present it is completely unknown whether the latter applies, some data concerning the transmission of bitter information into the brain exist. Indeed, single fiber recordings of rat chorda tympani and glossopharyngeal nerve fibers upon stimulation with various bitter compounds revealed individual

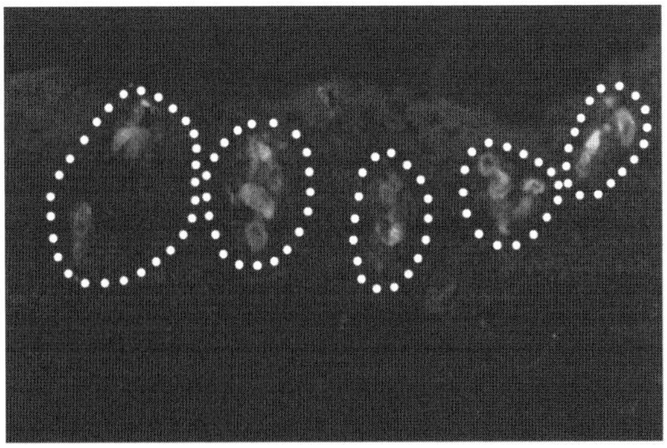

Figure 2.2 *Bitter taste receptor expression in human circumvallate papillae.* Cross-section through a human circumvallate papillae stained with antibodies specific for the human bitter taste receptor TAS2R38. Note that within each taste bud several cells express TAS2R38 (white). The taste buds are circled for easier visibility.

response patterns (Dahl *et al.*, 1997). The differential activation of these peripheral neurons by different oral bitter stimuli seems to be relayed to the brain stem as also NTS neurons respond non-uniformly to oral stimulation with different bitter compounds (Geran and Travers, 2006; Wilson *et al.*, 2012).

Also the next station of gustatory processing in rodents, the parabrachial nucleus (PBN), contains neurons responding to oral stimulation with different bitter substances in a non-uniform fashion (Geran and Travers, 2009). Although the response pattern of thalamic neurons upon gustatory stimulation has not been analyzed using a variety of different bitter stimuli and only few neurons (9%) exhibit unimodal activation by tastants (Verhagen *et al.*, 2003), the presence of a gustotopic map of taste qualities in the primary taste cortex of mice allows to pick up the trace of bitter response patterns again (Chen *et al.*, 2011). Using an in vivo calcium imaging approach to monitor cortical fields receiving gustatory input from thalamic neurons it was demonstrated that separate areas of the primary gustatory cortex are devoted to the processing of sweet, umami, salty, and bitter information. Sequential oral stimulations with the three bitter substances cycloheximide, denatonium, and quinine resulted in largely overlapping, yet somewhat distinct, activation patterns. As similar results were seen upon three repetitive stimulations with always the same bitter substance, these experiments remain inconclusive with respect to a possible discrimination among different bitter compounds.

A conclusive answer to the question of whether discrimination of bitter substances is possible or not must ultimately come from human psychophysical experiments and behavioral animal studies; however, so far, controversial results have been obtained. Whereas one study reported that rats fail to discriminate between the two intensity-matched ionic bitter stimuli denatonium and quinine (Spector and Kopka, 2002) other studies suggested that rodents respond differently to ionic bitter stimuli compared with non-ionic bitter compounds (Brasser *et al.*, 2005; Frank *et al.*, 2004).

2.5 EXTRAGUSTATORY BITTER TASTE RECEPTORS

An interesting development in taste research during the recent years has been the identification of G protein-coupled taste receptor gene expression in a steadily increasing number of non-gustatory tissues. This applies to bitter taste receptors as well (Fig. 2.3), which have been identified in brain (Ansoleaga *et al.*, 2013; Dehkordi *et al.*, 2012; Garcia-Esparcia *et al.*, 2013; Singh *et al.*, 2011), respiratory tract (for a recent review see (Tizzano and Finger, 2013)), heart (Foster *et al.*, 2013), gastrointestinal tract (for a recent review see (Behrens and Meyerhof, 2011)), male reproductive system (Li and Zhou, 2012; Voigt *et al.*, 2012; Xu *et al.*, 2013), bone marrow stromal and vascular smooth muscle cells (Lund *et al.*, 2013).

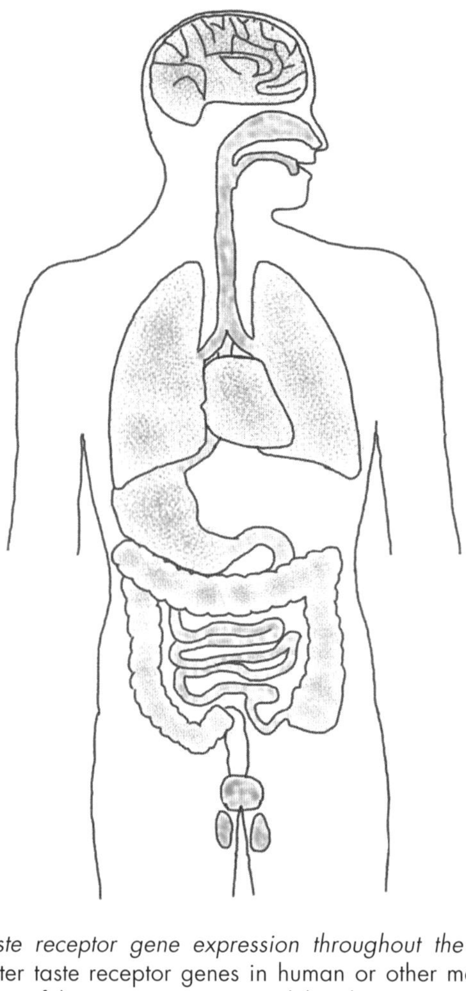

Figure 2.3 *Bitter taste receptor gene expression throughout the human body.* Tissues reported to express bitter taste receptor genes in human or other mammals are labeled in red. Note that large parts of the respiratory tract and the alimentary canal express bitter taste receptor genes. Additional sites of expression are brain, heart, urinary bladder and testes.

While the putative physiological roles of bitter taste receptors in some of the extraoral tissues are currently unknown, for the respiratory and gastrointestinal systems different functions have been proposed. Within the nasal epithelium and the upper respiratory tract bitter taste receptor gene expression is restricted to solitary chemosensory cells (SCC) (Finger *et al.*, 2003). Stimulation of mouse nasal SCCs with bitter substances mediates a depression of the respiratory rate, which may be important to minimize inhalation of potentially harmful xenobiotics. Recently, it was shown that bacterial quorum sensing molecules such as acyl-homoserine lactones are capable to activate nasal SCCs suggesting a role in anti-bacterial defense mechanisms (Sbarbati *et al.*, 2009; Tizzano *et al.*, 2010). This function of nasally expressed bitter taste receptors seems to be conserved as the presence of TAS2R expressing SCCs has recently been confirmed in human sinonasal mucosa (Barham *et al.*, 2013). However, bitter taste receptor expression is not restricted to SCCs also ciliated cells in the upper (Lee *et al.*, 2012) and lower (Shah *et al.*, 2009) respiratory tract have been shown to possess these receptors. Intriguingly, stimulation of these cells with bitter compounds resulted in changes in their ciliary beat frequency indicating a potential involvement in the clearing of airway epithelia (Shah *et al.*, 2009). Lee and colleagues suggested the involvement of human TAS2R38 expressed in ciliated cells as sensor for bacterial quorum sensing molecules. Indeed, it was shown that this receptor is activated by acyl-homoserine lactones and that the activation led to elevated ciliary beat frequency as well as direct antibacterial effects. Strikingly, individuals harboring the non-functional variant of the TAS2R38 gene, which occurs with high frequency in the human population, show a higher incidence of sinonasal bacterial infections (Lee *et al.*, 2012). In agreement with the human data it was shown that also the mouse nasal immune system responds to acyl-homoserine lactones and that this response is dependent on the taste signaling components PLCβ2 and TRPM5, but not α-gustducin (Lee *et al.*, 2014). Somewhat counterintuitively, bitter substances were shown to act on smooth muscle cells of human airways as powerful dilators of the bronchi and may hence represent potential drugs for the treatment of obstructive lung diseases (Deshpande *et al.*, 2010). Further studies confirmed the bronchodilatory effect of bitter compounds although contrasting evidence on the signal transduction mechanism was presented (Zhang *et al.*, 2013). The recent finding that *TAS2R* gene expression is upregulated in children with therapy-resistant asthma (Orsmark-Pietras *et al.*, 2013) suggests that bitter substance may indeed represent putative alternative treatment options for some patients.

Along with members of the TAS1R gene family (e.g., Dyer *et al.*, 2005) bitter taste receptor genes were detected in various tissues of the gastrointestinal system (Wu *et al.*, 2002). With respect to putative functional roles bitter taste receptor may fulfill in the alimentary canal, the absence of critical data has so far prevented that a conclusive picture emerged: 1.) Since many bitter substances are able to evoke profound pharmacological effects in the absence of bitter taste receptors, physiological activities observed in animal experiments per se are not a proof of bitter receptor signaling. 2.) For obvious reasons, most intervention studies were performed in rodent models, however, in contrast to human TAS2Rs the pharmacological properties of rodent bitter taste receptors are largely unknown. 3.) The lack of specific antisera raised against bitter taste receptors has for a long time prevented the identification of gastrointestinal cell types which would have been necessary to extrapolate putative functions from known cellular properties. Thus far in situ evidence for bitter taste expression was only provided for enteroendocrine L-cells using a non-validated antiserum against mouse Tas2r138 (Jeon *et al.*, 2008) as well as for a subset of colonic goblet cells, which were visualized using a Tas2r131 knock-in mouse model (Prandi *et al.*, 2013). Thus far, the evidence concerning bitter taste receptor functions in gastrointestinal tissues point in two somewhat different directions: The modulation of metabolic parameters via the activation of endocrine cells and physiological reactions directed to limit ingestion/exposure to potentially harmful bitter substances consistent with the role of bitter taste receptors in the oral cavity. A considerable number of studies on bitter substance mediated signaling in gastrointestinal tissues have used immortalized human and rodent cell lines of gastrointestinal origins. These, mostly enteroendocrine cell lines were shown by PCR only to express several bitter taste receptors as well as taste-related signaling molecules. Stimulation of such cell lines with bitter compounds resulted in the release of peptide hormones which are expected to directly or indirectly modulate the metabolism of the organism if a similar activation occurs in vivo. One of these cell lines that frequently were used to investigate bitter substance mediated signaling are NCI-H716 cells, a model for human enteroendocrine L-cells. These cells produce the incretin hormone GLP-1 that triggers pancreatic insulin secretion to facilitate reduction of circulating blood glucose. It was shown that these cells indeed respond to stimulation with bitter compounds with elevated intracellular calcium ion levels (Rozengurt *et al.*, 2006) as well as with GLP-1 release (Dotson *et al.*, 2008). It should, however, been noted that these cells are also equipped with the sweet taste receptor and that stimulation with sweet substances similarly triggers GLP-1 secretion (Jang *et al.*, 2007). Nevertheless, Dotson and colleagues reported an association between TAS2R9 genotype and glucose/insulin homeostasis in a human cohort study (Dotson *et al.*, 2008). The majority of studies on the gastrointestinal activity of bitter substances rather point to defense mechanisms against the (over-)ingestion of potentially harmful bitter compounds. One such mechanism would be a delayed emptying of the gastric content, which would reduce the total amount of putatively harmful food being ingested. Whereas in human studies contrasting evidence was revealed (Little *et al.*, 2009; Wicks *et al.*, 2005), Glendinning and colleagues demonstrated that intragastric infusion of the bitter substance denatonium reduced the speed of gastric emptying in rats significantly. Similarly, the administration of a mixture of bitter substances into the stomach of mice induced a delayed reduction in the speed of gastric emptying as well as a reduced food intake, although the short-term effect, an elevated food intake, which was likely caused by ghrelin secretion, appears counterintuitive

(Janssen *et al.*, 2011). Further distal, in the small intestine evidence has been provided that the activation of mouse bitter taste receptors regulate the expression of the xenobiotic transporter ABCB1 via a mechanism that involves the secretion of cholecystokinin (CCK) (Jeon *et al.*, 2011). Finally, bitter signaling in human and rodent colon is believed to elicit ion and fluid secretion into the lumen, a process that would help to flush out harmful colonic content (Kaji *et al.*, 2009).

2.6 OUTLOOK

The identification of bitter taste receptors in non-gustatory tissues has created some confusion about the general physiological role of these proteins. While some of the new information that have arisen from studies in non-gustatory TAS2R gene expression sites are fully compatible with a role in defense mechanisms against the uptake of toxic substances and hence, fit with their predicted gustatory function, other data appear to contradict such "unifying" hypotheses. The fact that bitter taste receptors protect the airways of mammals against inhalation of toxins or bacterial infections is difficult to reconcile with observations indicating that some bitter compounds may cause bronchodilation and therefore, could help, for example, asthma patients to improve breathing. Similarly, why should a bitter receptor on the tongue elicit rejective responses, but if activated further down along the alimentary canal facilitate important metabolic adjustments? What are the ligands for bitter taste receptors expressed in tissues that are not even exposed to the outside world such as brain, heart or testes?

It seems likely that convincing answers to these questions will require a lot of work, will take a long time before becoming available, and will likely associate individual bitter taste receptor genes with specific functions which cannot be generalized to the entire receptor family.

ACKNOWLEDGEMENTS

The authors thank Nadine Voigt for providing Figure 2.2 and the German Ministry of Education and Research (BMBF, #0315669) for financial support.

REFERENCES

Adler, E., Hoon, M. A., Mueller, K. L., Chandrashekar, J., Ryba, N. J. & Zuker, C. S. (2000). A novel family of mammalian taste receptors. *Cell* **100**, 693–702.

Akabas, M. H., Dodd, J. & Al-Awqati, Q. (1988). A bitter substance induces a rise in intracellular calcium in a subpopulation of rat taste cells. *Science* **242**, 1047–1050.

Ansoleaga, B., Garcia-Esparcia, P., Llorens, F., Moreno, J., Aso, E. & Ferrer, I. (2013). Dysregulation of brain olfactory and taste receptors in AD, PSP and CJD, and AD-related model. *Neuroscience* **248C**, 369–382.

Asano-Miyoshi, M., Abe, K. & Emori, Y. (2000). Co-expression of calcium signaling components in vertebrate taste bud cells. *Neurosci Lett* **283**, 61–64.

Barham, H. P., Cooper, S. E., Anderson, C. B., Tizzano, M., Kingdom, T. T., Finger, T. E., Kinnamon, S. C. & Ramakrishnan, V. R. (2013). Solitary chemosensory cells and bitter taste receptor signaling in human sinonasal mucosa. *Int Forum Allergy Rhinol* **3**, 450–457.

Behrens, M., Born, S., Redel, U., Voigt, N., Schuh, V., Raguse, J. D. & Meyerhof, W. (2012). Immunohistochemical detection of TAS2R38 protein in human taste cells. *PLoS One* **7**, e40304.

Behrens, M., Foerster, S., Staehler, F., Raguse, J. D. & Meyerhof, W. (2007). Gustatory expression pattern of the human TAS2R bitter receptor gene family reveals a heterogenous population of bitter responsive taste receptor cells. *J Neurosci* **27**, 12630–12640.

Behrens, M. & Meyerhof, W. (2011). Gustatory and extragustatory functions of mammalian taste receptors. *Physiol Behav* **105**, 4–13.

Beidler, L. M. & Smallman, R. L. (1965). Renewal of cells within taste buds. *J Cell Biol* **27**, 263–272.

Brasser, S. M., Mozhui, K. & Smith, D. V. (2005). Differential covariation in taste responsiveness to bitter stimuli in rats. *Chem Senses* **30**, 793–799.

Caicedo, A. & Roper, S. D. (2001). Taste receptor cells that discriminate between bitter stimuli. *Science* **291**, 1557–1560.

Chandrashekar, J., Kuhn, C., Oka, Y., Yarmolinsky, D. A., Hummler, E., Ryba, N. J. & Zuker, C. S. (2010). The cells and peripheral representation of sodium taste in mice. *Nature* **464**, 297–301.

Chaudhari, N. & Roper, S. D. (2010). The cell biology of taste. *J Cell Biol* **190**, 285–296.

Chen, X., Gabitto, M., Peng, Y., Ryba, N. J. & Zuker, C. S. (2011). A gustotopic map of taste qualities in the mammalian brain. *Science* **333**, 1262–1266.

Clapp, T. R., Stone, L. M., Margolskee, R. F. & Kinnamon, S. C. (2001). Immunocytochemical evidence for co-expression of Type III IP3 receptor with signaling components of bitter taste transduction. *BMC Neurosci* **2**, 6.

Dahl, M., Erickson, R. P. & Simon, S. A. (1997). Neural responses to bitter compounds in rats. *Brain Res* **756**, 22–34.

Defazio, R. A., Dvoryanchikov, G., Maruyama, Y., Kim, J. W., Pereira, E., Roper, S. D. & Chaudhari, N. (2006). Separate populations of receptor cells and presynaptic cells in mouse taste buds. *J Neurosci* **26**, 3971–3980.

Dehkordi, O., Rose, J. E., Fatemi, M., Allard, J. S., Balan, K. V., Young, J. K., Fatima, S., Millis, R. M. & Jayam-Trouth, A. (2012). Neuronal expression of bitter taste receptors and downstream signaling molecules in the rat brainstem. *Brain Res* **1475**, 1–10.

Deshpande, D. A., Wang, W. C., Mcilmoyle, E. L., Robinett, K. S., Schillinger, R. M., An, S. S., Sham, J. S. & Liggett, S. B. (2010). Bitter taste receptors on airway smooth muscle bronchodilate by localized calcium signaling and reverse obstruction. *Nat Med* **16**, 1299–1304.

Dotson, C. D., Zhang, L., Xu, H., Shin, Y. K., Vigues, S., Ott, S. H., Elson, A. E., Choi, H. J., Shaw, H., Egan, J. M., Mitchell, B. D., Li, X., Steinle, N. I. & Munger, S. D. (2008). Bitter taste receptors influence glucose homeostasis. *PLoS ONE* **3**, e3974.

Dyer, J., Salmon, K. S., Zibrik, L. & Shirazi-Beechey, S. P. (2005). Expression of sweet taste receptors of the T1R family in the intestinal tract and enteroendocrine cells. *Biochem Soc Trans* **33**, 302–205.

Finger, T. E., Bottger, B., Hansen, A., Anderson, K. T., Alimohammadi, H. & Silver, W. L. (2003). Solitary chemoreceptor cells in the nasal cavity serve as sentinels of respiration. *Proc Natl Acad Sci U S A* **100**, 8981–8986.

Foster, S. R., Porrello, E. R., Purdue, B., Chan, H. W., Voigt, A., Frenzel, S., Hannan, R. D., Moritz, K. M., Simmons, D. G., Molenaar, P., Roura, E., Boehm, U., Meyerhof, W. &

Thomas, W. G. (2013). Expression, regulation and putative nutrient-sensing function of taste GPCRs in the heart. *PLoS One* **8**, e64579.

Frank, M. E., Bouverat, B. P., Mackinnon, B. I. & Hettinger, T. P. (2004). The distinctiveness of ionic and nonionic bitter stimuli. *Physiol Behav* **80**, 421–431.

Garcia-Esparcia, P., Schluter, A., Carmona, M., Moreno, J., Ansoleaga, B., Torrejon-Escribano, B., Gustincich, S., Pujol, A. & Ferrer, I. (2013). Functional genomics reveals dysregulation of cortical olfactory receptors in Parkinson disease: novel putative chemoreceptors in the human brain. *J Neuropathol Exp Neurol* **72**, 524–539.

Geran, L. C. & Travers, S. P. (2006). Single neurons in the nucleus of the solitary tract respond selectively to bitter taste stimuli. *J Neurophysiol* **96**, 2513–2527.

Geran, L. C. & Travers, S. P. (2009). Bitter-responsive gustatory neurons in the rat parabrachial nucleus. *J Neurophysiol* **101**, 1598–1612.

Hisatsune, C., Yasumatsu, K., Takahashi-Iwanaga, H., Ogawa, N., Kuroda, Y., Yoshida, R., Ninomiya, Y. & Mikoshiba, K. (2007). Abnormal taste perception in mice lacking the type 3 inositol 1, 4, 5-trisphosphate receptor. *J Biol Chem* **282**(51), 7225–7231.

Huang, A. L., Chen, X., Hoon, M. A., Chandrashekar, J., Guo, W., Trankner, D., Ryba, N. J. & Zuker, C. S. (2006). The cells and logic for mammalian sour taste detection. *Nature* **442**, 934–938.

Huang, L., Shanker, Y. G., Dubauskaite, J., Zheng, J. Z., Yan, W., Rosenzweig, S., Spielman, A. I., Max, M. & Margolskee, R. F. (1999). Ggamma13 colocalizes with gustducin in taste receptor cells and mediates IP3 responses to bitter denatonium. *Nat Neurosci* **2**, 1055–1062.

Huang, Y. A., Maruyama, Y., Stimac, R. & Roper, S. D. (2008). Presynaptic (Type III) cells in mouse taste buds sense sour (acid) taste. *J Physiol* **586**, 2903–2912.

Huang, Y. J., Maruyama, Y., Dvoryanchikov, G., Pereira, E., Chaudhari, N. & Roper, S. D. (2007). The role of pannexin 1 hemichannels in ATP release and cell-cell communication in mouse taste buds. *Proc Natl Acad Sci U S A* **104**, 6436–6441.

Hwang, P. M., Verma, A., Bredt, D. S. & Snyder, S. H. (1990). Localization of phosphatidylinositol signaling components in rat taste cells: role in bitter taste transduction. *Proc Natl Acad Sci U S A* **87**, 7395–7399.

Ishimaru, Y., Inada, H., Kubota, M., Zhuang, H., Tominaga, M. & Matsunami, H. (2006). Transient receptor potential family members PKD1L3 and PKD2L1 form a candidate sour taste receptor. *Proc Natl Acad Sci U S A* **103**, 12569–12574.

Jang, H. J., Kokrashvili, Z., Theodorakis, M. J., Carlson, O. D., Kim, B. J., Zhou, J., Kim, H. H., Xu, X., Chan, S. L., Juhaszova, M., Bernier, M., Mosinger, B., Margolskee, R. F. & Egan, J. M. (2007). Gut-expressed gustducin and taste receptors regulate secretion of glucagon-like peptide-1. *Proc Natl Acad Sci U S A* **104**(38), 15069–15074.

Janssen, S., Laermans, J., Verhulst, P. J., Thijs, T., Tack, J. & Depoortere, I. (2011). Bitter taste receptors and alpha-gustducin regulate the secretion of ghrelin with functional effects on food intake and gastric emptying. *Proc Natl Acad Sci U S A* **108**, 2094–2099.

Jeon, T. I., Seo, Y. K. & Osborne, T. F. (2011). Gut bitter taste receptor signalling induces ABCB1 through a mechanism involving CCK. *Biochem J* **438**, 33–37.

Jeon, T. I., Zhu, B., Larson, J. L. & Osborne, T. F. (2008). SREBP-2 regulates gut peptide secretion through intestinal bitter taste receptor signaling in mice. *J Clin Invest* **118**, 3693–3700.

Kaji, I., Karaki, S., Fukami, Y., Terasaki, M. & Kuwahara, A. (2009). Secretory effects of a luminal bitter tastant and expressions of bitter taste receptors, T2Rs, in the human and rat large intestine. *Am J Physiol Gastrointest Liver Physiol* **296**, G971–G981.

Kusakabe, Y., Yamaguchi, E., Tanemura, K., Kameyama, K., Chiba, N., Arai, S., Emori, Y. & Abe, K. (1998). Identification of two alpha-subunit species of GTP-binding proteins, Galpha15 and Galphaq, expressed in rat taste buds. *Biochim Biophys Acta* **1403**, 265–272.

Kusakabe, Y., Yasuoka, A., Asano-Miyoshi, M., Iwabuchi, K., Matsumoto, I., Arai, S., Emori, Y. & Abe, K. (2000). Comprehensive study on G protein alpha-subunits in taste bud cells, with special reference to the occurrence of Galphai2 as a major Galpha species. *Chem Senses* **25**, 525–531.

Lee, R. J., Chen, B., Redding, K. M., Margolskee, R. F. & Cohen, N. A. (2014). Mouse nasal epithelial innate immune responses to Pseudomonas aeruginosa quorum-sensing molecules require taste signaling components. *Innate Immun* **20**(6), 606–617.

Lee, R. J., Xiong, G., Kofonow, J. M., Chen, B., Lysenko, A., Jiang, P., Abraham, V., Doghramji, L., Adappa, N. D., Palmer, J. N., Kennedy, D. W., Beauchamp, G. K., Doulias, P. T., Ischiropoulos, H., Kreindler, J. L., Reed, D. R. & Cohen, N. A. (2012). T2R38 taste receptor polymorphisms underlie susceptibility to upper respiratory infection. *J Clin Invest* **122**, 4145–4159.

Li, F. & Zhou, M. (2012). Depletion of bitter taste transduction leads to massive spermatid loss in transgenic mice. *Mol Hum Reprod* **18**, 289–297.

Little, T. J., Gupta, N., Case, R. M., Thompson, D. G. & Mclaughlin, J. T. (2009). Sweetness and bitterness taste of meals per se does not mediate gastric emptying in humans. *Am J Physiol Regul Integr Comp Physiol* **297**, R632–R639.

LopezJimenez, N. D., Cavenagh, M. M., Sainz, E., Cruz-Ithier, M. A., Battey, J. F. & Sullivan, S. L. (2006). Two members of the TRPP family of ion channels, Pkd1l3 and Pkd2l1, are co-expressed in a subset of taste receptor cells. *J Neurochem* **98**, 68–77.

Lund, T. C., Kobs, A. J., Kramer, A., Nyquist, M., Kuroki, M. T., Osborn, J., Lidke, D. S., Low-Nam, S. T., Blazar, B. R. & Tolar, J. (2013). Bone marrow stromal and vascular smooth muscle cells have chemosensory capacity via bitter taste receptor expression. *PLoS One* **8**, e58945.

Matsunami, H., Montmayeur, J. P. & Buck, L. B. (2000). A family of candidate taste receptors in human and mouse. *Nature* **404**, 601–604.

McLaughlin, S. K., Mckinnon, P. J. & Margolskee, R. F. (1992). Gustducin is a taste-cell-specific G protein closely related to the transducins. *Nature* **357**, 563–569.

Miller, I. J., JR. (ed.) (1995). *Anatomy of the peripheral taste system*, New York: Dekker.

Miyoshi, M. A., Abe, K. & Emori, Y. (2001). IP(3) receptor type 3 and PLCbeta2 are co-expressed with taste receptors T1R and T2R in rat taste bud cells. *Chem Senses* **26**, 259–265.

Mueller, K. L., Hoon, M. A., Erlenbach, I., Chandrashekar, J., Zuker, C. S. & Ryba, N. J. (2005). The receptors and coding logic for bitter taste. *Nature* **434**, 225–229.

Orsmark-Pietras, C., James, A., Konradsen, J. R., Nordlund, B., Soderhall, C., Pulkkinen, V., Pedroletti, C., Daham, K., Kupczyk, M., Dahlen, B., Kere, J., Dahlen, S. E., Hedlin, G. & Melen, E. (2013). Transcriptome analysis reveals upregulation of bitter taste receptors in severe asthmatics. *Eur Respir J* **42**, 65–78.

Perez, C. A., Huang, L., Rong, M., Kozak, J. A., Preuss, A. K., Zhang, H., Max, M. & Margolskee, R. F. (2002). A transient receptor potential channel expressed in taste receptor cells. *Nat Neurosci* **5**, 1169–1176.

Prandi, S., Bromke, M., Hubner, S., Voigt, A., Boehm, U., Meyerhof, W. & Behrens, M. (2013). A subset of mouse colonic goblet cells expresses the bitter taste receptor tas2r131. *PLoS One* **8**, e82820.

Romanov, R. A., Rogachevskaja, O. A., Bystrova, M. F., Jiang, P., Margolskee, R. F. & Kolesnikov, S. S. (2007). Afferent neurotransmission mediated by hemichannels in mammalian taste cells. *Embo J* **26**, 657–667.

Rossler, P., Boekhoff, I., Tareilus, E., Beck, S., Breer, H. & Freitag, J. (2000). G protein betagamma complexes in circumvallate taste cells involved in bitter transduction. *Chem Senses* **25**, 413–421.

Rozengurt, N., Wu, S. V., Chen, M. C., Huang, C., Sternini, C. & Rozengurt, E. (2006). Colocalization of the alpha-subunit of gustducin with PYY and GLP-1 in L cells of human colon. *Am J Physiol Gastrointest Liver Physiol* **291**, G792–G802.

Ruiz-Avila, L., Mclaughlin, S. K., Wildman, D., Mckinnon, P. J., Robichon, A., Spickofsky, N. & Margolskee, R. F. (1995). Coupling of bitter receptor to phosphodiesterase through transducin in taste receptor cells. *Nature* **376**, 80–85.

Sbarbati, A., Tizzano, M., Merigo, F., Benati, D., Nicolato, E., Boschi, F., Cecchini, M. P., Scambi, I. & Osculati, F. (2009). Acyl homoserine lactones induce early response in the airway. *Anat Rec (Hoboken)* **292**, 439–448.

Shah, A. S., Ben-Shahar, Y., Moninger, T. O., Kline, J. N. & Welsh, M. J. (2009). Motile cilia of human airway epithelia are chemosensory. *Science* **325**, 1131–1134.

Singh, N., Vrontakis, M., Parkinson, F. & Chelikani, P. (2011). Functional bitter taste receptors are expressed in brain cells. *Biochem Biophys Res Commun* **406**, 146–151.

Spector, A. C. & Kopka, S. L. (2002). Rats fail to discriminate quinine from denatonium: implications for the neural coding of bitter-tasting compounds. *J Neurosci* **22**, 1937–1941.

Taruno, A., Vingtdeux, V., Ohmoto, M., Ma, Z., Dvoryanchikov, G., Li, A., Adrien, L., Zhao, H., Leung, S., Abernethy, M., Koppel, J., Davies, P., Civan, M. M., Chaudhari, N., Matsumoto, I., Hellekant, G., Tordoff, M. G., Marambaud, P. & Foskett, J. K. (2013). CALHM1 ion channel mediates purinergic neurotransmission of sweet, bitter and umami tastes. *Nature* **495**, 223–226.

Tizzano, M., Dvoryanchikov, G., Barrows, J. K., Kim, S., Chaudhari, N. & Finger, T. E. (2008). Expression of Galpha14 in sweet-transducing taste cells of the posterior tongue. *BMC Neurosci* **9**, 110.

Tizzano, M. & Finger, T. E. (2013). Chemosensors in the nose: guardians of the airways. *Physiology (Bethesda)* **28**, 51–60.

Tizzano, M., Gulbransen, B. D., Vandenbeuch, A., Clapp, T. R., Herman, J. P., Sibhatu, H. M., Churchill, M. E., Silver, W. L., Kinnamon, S. C. & Finger, T. E. (2010). Nasal chemosensory cells use bitter taste signaling to detect irritants and bacterial signals. *Proc Natl Acad Sci U S A* **107**, 3210–3215.

Verhagen, J. V., Giza, B. K. & Scott, T. R. (2003). Responses to taste stimulation in the ventroposteromedial nucleus of the thalamus in rats. *J Neurophysiol* **89**, 265–275.

Voigt, A., Hubner, S., Lossow, K., Hermans-Borgmeyer, I., Boehm, U. & Meyerhof, W. (2012). Genetic labeling of Tas1r1 and Tas2r131 taste receptor cells in mice. *Chem Senses* **37**, 897–911.

Wicks, D., Wright, J., Rayment, P. & Spiller, R. (2005). Impact of bitter taste on gastric motility. *Eur J Gastroenterol Hepatol* **17**, 961–965.

Wilson, D. M., Boughter, J. D., Jr. & Lemon, C. H. (2012). Bitter taste stimuli induce differential neural codes in mouse brain. *PLoS One* **7**, e41597.

Wong, G. T., Gannon, K. S. & Margolskee, R. F. (1996). Transduction of bitter and sweet taste by gustducin. *Nature* **381**, 796–800.

Wu, S. V., Rozengurt, N., Yang, M., Young, S. H., Sinnett-Smith, J. & Rozengurt, E. (2002). Expression of bitter taste receptors of the T2R family in the gastrointestinal tract and enteroendocrine STC-1 cells. *Proc Natl Acad Sci U S A* **99**, 2392–2397.

Xu, J., Cao, J., Iguchi, N., Riethmacher, D. & Huang, L. (2013). Functional characterization of bitter-taste receptors expressed in mammalian testis. *Mol Hum Reprod* **19**, 17–28.

Yoshida, R., Miyauchi, A., Yasuo, T., Jyotaki, M., Murata, Y., Yasumatsu, K., Shigemura, N., Yanagawa, Y., Obata, K., Ueno, H., Margolskee, R. F. & Ninomiya, Y. (2009). Discrimination of taste qualities among mouse fungiform taste bud cells. *J Physiol* **587**, 4425–4439.

Zhang, C. H., Lifshitz, L. M., Uy, K. F., Ikebe, M., Fogarty, K. E. & ZHUGE, R. (2013). The cellular and molecular basis of bitter tastant-induced bronchodilation. *PLoS Biol* **11**, e1001501.

Zhang, Y., Hoon, M. A., Chandrashekar, J., Mueller, K. L., Cook, B., Wu, D., Zuker, C. S. & Ryba, N. J. (2003). Coding of sweet, bitter, and umami tastes: different receptor cells sharing similar signaling pathways. *Cell* **112**, 293–301.

3 Bitterness Perception in Humans: An Evolutionary Perspective

Hui Yang and Peng Shi

Humans have five primary taste sensations: sweetness, sourness, saltiness, bitterness and umami (Kinnamon & Cummings, 1992; Lindemann, 1996; Chaudhari *et al.*, 2000). Among them, bitterness perception is regarded as a warning sensation to avoid poison intake, because poisonous compounds existing in nature usually taste bitter to humans (Glendinning, 1994). Bitterness perception induces aversive reactions thus helps prevent the ingestion of potential toxins. In vertebrates, bitterness perception is initiated by bitter tastants binding to a group of G protein-coupled receptors called bitter taste receptors (Tas2Rs). The number of Tas2Rs varies enormously among different species, ranging from no functional gene in dolphin to ~50 in frog (Li & Zhang, 2013). The variability of *Tas2R* gene numbers among species is not well explained, though various hypotheses have been proposed. The major theory hypothesized that vertebrates *Tas2R* gene repertoires were shaped by their food preferences, for example, herbivores tend to have more *Tas2R* genes because there are more potential toxic compounds in plants tissues than in animal tissues (Shi & Zhang, 2006; Li & Zhang, 2013). Meanwhile, recent functional data (Meyerhof *et al.*, 2010) raised the possibility that broader tuning property of each receptor may compensate for the small repertoire of Tas2Rs. Despite the interspecific gene number variation, individuals within a population could exhibit sensitivity discrepancy to certain bitter tastants. For example, it has long been observed that some people taste a chemical compound called phenylthiocarbamide (PTC) as bitter, while others observed no taste (Fox, 1932). In the last decade, knowledge about the genetics and mechanisms of bitter taste perception has been extended rapidly. Here, we review the general features of the evolution of *Tas2R* genes and the advances in their functional studies. We discuss about the factors that may impact on bitter taste ability and the genetical basis of tasteblindness and its origination. With those, we try to provide a more comprehensive understanding of the bitter taste perception from an evolutionary view.

Bitterness: Perception, Chemistry and Food Processing, First Edition. Edited by Michel Aliani & Michael N. A. Eskin.

3.1 BITTER TASTE RECEPTORS - A GROUP OF G PROTEIN-COUPLED RECEPTOR (GPCR) MEMBERS

Bitter taste transduction starts with the interaction between bitter tastants and bitter taste receptors which are expressed by the cells reside in the taste buds on the papillae of the tongue (Hoon *et al.*, 1999). Taste information is then passed through via a series of signal transduction cascades and finally reaches the brain. Therefore, bitter taste receptors act as the first and the most crucial component in bitter signal transduction pathway. In vertebrates, bitter taste receptors are ~300 amino acids in length and encoded by a single-exon coding region of *Tas2R* genes (Adler *et al.*, 2000; Matsunami *et al.*, 2000). They belong to the G protein-coupled receptors (GPCRs), which are characterized by seven transmembrane α-helical regions, a short extracellular N-terminus, and an intracellular C-terminus. Although being structurally like the vomeronasal receptor type 1 (V1R), Tas2Rs share no sequence similarity with V1Rs. Furthermore, the classification of Tas2Rs in GPCR superfamily is still unclear. Some placed them in the Class A GPCRs (Temussi, 2009), whereas others described them as non-Class A members (Singh *et al.*, 2011), either as a putative separate family (Horn *et al.*, 2003; Isberg *et al.*, 2014) or grouped with the frizzled receptors (Fredriksson *et al.*, 2003).

Tas2R genes form a diverse gene family. They share 30–70% overall sequence identity among themselves. Meanwhile, they contain highly conserved motifs in the first, second, third and last transmembrane regions (Adler *et al.*, 2000). Detailed investigation of the role of 15 conserved residues in the transmembrane region by amino acid replacement revealed that replacement of Asn–$66^{2.65}$ and Asn–$24^{1.50}$ led to significant agonist-induced signaling loss (Singh *et al.*, 2011), suggesting the essential roles in receptor activation played by the transmembrane domains. On the other hand, the most divergent segments of Tas2Rs locate in the extracellular regions, where the amino acid residues are more variable. It is presumed to reflect the wide spectrum of Tas2Rs to recognize bitter substances because extracellular regions are supposed to be the potential sites where bitter ligands bind to and interact with the receptors (Adler *et al.*, 2000; Gilbertson *et al.*, 2000).

3.2 *Tas2R* GENE FAMILY - A HIGHLY DIVERSE FAMILY IN VERTEBRATES

Since first identified in year 2000, *Tas2R* gene repertoires have been described in more than 50 vertebrate genomes, including mammals, birds, reptiles, amphibians and fishes (Shi & Zhang, 2006; Dong *et al.*, 2009; Li & Zhang, 2013). Interestingly, the number of *Tas2R* genes varies greatly among these species, with the smallest of zero functional gene in dolphin and the largest of 51 in frog (Fig. 3.1) (Li & Zhang, 2013). Except these two extrema, most mammals investigated have 15–30 intact *Tas2R* genes and non-mammal vertebrates usually possess less than 10 intact *Tas2R* genes. Because bitter taste receptors detect potential toxins, the number of *Tas2R* genes is generally regarded to reflect the importance to recognize bitter substances for each species, which in turn correlates with species-specific diet preferences. Based on current comparative genomics data, the *Tas2R* gene number distribution in vertebrates appears to be positively correlated with the variety of bitter poisons each species may encounter in its living environment (Shi & Zhang, 2006; Li & Zhang, 2013). Herbivores and omnivores tend to have larger *Tas2R* gene family size, probably because their diets contain larger fraction of plants. In contrast, carnivores possess smaller number of *Tas2R* genes, which is consistent with the fact that animal tissues contain less toxic compounds than plant tissues do. These observations support the hypothesis that dietary preferences shaped the diverse *Tas2R* gene family repertoires (Shi & Zhang, 2006; Li & Zhang, 2013).

Although the diet-driven hypothesis can explain *Tas2R* diversity in general, unmatched observations still exist. For instance, birds belong to omnivores, whereas they have an extremely small size of two to three intact *Tas2R* genes (Fig. 3.1). The question then emerges whether bitter taste is not so important in birds or smaller size of *Tas2R* gene family can be compensated by broader tuning property of individual receptor. Functional screening of human Tas2Rs with various natural or synthetic bitter chemicals revealed that bitter taste receptors could be roughly divided into broadly, narrowly and intermediately tuned groups. For example, hTas2R3 and hTas2R5 responded only to single compound while hTas2R14 responded to 33 out of 104 tested chemicals. Other hTas2Rs could be activated by a variety *of* bitter compounds in-between (Meyerhof *et al.*, 2010). Such tuning property raises the possibility that broader tuning of individual receptors might compensate for the small size of *Tas2R* gene repertoires, that is, if birds *Tas2Rs* exhibit wider tuning spectra of each member, it is still possible for them to recognize equal or similar size of bitter compounds as other animals do. Indeed, recent functional assay testing with birds *Tas2Rs* demonstrated the broad tuning property of chicken *Tas2Rs*, supporting the compensation theory (Behrens et al., 2014). This would become a good supplement to the diet-driven hypothesis. When more tuning breadth data are collected in the future, we could expect a better understanding of *Tas2R* function.

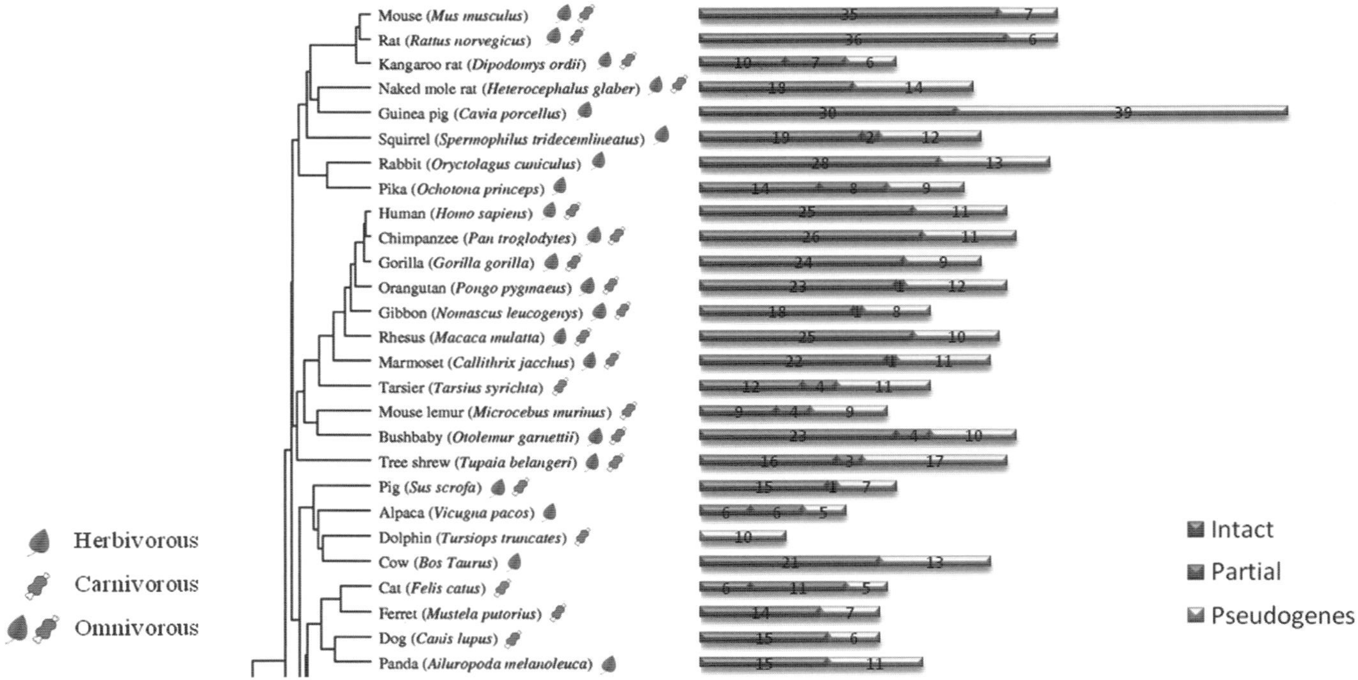

Figure 3.1 The *Tas2R* gene repertoires in vertebrates. Dietary information and the number of *Tas2R* genes are shown after each species name. The scale beneath the phylogeny indicates the divergence time and the one below the bars indicates the total number of *Tas2R* genes (intact, partial and pseudogenes) in each species. (Modified from Li & Zhang, 2013).

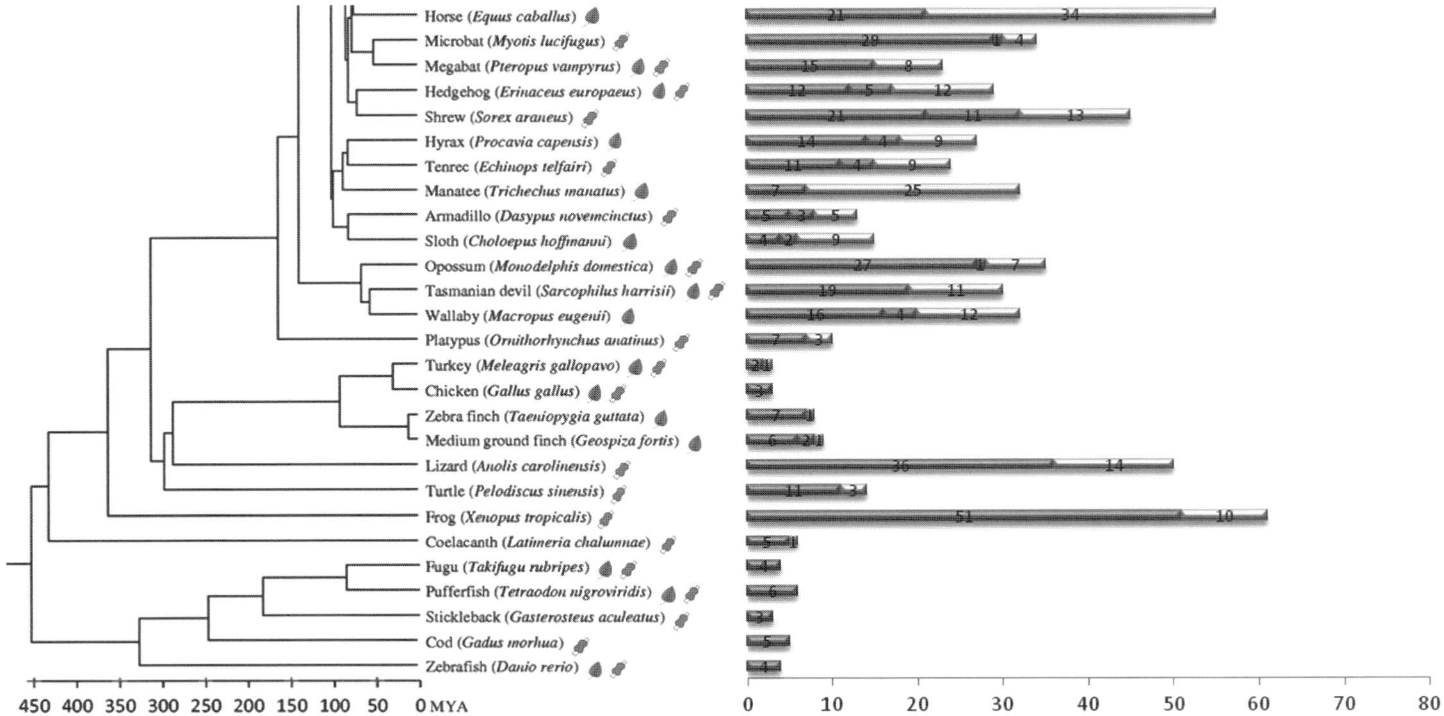

Figure 3.1 (*Continued*)

3.3 THE EVOLUTION OF *Tas2R* GENE FAMILY IN VERTEBRATES

Tas2R repertoires vary enormously among species, it is thus intriguing to find out whether they had common ancestors and how they evolved over time. A study on human and mouse *Tas2R*s revealed that a portion of *Tas2R* genes show one-to-one orthologous pairing, while the others group as species-specific clusters (Shi *et al.*, 2003). Evolutionary analyses found that the one-to-one orthologous genes were subject to strong selective constraints, suggesting that these genes play a basic role in bitter perception, probably detecting bitter compounds that are common to a wide range of animals. On the contrary, the species-specific gene clusters evolved rapidly and exhibit high diversity, indicating their roles in recognizing special bitter compounds each species encountered in its given environment (Shi *et al.*, 2003). Subsequent researches with extensive species also support the common-and-specialized structure in *Tas2R* gene phylogeny (Go, 2006; Shi & Zhang, 2006; Li & Zhang, 2013), agreeing with previous hypothesis.

Recent evolutionary analysis of *Tas2R* genes in 54 vertebrates suggests that the major division of *Tas2R* genes occurred between fishes and tetrapods. While fishes, frog and lizard possess several basal lineages, birds, turtle and all mammalian *Tas2R* genes seem to originate from only one basal lineage (Li & Zhang, 2013). Gene number estimation in ancestral species using the reconciled-tree method inferred that the ancestral size of *Tas2R* gene repertoire was less than 10, no matter in the common ancestor of tetrapods or that of mammals (Li & Zhang, 2013). Comparing to contemporary *Tas2R* gene repertoires, in addition to the considerably high percentage of pseudogenes (Fig. 3.1), it is obvious that birth-and-death process was prevalent throughout vertebrate *Tas2R* gene evolution, resembling the evolutionary pattern in other chemosensory gene families (Nei *et al.*, 2008; Shi & Zhang, 2009). Notably, dramatic gene expansion occurred in the branches leading to frog, lizard and several mammals including guinea pig, microbat and the common ancestor of mouse and rat; meanwhile, massive gene losses were observed in the branches leading to dolphin and manatee (Li & Zhang, 2013). Overall, gene gain and loss predominated *Tas2R* gene evolution and shaped the contemporary gene repertoires. However, the reasons underlying such expansions and contractions are not fully understood. Except for factors mentioned above, such as diet preference and receptor tuning breadth, other factors may also impact the evolution of *Tas2R* gene repertoires. Recently, a study raised the possibility that feeding behavior may influence *Tas2R* repertoire as well. The way dolphins swallow food without chewing is probably related to their *Tas2R* genes loss because they need no taste under such specialized feeding approach (Jiang *et al.*, 2012). Thus, factors impacting *Tas2R* gene evolution seem more complex than previously thought. Extensive efforts are needed in future works to provide a comprehensive understanding of *Tas2R* diversity.

3.4 DIVERSE SELECTIVE FORCES DROVE THE EVOLUTION OF *Tas2R* GENES IN PRIMATES

As the close relatives of humans, nonhuman primates attracted much attention in biological researches, including bitter perception studies. Several groups investigated the patterns of molecular evolution of *Tas2R* genes in a variety of primate species and revealed that these genes were under more relaxed functional constraints in primates than in rodents, despite less gene number in primates (Parry *et al.*, 2004; Wang *et al.*, 2004; Fischer *et al.*, 2005; Go *et al.*, 2005). It was reflected by the higher proportion of pseudogenes in primates (27–41%) than in mice (17%) (Fischer *et al.*, 2005; Go *et al.*, 2005; Li & Zhang, 2013). In addition, repeated lineage-specific pseudogenizations were observed among primate species (Go *et al.*, 2005), consistent with the functional constraints relaxation. Estimates of per site nonsynonymous to synonymous substitution rates (dN/dS) in *Tas2R* genes, which reflect sequence variability driven by selective pressures, exhibited higher values in primate species than in rodents (Wang *et al.*, 2004; Fischer *et al.*, 2005; Go *et al.*, 2005), indicating primates *Tas2R* gene sequences could change more freely than rodents *Tas2R*s do. Even within primates, the mean dN/dS ratio for *Tas2R* genes (0.93) is higher than the average for genes in primates (0.21) (Fischer *et al.*, 2005), showing higher variability of *Tas2R* genes. These evidences further support that looser selective forces acted on primates *Tas2R* genes.

Beside the interspecific variation survey, *Tas2R* gene polymorphism was also investigated in human and chimpanzee populations (Wang *et al.*, 2004; Sugawara *et al.*, 2011; Hayakawa *et al.*, 2012). Analyses with all 25 functional *Tas2R* genes in 22 human individuals from diverse geographic origins show signals of neutral evolution, for example, equal levels of synonymous and nonsynonymous polymorphisms, equal rates of synonymous and nonsynonymous substitutions irrespective of functional domains division. Moreover, segregation of non-functional alleles in populations and fixation of pseudogenes in the species were observed. These all suggest that human *Tas2R* genes lacked selective constraints during their evolution (Wang *et al.*, 2004). Later survey of the sequence variations in all 28 putative functional *Tas2R* genes in 46 western chimpanzees showed high nucleotide diversity as well (Sugawara *et al.*, 2011). In general, nucleotide diversity per site (π) is 0.8×10^{-3} and Watterson's θ per site is 0.6×10^{-3} in chimpanzees (Sugawara *et al.*, 2011), alike those of 1.2×10^{-3} and 1.1×10^{-3} in humans, respectively (Wang *et al.*, 2004), suggesting similar selective constraints relaxation occurred in chimpanzee evolution. However, most western chimpanzees had two or three more functional genes than humans, though the number of functional genes varied among chimpanzee individuals (Sugawara *et al.*, 2011), indicating relatively weaker relaxation acted on chimpanzees than on humans. Combined all above data, it is most likely that the functional relaxation operated in two steps: it first started early in the ancestry of humans and chimpanzees and then strengthened in the human lineage along recently (Wang *et al.*, 2004). The most direct explanation for this is that humans and chimpanzees reduced their needs for bitter taste because of the diet change. Consistently, human diet changed about 2 million years ago (MYA) with increase of meat and decrease of plants, which reduced the number of bitter poisons humans might intake. On the other hand, meat accounts for 2–13% of chimpanzee diet but is never found in other great apes' diets, which also reduces the potential of toxins intake in chimpanzees (Wang *et al.*, 2004). This explains the common selective pressure relaxation in both humans and chimpanzees. Specifically for humans, a computer simulation estimated the rate of pseudogenization in human *Tas2R* genes and deduced that the human-specific fixation of pseudogenes occurred 0.75 MY ago, which is in accordance with the time when controlled use of fire was started (~0.8 MYA). Because cooking significantly detoxifies poisonous food, it probably triggered the reinforced relaxation on human *Tas2R* genes and led to current observations (Wang *et al.*, 2004).

Relaxation of selective constraints acts as a double-edged sword. Lesser constraints enable more nonsynonymous substitutions in *Tas2R* genes to accumulate, which in turn could diminish or lose their original function. On the other hand, diversified amino acid sequences increase the potential to recognize novel bitter ligands, extending the ability to taste a broader range of bitter substances (Sugawara *et al.*, 2011). The observed high level of sequence diversity of *Tas2R* genes in human and chimpanzee populations therefore might lead to correspondingly high polymorphisms in bitter taste perception. Indeed, differential bitter taste sensations among individuals have long been observed and reported (Fox, 1932; Fischer *et al.*, 1961; Glanville & Kaplan, 1965), and associations between *Tas2R* variants and bitter sensations have been identified (Hayes *et al.*, 2011). As a result, the taster-and-nontaster phenotypes were connected with the *Tas2R* molecular genotypes, which enables us to further dissect the mechanism of bitter taste perception at the molecular level.

Although the relaxation was observed in general among *Tas2R* genes in primates, refined inspections on certain gene member discovered various forces played a role during *Tas2R* evolution. An examination of selective signatures on *Tas2R38* gene among 40 primate species uncovered relatively high nonsynonymous substitution rate (ω=dN/dS=0.6) but significantly lower than expected under neutral evolution, suggesting that purifying selection was the major constraint shaping the structure of this gene (Wooding, 2011). Furthermore, investigations in African populations showed high level of rare nonsynonymous variants in *Tas2R38* gene, which were probably arisen by recent diversifying selection (Campbell *et al.*, 2012), and positive selection in *Tas2R16* gene, indicating local adaptation (Campbell *et al.*, 2013). These results combined to suggest that *Tas2R* genes have undergone complex selective forces during primates evolution. Subsequent analyses with more *Tas2R* gene members in larger sample size would scrutinize the various selective pressures and their roles in *Tas2R* gene evolution.

3.5 GENETICAL BASIS OF TASTEBLINDNESS – HUMAN PTC PERCEPTION AS AN EXAMPLE

The discovery of individual bitter taste perceptive variations in human could be traced back to as early as year 1931 in a laboratory incident. When Dr. A. L. Fox was pouring phenylthiocarbamide (PTC, a fine crystalline powder) into a bottle he accidentally released some in the air. A colleague nearby complained about the bitter taste of the dust, but Dr. Fox could taste nothing. From this starting point, Dr. Fox investigated a large number of people and found that PTC taste sensitivity variance did exist among individuals regardless of age, gender or race. Most people fall into one of the two categories: tasters who taste PTC as extensive bitter even at very low concentrations, and nontasters who observe no taste of the compound unless at extremely high concentrations (Fox, 1932). The finding caught much attention in the academic community. In the following decade, thousands of samples were tested by several groups and the nontaster frequency was estimated ranging from 13% to 63% with an average of ~50% (reviewed in Wooding, 2006). In addition, intensive efforts were made, trying to explain the phenomenon and find out the mechanism underlying the distinct sensitivities. However, little progress was made on this aspect until the identification of bitter taste receptors (Hoon *et al.*, 1999; Adler *et al.*, 2000; Chandrashekar *et al.*, 2000). The discovery of *Tas2R* gene family provided targets to study the genetics of PTC sensation. Association studies conducted by two groups separately revealed that the variation of *Tas2R38* gene accounts for 50–80% PTC perceptive variance (Drayna *et al.*, 2003; Kim *et al.*, 2003). In addition, both *in vitro* functional assay (Bufe *et al.*, 2005) and 3D structural modeling (Tan *et al.*, 2012) confirmed the binding of PTC and activation of Tas2R38. Consistently, two major haplotypes are responsible for the taster and nontaster phenotypes, respectively. Notably, the frequencies of these two haplotypes in human populations fit well with the frequencies estimated from phenotype data (Guo & Reed, 2001; Wooding *et al.*, 2004). Thus, the PTC tasteblindness seems to be well explained by the nontaster haplotype of *Tas2R38* gene.

Close comparison of the major PTC taster and nontaster haplotypes showed only three amino acids differences between each other (Drayna *et al.*, 2003; Kim *et al.*, 2003; Wooding *et al.*, 2004). Particularly, there are no premature stop codons or frameshifts to disrupt the ORF in the nontaster allele. In addition, the responses to PTC of those haplotypes intermediate to the taster and nontaster haplotypes were attenuated but not completely eliminated (Bufe *et al.*, 2005), suggesting that the nontaster allele is not a pseudogene. Further, a study found divergent taste responses to the fruits of *Antidesma bunius* and what fascinated people was that all subjects who tasted antidesma berries as bitter were PTC nontasters, whereas no antidesma responders were found in all PTC tasters (Henkin & Gillis, 1977). A possibility is therefore emerging based on these evidences that the PTC nontaster allele is not null and may respond to some compounds other than PTC (Wooding, 2006). Recent screening of all human Tas2Rs with 104 natural and synthetic bitter chemicals confirmed that hTas2R38 responds to a series of agonists (Meyerhof *et al.*, 2010). However, the corresponding ligands repertoires between taster and nontaster haplotypes remain largely unknown. As functional data accumulated, it would be answered in the future.

3.6 PTC TASTEBLINDNESS IN HUMANS AND CHIMPANZEES - SHARED PHENOTYPE RESULTED FROM UNSHARED GENOTYPES

Interestingly, behavioral test performed in chimpanzees showed PTC taster and nontaster variations as well, and the proportions are similar to those in human populations, respectively (Fisher *et al.*, 1939). It is curious to find out whether the perceptive variances originated before human and chimpanzee split. With the identification of Tas2R38 as the major determinant of PTC perception, it is possible to attack this problem by examining the evolutionary pattern of this molecule.

The patterns of *Tas2R38* DNA sequence variation have been surveyed in several geographically diverse human populations from Africa, Asia, Europe and North America (Wooding *et al.*, 2004; Campbell *et al.*, 2012). The results revealed that nucleotide diversity of *Tas2R38* gene is strikingly higher than expected while levels of differentiation are lower than average for humans, indicating balancing selection has maintained the common haplotype variation in human populations (Wooding *et al.*, 2004). Additionally, novel rare nonsynonymous polymorphisms recently arisen only in Africans were identified, suggesting recent selective pressures also shaped the unusually high level of rare nonsynonymous variants in Africans. These data indicate that human *Tas2R38* gene underwent complex evolution that while ancient balancing selection maintained common haplotype variation across global populations, recent selection raised the frequencies of rare variants in Africans (Campbell *et al.*, 2012).

Tas2R38 gene sequence variations were also investigated in various non-human primates. A study with 40 species representing all major primate taxa revealed extensive variation at *Tas2R38* locus (Wooding, 2011), indicating high level of polymorphism of this gene in general. Specifically, sequence analyses in a chimpanzee population discovered two common *Tas2R38* alleles that associated with PTC sensitivity, which is in accordance with the behavioral experiments (Wooding, 2006). However, unlike human taster/nontaster alleles which are distinguished by three nonsynonymous substitutions, the chimpanzee taster/nontaster alleles differ by a mutation at the initiation codon, making a change from ATG to AGG in the nontaster allele. The alteration results in a truncated polypeptide, which does not respond to PTC *in vitro* (Wooding, 2006). Taken together, these findings demonstrate that although humans and chimpanzees show similar taste responses to PTC as tasters and nontasters, and even they both maintained two common alleles responsible for the distinct sensitivities, respectively, the underlying molecular mechanisms are totally different in these two species. The nontaster alleles seem to evolve independently in human and chimpanzee, that is, while human allele exhibits three amino acids substitutions, the chimpanzee allele appears to be a null (Wooding, 2006). Recently, PTC nontaster Japanese macaques were observed and their *Tas2R38* nontaster allele was identified with an ATG to ACG change at the initiation codon, which is like neither that in humans nor that in chimpanzees, suggesting that the nontaster allele variations arose at least three times independently in primates (Suzuki *et al.*, 2010).

3.7 CLOSING REMARKS

Since the behavioral bitter taste sensitivity variants were observed eight decades ago, especially since the bitter taste receptor genes were identified, rapid progresses have been made to understand the physiology and genetics of bitter perception. Nonetheless, the details remain largely unknown. For example, distinct bitter sensitivity variation and highly diverse *Tas2R* gene repertoires were observed but the precise mechanism underlying it keeps elusive. Beneficial from the explosion of genome sequencing data, extensive *Tas2R* gene sequences from widely distributed non-model species could be compared and analyzed. Wider species selection provides more chances to find sequence alterations and their possible impact on the receptor function could be speculated when the evolutionary history and eco-factors of each species are taken into account. As an example, the evolutionary view was proved to be useful in explaining PTC tasteblindness. With this guidance and the improved *in vitro* functional assay, scientific hypotheses could be proposed and tested feasibly. As such data accumulated in the future, it is expected that more underlying the diversification of bitter taste receptors would be disclosed, and we will step further in exploring the mysterious field of taste perception field.

ACKNOWLEDGEMENT

This work was supported by grants from National Natural Science Foundation of China (31271329), Ministry of Human Resources and Social Security of the People's Republic of China to HY, and funds from National Natural Science Foundation of China (30930015) to PS.

REFERENCES

Adler, E., Hoon, M. A., Mueller, K. L., Chandrashekar, J., Ryba, N. J. & Zuker, C. S. (2000). A novel family of mammalian taste receptors. *Cell* **100**(6), 693–702.

Behrens, M., Korsching, S. I. & Meyerhof, W. (2014). Tuning properties of avian and frog bitter taste receptors dynamically fit gene repertoire sizes. *Mol Biol Evol* **31**(12), 3216–3227.

Bufe, B., Breslin, P. A. S., Kuhn, C., Reed, D. R., Tharp, C. D., Slack, J. P., Kim, U.-K., Drayna, D. & Meyerhof, W. (2005). The Molecular Basis of Individual Differences in Phenylthiocarbamide and Propylthiouracil Bitterness Perception. *Curr Biol* **15**(4), 322–327.

Campbell, M. C., Ranciaro, A., Froment, A., Hirbo, J., Omar, S., Bodo, J.-M., Nyambo, T., Lema, G., Zinshteyn, D., Drayna, D., Breslin, P. A. S. & Tishkoff, S. A. (2012). Evolution of Functionally Diverse Alleles Associated with PTC Bitter Taste Sensitivity in Africa. *Mol Biol Evol* **29**(4), 1141–1153.

Campbell, M. C., Ranciaro, A., Zinshteyn, D., Rawlings-Goss, R., Hirbo, J., Thompson, S., Woldemeskel, D., Froment, A., Rucker, J. B., Omar, S., Bodo, J.-M., Nyambo, T., Belay, G., Drayna, D., Breslin, P. A. S. & Tishkoff, S. A. (2013). Origin and Differential Selection of Allelic Variation at TAS2R16 Associated with Salicin Bitter Taste Sensitivity in Africa. *Mol Biol Evol* **31**(2), 288–302.

Chandrashekar, J., Mueller, K. L., Hoon, M. A., Adler, E., Feng, L., Guo, W., Zuker, C. S. & Ryba, N. J. (2000). T2Rs function as bitter taste receptors. *Cell* **100**(6), 703–711.

Chaudhari, N., Landin, M. A. & Roper, S. D. (2000). A metabotropic glutamate receptor variant functions as a taste receptor. *Nat Neurosci* **3**(2): 113–119.

Dong, D., Jones, G. & Zhang, S. (2009). Dynamic evolution of bitter taste receptor genes in vertebrates. *BMC Evol Biol* **9**, 12.

Drayna, D., Coon, H., Kim, U. K., Elsner, T., Cromer, K., Otterud, B., Baird, L., Peiffer, A. P. & Leppert, M. (2003). Genetic analysis of a complex trait in the Utah Genetic Reference Project: a major locus for PTC taste ability on chromosome 7q and a secondary locus on chromosome 16p. *Hum Genet* **112**(5–6), 567–572.

Fischer, A., Gilad, Y., Man, O. & Paabo, S. (2005). Evolution of bitter taste receptors in humans and apes. *Mol Biol Evol* **22**(3), 432–436.

Fischer, R., Griffin, F., England, S. & Garn, S. M. (1961). Taste thresholds and food dislikes. *Nature* **191**, 1328.

Fisher, R. A., Ford, E. B. & Huxley, J. (1939). Taste-testing the anthropoid apes. *Nature* **144**, 750–750.

Fox, A. L. (1932). The Relationship between Chemical Constitution and Taste. *P Natl Acad Sci USA* **18**(1), 115–120.

Fredriksson, R., Lagerstrom, M. C., Lundin, L. G. & Schioth, H. B. (2003). The G-protein-coupled receptors in the human genome form five main families. Phylogenetic analysis, paralogon groups, and fingerprints. *Mol Pharmacol* **63**(6), 1256–1272.

Gilbertson, T. A., Damak, S. & Margolskee, R. F. (2000). The molecular physiology of taste transduction. *Curr Opin Neurobiol* **10**(4), 519–527.

Glanville, E. V. & Kaplan, A. R. (1965). Food Preference and Sensitivity of Taste for Bitter Compounds. *Nature* **205**(4974), 851–853.

Glendinning, J. I. (1994). Is the Bitter Rejection Response Always Adaptive. *Physiol Behav* **56**(6), 1217–1227.

Go, Y. (2006). Lineage-specific expansions and contractions of the bitter taste receptor gene repertoire in vertebrates. *Mol Biol Evol* **23**(5), 964–972.

Go, Y., Satta, Y., Takenaka, O. & Takahata, N. (2005). Lineage-specific loss of function of bitter taste receptor genes in humans and nonhuman primates. *Genetics* **170**(1), 313–326.

Guo, S. W. & Reed, D. R. (2001). The genetics of phenylthiocarbamide perception. *Ann Hum Biol* **28**(2), 111–142.

Hayakawa, T., Sugawara, T., Go, Y., Udono, T., Hirai, H. & Imai, H. (2012). Eco-geographical diversification of bitter taste receptor genes (TAS2Rs) among subspecies of chimpanzees. *Pan troglodytes PLoS One* **7**(8), e43277.

Hayes, J. E., Wallace, M. R., Knopik, V. S., Herbstman, D. M., Bartoshuk, L. M. & Duffy, V. B. (2011). Allelic variation in TAS2R bitter receptor genes associates with variation in sensations from and ingestive behaviors toward common bitter beverages in adults. *Chem Senses* **36**(3), 311–319.

Henkin, R. I. & Gillis, W. T. (1977). Divergent taste responsiveness to fruit of the tree Antidesma bunius. *Nature* **265**(5594), 536–537.

Hoon, M. A., Adler, E., Lindemeier, J., Battey, J. F., Ryba, N. J. & Zuker, C. S. (1999). Putative mammalian taste receptors: a class of taste-specific GPCRs with distinct topographic selectivity. *Cell* **96**(4), 541–551.

Horn, F., Bettler, E., Oliveira, L., Campagne, F., Cohen, F. E. & Vriend, G. (2003). GPCRDB information system for G protein-coupled receptors. *Nucleic Acids Res* **31**(1), 294–297.

Isberg, V., Vroling, B., van der Kant, R., Li, K., Vriend, G. & Gloriam, D. (2014). GPCRDB: an information system for G protein-coupled receptors. *Nucleic Acids Research* **42**(D1), D422–D425.

Jiang, P., Josue, J., Li, X., Glaser, D., Li, W., Brand, J. G., Margolskee, R. F., Reed, D. R. & Beauchamp, G. K. (2012). Major taste loss in carnivorous mammals. *P Natl Acad Sci USA* **109**(13), 4956–4961.

Kim, U. K., Jorgenson, E., Coon, H., Leppert, M., Risch, N. & Drayna, D. (2003). Positional cloning of the human quantitative trait locus underlying taste sensitivity to phenylthiocarbamide. *Science* **299**(5610), 1221–1225.

Kinnamon, S. C. & Cummings, T. A. (1992). Chemosensory transduction mechanisms in taste. *Annu Rev Physiol* **54**, 715–731.

Li, D. & Zhang, J. (2014). Diet Shapes the Evolution of the Vertebrate Bitter Taste Receptor Gene Repertoire. *Mol Biol Evol*, **31**(2), 303–309.

Lindemann, B. (1996). Taste reception. *Physiol Rev* **76**(3), 718–766.

Matsunami, H., Montmayeur, J. P. & Buck, L. B. (2000). A family of candidate taste receptors in human and mouse. *Nature* **404**(6778), 601–604.

Meyerhof, W., Batram, C., Kuhn, C., Brockhoff, A., Chudoba, E., Bufe, B., Appendino, G. & Behrens, M. (2010). The molecular receptive ranges of human TAS2R bitter taste receptors. *Chem Senses* **35**(2), 157–170.

Nei, M., Niimura, Y. & Nozawa, M. (2008). The evolution of animal chemosensory receptor gene repertoires: roles of chance and necessity. *Nat Rev Genet* **9**(12), 951–963.

Parry, C. M., Erkner, A. & le Coutre, J. (2004). Divergence of T2R chemosensory receptor families in humans, bonobos, and chimpanzees. *P Natl Acad Sci USA* **101**(41), 14830–14834.

Shi, P. & Zhang, J. (2009). Extraordinary Diversity of Chemosensory Receptor Gene Repertoires Among Vertebrates. In: *Chemosensory Systems in Mammals, Fishes, and Insects*. (Korsching S. & Meyerhof, W., eds), pp. 57–75. Springer Berlin Heidelberg.

Shi, P., Zhang, J., Yang, H. & Zhang, Y. P. (2003). Adaptive diversification of bitter taste receptor genes in Mammalian evolution. *Mol Biol Evol* **20**(5), 805–814.

Shi, P. & Zhang, J. Z. (2006). Contrasting modes of evolution between vertebrate sweet/umami receptor genes and bitter receptor genes. *Mol Biol Evol* **23**(2), 292–300.

Singh, N., Pydi, S. P., Upadhyaya, J. & Chelikani, P. (2011). Structural basis of activation of bitter taste receptor T2R1 and comparison with Class A G-protein-coupled receptors (GPCRs). *J Biol Chem* **286**(41), 36032–36041.

Sugawara, T., Go, Y., Udono, T., Morimura, N., Tomonaga, M., Hirai, H. & Imai, H. (2011). Diversification of bitter taste receptor gene family in western chimpanzees. *Mol Biol Evol* **28**(2), 921–931.

Suzuki, N., Sugawara, T., Matsui, A., Go, Y., Hirai, H. & Imai, H. (2010). Identification of non-taster Japanese macaques for a specific bitter taste. *Primates* **51**(4), 285–289.

Tan, J., Abrol, R., Trzaskowski, B. & Goddard, W. A., 3rd (2012). 3D structure prediction of TAS2R38 bitter receptors bound to agonists phenylthiocarbamide (PTC) and 6-n-propylthiouracil (PROP). *J Chem Inf Model* **52**(7), 1875–1885.

Temussi, P. A. (2009). Sweet, bitter and umami receptors: a complex relationship. *Trends Biochem Sci* **34**(6), 296–302.

Wang, X., Thomas, S. D. & Zhang, J. (2004). Relaxation of selective constraint and loss of function in the evolution of human bitter taste receptor genes. *Hum Mol Genet* **13**(21), 2671–2678.

Wooding, S. (2006). Phenylthiocarbamide: a 75-year adventure in genetics and natural selection. *Genetics* **172**(4), 2015–2023.

Wooding, S. (2011). Signatures of natural selection in a primate bitter taste receptor. *J Mol Evol* **73**(5-6), 257–265.

Wooding, S., Kim, U. K., Bamshad, M. J., Larsen, J., Jorde, L. B. & Drayna, D. (2004). Natural selection and molecular evolution in PTC, a bitter-taste receptor gene. *Am J Hum Genet* **74**(4), 637–646.

SECTION II

THE CHEMISTRY OF BITTERNESS

4 Fruits and Vegetables

Ernst Hoehn and Daniel Baumgartner

4.1 INTRODUCTION

Fruits and vegetables are essential for balanced diets and are valuable sources of nutrients as well as secondary plant metabolites referred to as phytonutriceuticals or phytonutrients (Barratt-Fornell & Drewnowski, 2002; Hounsome *et al.*, 2008; Dias & Ortiz, 2012). Plants produce a vast and diverse assortment of over 45,000 secondary metabolites that can be divided into four major groups: phenolic and polyphenolic compounds, terpenes, alkaloids and sulfur compounds (Crozier, 2003; Saltmarsh *et al.*, 2003; Hounsome *et al.*, 2008). Diets rich in vegetables and fruits have been linked with lower rates of chronic diseases (Beecher, 1999; Dinehart *et al.*, 2006; Steevens *et al.*, 2011). In many countries consumption of at least five portions of fruits and vegetables a day are recommended in programs such as "five a day" following the recommendations of the World Health Organization (WHO). In spite of these recommendations consumption of vegetables and fruit often remains insufficient (Drewnowski & Gomez-Carneros, 2000; Guenther *et al.*, 2006; Peltzer & Pengpid, 2012). Many people dislike eating vegetables or some fruits because of their bitterness (Duffy *et al.*, 2010; Andreeva *et al.*, 2013; Sharafi *et al.*, 2013; Garcia-Burgos & Zamora, 2015). In particular, children show preferences for sweeter foods and aversions to bitter vegetables (Drewnowski, 1997; Steiner *et al.*, 2001; Zeinstra *et al.*, 2010; Bai *et al.*, 2014; Feeney *et al.*, 2014). The taste of fruits and vegetables determines how well they are liked or accepted as part of the diet (Cox *et al.*, 2012). One particular concern is that bitterness is generally linked with dietary danger (Glendinning, 1994; Barratt-Fornell & Drewnowski, 2002). Thus the sense of taste has evolved in which humans and animals have become alerted to the bitter taste of food toxins and as a result carefully avoid ingestion of these harmful foods (Tepper, 2008; Reed *et al.*, 2010). Bitter threshold concentrations of toxic compounds, however, are not always lower than their

Bitterness: Perception, Chemistry and Food Processing, First Edition. Edited by Michel Aliani & Michael N. A. Eskin.

toxicity thresholds while non-toxic compounds may taste bitter but are harmless (Glendinning, 1994). Furthermore, not all poisons are bitter (Reed *et al.*, 2010).

Humans are generally very sensitive to bitterness and can detect bitter compounds at much lower concentrations compared to compounds with other basic tastes. Furthermore, bitter perception is highly individual (Meyerhof, 2005; Garcia-Burgos & Zamora, 2015). More than eight decades ago, Fox (1932) discovered that phenylthiocarbamide (PTC) was tasteless to about 30% of individuals and moderate-to-intensely bitter to the majority of individuals. Since then many studies have been carried out using PTC and since the 1960s investigated perception of bitterness and its heritability in a related compound 6-n-propylthiouracil (PROP) (Bufe *et al.*, 2005). Furthermore, an investigation of heritability and genetic covariation of sensitivity to PROP and other bitter compounds suggested that their perception is influenced by several sets of genes (Delwiche *et al.*, 2001; Drayna, 2005; Hansen *et al.*, 2006; Tepper, 2008). PROP sensitivity explained that the variability in vegetable preference and intake was due on the different sensitivities to bitterness. In addition a quinine marker explained variability in vegetable preference and intake via vegetable bitterness and sweetness (Dinehart *et al.*, 2006; Mennella *et al.*, 2005). Some experiments showed that PROP bitterness ratings were specific markers for vegetable glucosinolates but did not correlate with bitterness for other bitter compounds (Bufe *et al.*, 2005; Hansen *et al.*, 2006).

For many compounds of plant origin, bitterness prediction based on their molecular structure is limited. Models were proposed based on the Shallenberger-Acree-Kier nomenclature (Shallenberger & Acree, 1971; Belitz & Wieser, 1985; Roy, 1992; Kubo, 1994; Acree & Lindley, 2008) which gave insight into possible interactions of compounds with bitter taste receptors but showed that sensory evaluation was required to determine bitterness of a compound. However, relationships between chemical structure and taste indicated that sweet and bitter taste were closely related (Maga, 1990; Roy, 1992; Walters, 1996). Based on this, a number of studies were conducted to find ways to inhibit or reduce bitterness. It has long been known that the use of sugar, non-nutritive sweeteners or fat and other ingredients such as sodium salts or trisodium citrate can reduce or mask bitterness in a variety of foods (Roy, 1990; 1992; Breslin & Beauchamp, 1997; Napoleone *et al.*, 2007; Capaldi & Privitera, 2008; Ley, 2008; Wilkie *et al.*, 2013). A recent study by Sharafi *et al.* (2013) investigated the masking effect of sodium acetate, sodium chloride and aspartame on the bitterness of asparagus, Brussel sprouts and kale. They found that masking vegetable bitterness depended on vegetable type and taste phenotype. It appeared that masking bitterness with mainly sweeteners could suppress bitterness and increase acceptance if they were matched to perceived vegetable bitterness or to self-reported vegetable disliking. Similar findings were reported in a study determining suppression of bitterness in vegetables by sodium chloride (Wilkie *et al.*, 2014). Furthermore, the authors pointed out that sodium chloride might interact fundamentally differently with a full food matrix than it does with quinine hydrochloride or other bitter compounds studied in water, thus limiting generalizability of chemical suppression studies.

Many secondary bitter plant metabolites provide resistance against insects, fungi, bacteria, thus they may be considered 'natural pesticides'. In some cases they are present constitutively but some increase of their concentrations or even their formation can be observed in response to infections, wounding or environmental stress. Hence, the amounts found in vegetables and fruit are influenced by variety, growing conditions (temperature, water supply, fertilisation), development stage, harvesting date/ripeness, year, and geographical location as well as postharvest factors such as storage, preparation (peeling) and processing operations (Seljasen *et al.*, 2001; Kjellenberg *et al.*, 2010; Rouphael *et al.*, 2012; Tiwari & Cummins, 2013). For a long time debittering of plant foods was a major concern of agriculture and the food industry. Selective breeding out of bitter compounds or their removal during processing has resulted in low amounts of bitter compounds in current plant foods (Fenwick *et al.*, 1990; Drewnowski *et al.*, 2001; Dias & Ortiz, 2012). More recently, many positive health effects have been associated with bitter secondary metabolites which promoted engineering plant foods with enhanced concentrations of phytonutrients (Farnham *et al.*, 1999; Dias & Ortiz, 2012; Qian *et al.*, 2015). Drewnowski *et al.* (2000), however, pointed out that good taste and good health may be incompatible as it relates to the bitter secondary metabolites. Sun *et al.* (2006) proposed genetically modifying plant foods by functional expression of the taste modifying protein miraculin (Kurihara & Beidler, 1968; Theerasilp & Kurihara, 1988). Miraculin modifies a sour taste into a sweet taste. There are other naturally occurring taste modifying proteins as well as sweet proteins identified including thaumatin, monellin, mabinalin, pentadin, brazzein and curculin (Van der Wel *et al.*, 1989; Yamashita *et al.*, 1990; Liu *et al.*,1993; Ming & Hellekant, 1994). All of them have been cloned, sequenced and many of these proteins have been expressed in foreign hosts (Gibbs *et al.*,1996; Faus, 2000). Sweetness could be increased in plant foods, as shown by Sun *et al.* (2006) with transgenic lettuce. Whether perceived bitterness in plants containing bitter compounds could be suppressed or masked in this way remains to be determined.

One of the first reviews on the bitterness in foods was published in 1990 edited by Rouseff encompassing the current state of knowledge of bitterness in foods and beverages. A subsequent excellent review "*Bitter taste, phytonutrients, and the consumer*" published in 2000 by Drewnowski and Gomez-Carneros lists the bitter components in foods and points out the dilemma between the beneficial effects of bitter phytonutrients on health and their incompatibility with consumer acceptance. The data from both reviews are compiled in Table 4.1 together with data from studies covering years 2000 to 2015.

Table 4.1 Bitter phytonutrients in fruits and vegetables.

"Phytonutrient class"	Compound (trivial name)	Genus	Commodity	Source (Ref)
Cyanogenic glycosides	Amygdalin	*Rosacea*	Almond	Sanchez-Pérez *et al.* 2008
			Apricot	Cervellati *et al.* 2012
			Saskatoon berry	Mazza and Cotrell 2008
Glucosinolates	Sinigrin, Progoitrin	*Brassica*	Cabbage, Broccoli, Brussels sprouts, Cauliflower, Turnip/Swede, Collard, Kale, Chinese Kale	Chin *et al.* 1996, VanDorn *et al.* 1998, Baik *et al.* 2003, Zabaras *et al.* 2012 Qian *et al.* 2015
			Chinese cabbage	Lee *et al.* 2014
Glycoalkaloids	α-Solanine α-Chaconine β_2-Chaconine Solanidine Tomatine	*Solanum*	Potato	Valkonen *et al.* 1996, Zitnak and Filadelfi 1985 Zarzecka *et al.* 2013
			Tomato	Friedman and Levin 1995, Kozukue and Friedman 2003, Kozukue *et al.* 2004, Mulatu *et al.* 2006, Tohge *et al.* 2014
			Egg plant	Plazas *et al.* 2013, Sabolu *et al.* 2014
Phenolics	Flavonoids Phenolic acids Flavones Flavanols Quercitin			
		Daucus	Carrot	Alasalvar *et al.* 2001
		Brassica	Cauliflower	Llorach *et al.* 2003 Cartea *et al.* 2011
		Solanum	Egg plant	Plazas *et al.* 2013, Sabolu *et al.* 2014
		Citrus	Grapefruit, Orange	Khan *et al.* 2014
		Malus	Apple	Peleg *et al.* 1999
		Cichorium	Endive	Ferioli *et al.* 2015

Polyacetylenes	Falcarinol	*Daucus*	Carrot	Czepa and Hofmann 2003, Czepa and Hofmann 2004, Christensen and Brandt 2006, Kjellenberg *et al.* 2010, Agulió-Aguyayo *et al.* 2013
		Foeniculum	Fennel	Rawson *et al.* 2013a
		Pastinaca	Parsnip	Koidis *et al.* 2012
Triterpenes	Limonin, Nomilin	*Citrus*	Orange, Grapefruit, Lemon	Peterson *et al.* 2006a, Baldwin *et al.* 2010, Dea *et al.* 2013
	Cucurbitacin B and E	*Cucurbitacea*	Cucumber, Gherkin, Zucchini, Squash, Pumpkin, Melon	Zhang *et al.* 2013
Saponins	Asparasaponin, Protodicosin, Neoprotodicosin, Dicosin, Methyl protodicosin	*Asparagus*	Asparagus	Schwarzbach *et al.* 2006, Brückner *et al.* 2010, Dawid and Hofmann 2012a, 2012b, 2014, Vincken *et al.* 2007
Sesquiterpene lactones	Lactucin	*Lactuca*	Lettuce, Endive	Graziani 2015, Wulfkuehler *et al.* 2013b
	Lactucopicrin	*Cichorium*	Chicory	Price *et al.* 1990, François *et al.* 2008, Wulfkuehler *et al.* 2013a, 2014, Peters and van Amerongen 1996, Peters *et al.* 1997, Peters and van Amerongen 1998
	Cynaropicrin	*Cynara*	Artichoke	Samek *et al.* 1971

Compilation based on Rouseff (1990) encompassing data concerning bitterness in foods and beverages. Bitter taste, phytonutrients, and the consumer, an often cited excellent review published in 2000 by Drewnowski and Gomez-Carneros lists bitter components in foods and points out the dilemma between beneficial effects on health of bitter phytonutrients and its incompatibility with consumer acceptance. Data of both reviews are compiled in this table and supplemented with data from studies covering years 2000 to 2015.

4.2 FRUITS

Fruits are normally dominated by the sweetness of sugars and tartness of organic acids (Legua *et al.*, 2014). In some fruits, such as citrus, berries, apples and in products such as wine, cider and some fruit juices bitterness has been reported. Phenolic compounds are the most common bitterness causing components in fruits. Furthermore, phenolics also contribute to astringency in berries as well as in wine and apple ciders (Arnold *et al.*, 1980; Lea, 1990; Robichaud & Noble, 1990; Noble, 1994; Peleg *et al.*, 1999; Laaksonen *et al.*, 2010, 2013; Laaksonen, 2011; Jimenez-Garcia *et al.*, 2013). Astringency is a tactile sensation often confused with bitterness (Lea & Arnold, 1978; Arnold *et al.*, 1980; Cerf-Ducastel *et al.*, 2001; Bajec & Pickering, 2008). It includes, according to Singleton and Esau (1969), the "dry-mouth" feeling, puckering or roughing sensation attributed to the interactions of phenolics with proteins of the mouth. Phenolics such as gallic acid, catechin and tannic acid all show increasing bitterness as well as astringency with higher concentrations (Arnold *et al.*, 1980; Robichaud & Noble, 1990). Furthermore, astringency generally increases with molecular weight while bitterness peaks with tetramers (Arnold *et al.*, 1980; Lea, 1990). However, during ripening, fruit astringency decreases as flavonoids polymerize to sensory inactive compounds (Joslyn & Goldstein, 1964). In aged wine the loss of astringency results from polymerization and precipitation of the astringent flavonoids (Singleton & Noble, 1976; Lea, 1990). Most investigations on astringency are concerned with wine, apple cider and some berries and their juices (Laaksonen *et al.*, 2013). Astringency and bitterness are both considered negative sensory factors so that phenolics may therefore negatively affect pleasantness of food (Lesschaeve & Noble, 2005; Laaksonen *et al.*, 2013). In mature fruits, however, bitterness as well as astringency are rarely found as the highest concentrations of bitter and astringent compounds generally occur in immature fruits (Rouseff, 1990). Consequently in ripe desert apples bitterness and astringency are absent with the exception of those apples showing symptoms of bitter pit, a physiological disorder caused by Ca-deficiency (Dart, 2004). Exceptions are "bittersweet" cider apples which show a bitter taste and are especially grown in England and France for cider production (Lea, 1990).

4.2.1 Flavonoids, flavonols and limonoid aglycones in grapefruit, orange and lemons

Bitterness in citrus fruit is usually found in the juices and seldom, with a few exceptions, in intact, mature and healthy fruits. The compounds responsible for bitterness fall into two different classes and include limonoids and flavanone neohesperidosides (Rouseff & Matthews, 1984; Hasegawa & Maier, 1990; Rouseff, 1990). The limonoids include two bitter compounds limonin and nomilin (Dea *et al.*, 2013). Both of these components develop gradually in the juices of oranges, grapefruit, lemons and other citrus varieties after extraction (Fig. 4.1).

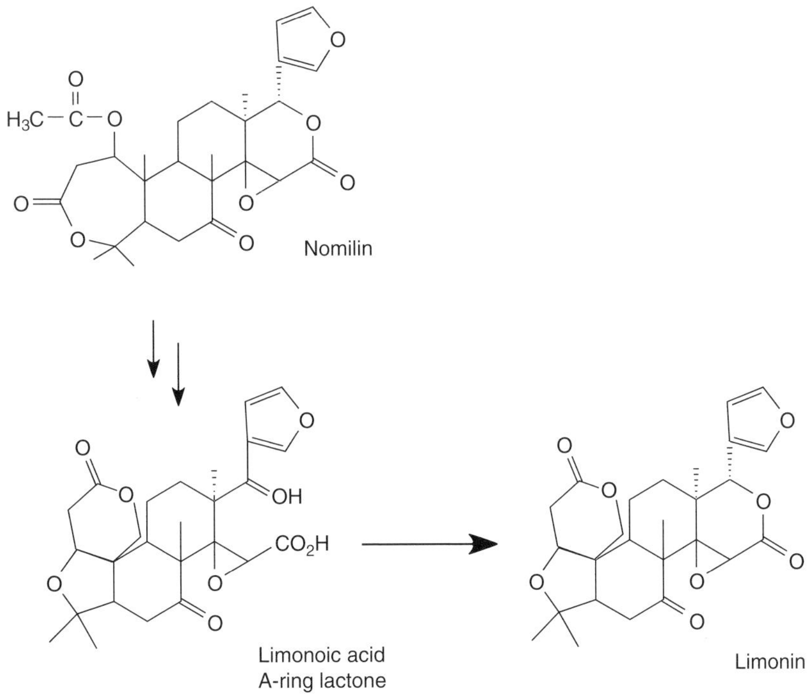

Figure 4.1 Delayed bitterness in citrus, formation of Limonin derived from Nomilin via Limoneic acid A-ring lactone (Hasegawa and Maier, 1990). Reproduced with permission of Elsevier.

The precursor of these compounds is the nonbitter limonoate A-ring lactone which is gradually converted to the bitter components. This process is called "delayed bitterness" (Hasegawa & Maier, 1990; Yusof *et al.*, 1990). The concentrations of limonin and nomilin in orange fruits of Hunaglongbin-infected trees can reach levels four times as high as the level normally present in healthy fruits (Baldwin *et al.*, 2010; Plotto *et al.*, 2010). The threshold levels of limonin and nomilin were first reported at concentrations of about 6 mg L^{-1} (Hasegawa & Maier, 1990). More recently, recognition thresholds in orange juice were reported by Dea *et al.* (2013) to be 4.7 and 2.6 mg L^{-1} for limonin and nomilin, respectively. The same authors found synergistic effects between limonin and nomilin in juices of fruits harvested from Hunaglongbin-infected trees. Adding nomilin at a subthreshold level of 2 mg L^{-1}, decreased limonin threshold to 2.6 mg L^{-1}. In the same study they showed that addition of sucrose, but not citric acid, decreased the perception of bitterness induced by limonin and nomilin in orange juice.

The flavanone neohesperidoside naringin (Fig. 4.2) imparts desirable bitterness to grapefruits but undesirable characteristics to such citrus cultivars as oranges and mandarins. Consequently bitterness has been under positive and negative selection and breeding for many years (Frydman *et al.*, 2013).

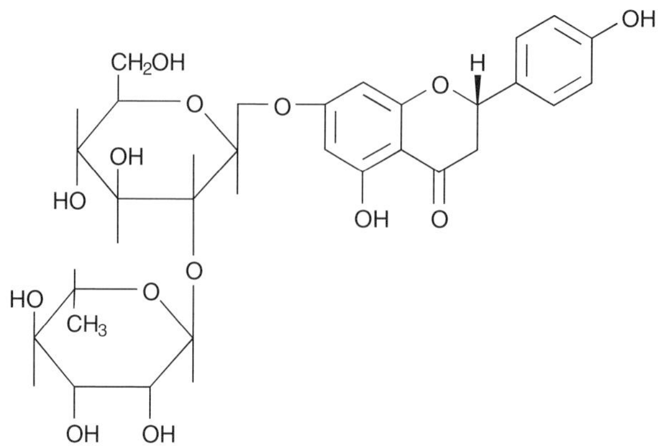

Figure 4.2 Flavanone neohesperidoside Naringin inducing bitterness in grapefruit and other citrus (Puri, 1990). Reproduced with permission of Elsevier.

Naringin levels found in grapefruits averaged 170 mg kg^{-1} edible fruit (Peterson *et al.*, 2006a). In lemons naringin levels were much lower at 1.8 mg kg^{-1} while in limes no naringin was detected. Sour oranges and tangelos contain 188 mg kg^{-1} and 56 mg kg^{-1} naringin, respectively and are slightly bitter. Sour oranges are usually used for production of marmalades and not eaten raw. In sweet oranges naringin amounted for only 1.7 mg kg^{-1} (Peterson *et al.*, 2006b). The detection threshold for naringin in water is 20 ppm but there is a wide range of sensitivity among tasters (Guadagni *et al.*, 1973). According to Puri (1990) grapefruit juice containing 300 to 500 ppm naringin has a characteristic, desirable bitterness. Juices containing more than 700 ppm naringin, however, are considered too bitter by most consumers.

4.2.2 Cyanogenic glycoside in apricot, almonds and other species

Cyanogenic glycosides are secondary plant metabolites composed of a cyanogenic aglycone (hydroxynitriles) and from one or up to three sugar moieties (mostly glucose). The hydroxynitriles are derived from the five amino acids tyrosine, phenylalanine, valine, leucine and isoleucine (Ballhorn, 2011). The number of cyanogenic glycoside containing plant species is over 3000 belonging to families such as *Fabaceae*, *Rosaceae*, *Linaceae*, *Compositae* and others (Vetter, 2000; Bak *et al.*, 2006; Mazza & Cottrell, 2008; Ballhorn, 2011; Cervellati *et al.*, 2012). Important in terms of bitterness are amygdalin (a disaccharide) and prunasin (a monosaccharide), they are cyanogenic glycosides found in bitter almonds, bitter apricot kernels, and kernels of apples, cherries, plums, and peaches as well as in seeds of flax, lima bean, cycas plants and in other plant foods (Mazza & Cottrell, 2008; Barceloux, 2009; Bolarinwa *et al.*, 2014). Cyanogenic glycosides are the chemical defense systems of plants against pathogens, insects and herbivores.

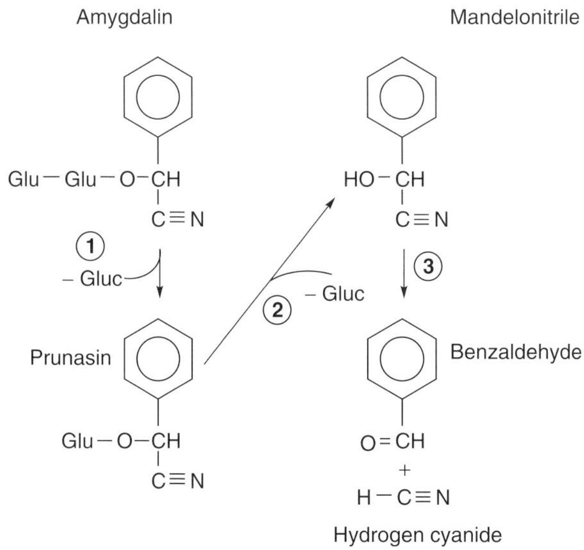

Figure 4.3 Cyanogenic glycosides amygdalin and prunasin and release of glucose, benzaldehyd and HCN (Balhorn, 2011). Reproduced with permission of Elsevier.

Intact amygdalin and purasin are not toxic but when plant tissues containing cyanogenic glycosides are damaged enzymatic hydrolysis occurs producing glucose, benzaldehyde and highly toxic hydrogen cyanide (HCN) (Fig. 4.3). Following consumption of food containing cyanogenic glycosides, HCN may also be generated by the action of enzymes from the gut microflora (Carter *et al.*, 1980). The toxicity of HCN is mainly due to its inhibition of the mitochondrial respiration pathway (Ballhorn *et al.*, 2009). Amygdalin levels in sweet, semi sweet and in bitter almonds range from 0 –900 mg kg^{-1}, 0 – 3000 mg kg^{-1} and 300 – 68500 mg kg^{-1}, respectively (Wirthensohn *et al.*, 2008; Bolarinwa *et al.*, 2014). Thus amygdalin and HCN can be used as indicators of the presence of bitter almonds in shipments of sweet almonds (Toomey *et al.*, 2012) or discrimination of almonds with respect to their bitterness (Borràs *et al.*, 2014). According to Ballhorn *et al.* (2009) cyanide toxicity levels for humans range from 0.5 to 3.5 mg kg^{-1} body weight. Bitter almonds can contain up to 5% amygdalin that corresponds to about 0.3% HCN or according to Ballhorn (2011) to about 1 mg HCN per seed. Based on this 10-15 bitter almond seeds are considered lethal for children or 50–60 seeds for adults. Sweet almonds contain much lower amounts of amygdalin, however, up to 2% of sweet almonds are bitter and contain amygdalin in comparable amounts to bitter almonds (Ballhorn, 2011). Other prunus species like apricot kernels contain on average 9100–36,600 mg kg^{-1} amygdalin depending on the cultivar. In bitter apricot kernels, amygdalin levels are very high and may reach 55 g kg^{-1} (Femenia *et al.*, 1995; Hayta & Alpaslan, 2011; Bolarinwa *et al.*, 2014). In a rather special berry, the Saskatoon berry (*Amelanchier alnifolia* Nutt.) native to the southern Yukon and Northwest Territories, the Canadian prairies and the northern plains of the United States, amygdalin contents range from 6.6–129.2 mg kg^{-1} (Mazza & Cottrell, 2008). The authors indicated that to suffer intoxication an adult person (80 kg) would have to ingest about 9.6 kg of fresh berries whereas a child (25 kg) would have to ingest about 3 kg of fresh berries of the varieties having the higher cyanogens content assuming of course that there is a 100% hydrolysis of the glycosides rather than any other metabolic fate. However, cyanide poisonings are rare but the correct diagnosis is often delayed until patients are hospitalized (Shragg *et al.*, 1982; Geller *et al.*, 2006). Health promoting effects of amygdalin were popularized as a cancer cure but clinical trials did not support these claims (Miller *et al.*, 1981; Milazzo *et al.*, 2011).

Bitterness of almonds or apricot kernels was highly and significantly correlated with their amygdalin content (Femenia *et al.*, 1995; Sánchez-Pérez *et al.*, 2008; Wirthensohn *et al.*, 2008; Borràs *et al.*, 2014). However, it is through the enzymatic breakdown of amygdalin that benzaldehyde and HCN are liberated producing the bitter taste (Wirthensohn *et al.*, 2008; Barceloux, 2009; Cervellati *et al.*, 2012; Bolarinwa *et al.*, 2014; Borràs *et al.*, 2014). Benzaldehyde is bitter but also elicits almond like aroma (Belitz *et al.*, 2009). The odor threshold level of benzaldehyde in water is reported to be 0.35 mg L^{-1} (Belitz *et al.*, 2009).

Sweet almonds and apricot kernels as well as bitter types are all nutrient-dense foods and potentially valuable sources of proteins and lipids (Femenia *et al.*, 1995; King *et al.*, 2008; Erdogan-Orhan & Kartal, 2011; Hayta & Alpaslan, 2011; Yada *et al.*, 2011). Utilization of bitter almonds as well as of bitter apricot kernels as food products is restricted because of bitterness and toxicity of HCN. Apricot pits and hence kernels the major by-products of fruit processing are discarded. Kernel bitterness or the content of amygdalin in almonds as well as in apricot kernels seems to be a monogenetic recessive trait (Sánchez-Pérez *et al.*, 2008, 2010, 2012; Negri *et al.*, 2008; Wirthensohn *et al.*, 2008). However, studies indicated that the cyanoglucoside content of seeds is linked to five genes involved in their biosynthesis, transport and catabolism (Negri *et al.*, 2008; Sanchez-Perez *et al.*, 2008; Cervellati *et al.*, 2012). Studies aimed at identifying the quantitative trait loci (QTL) in apricots (Cervellati *et al.*, 2012) may lead to the development and cultivation of apricot cultivars with sweet seeds and thus increase marketability of this by-product of fruit processing.

4.3 VEGETABLES

4.3.1 Brussels sprouts, cabbage, cauliflower, turnips/Swedes and collards and kale: glucosinolates/isothiocyanates and phenolics

Glucosinolates (Fig. 4.4) and their breakdown products have been associated with the bitterness in many vegetables of the *Brassica* genus such as broccoli, Brussels sprouts, cabbage, cauliflower, Chinese cabbage, turnips/Swedes, collards, kale and

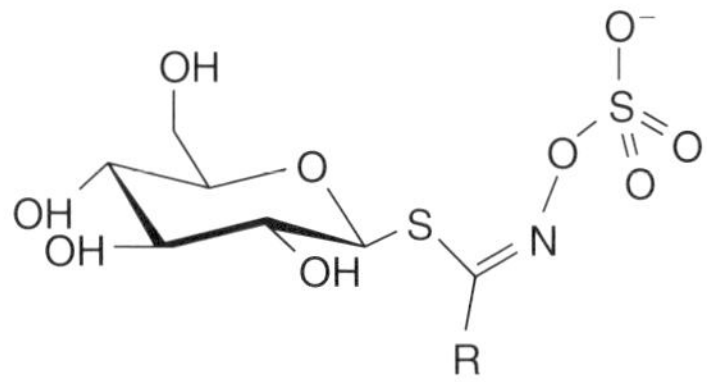

Basic glucosinolate structure, *R* = aliphatic, aromatic or indolic side chain.

Figure 4.4 Glucosinolates in Brassica (Fenwick *et al.*, 1990). Reproduced with permission of Elsevier.

others (VanEtten *et al.*, 1979; Fenwick *et al.*, 1983a; Bedford, 1989; van Doorn *et al.*, 1998; Mithen, 2001; Baik *et al.*, 2003; Schonhof *et al.*, 2004; Zabaras *et al.*, 2013; Lee *et al.*, 2014; Park *et al.*, 2014; Qian *et al.*, 2015). Bitter effects are attributed to sinigrin, gluconapin and progoitrin, respectively its breakdown product goitrin and also glucobrassicin and neoglucobrassicin (Fenwick *et al.*, 1983a,b; Zabaras *et al.*, 2013). Studies on Brussels sprouts by Fenwick *et al.* (1983a) indicated a strong association between bitterness and the presence of sinigrin and progoitrin. However, it was pointed out that sinigrin was bitter *per se*. The bitterness of progoitrin was associated with its decomposition product goitrin which is formed when plants are damaged (e.g., by chewing): progoitrin reacts with myrosinase and goitrin as well as other breakdown products are formed (Wooding *et al.*, 2010). Goitrin is structurally similar to PROP and PTC. Thus perception of goitrin and hence taste responses to vegetables containing goitrin seem to vary individually and might be associated to genetic mutations in the TAS2R38 gene (Sandell & Breslin, 2006; Wooding *et al.*, 2010; Behrens *et al.*, 2013). The distribution of taste thresholds for PTC, PROP and goitrin vary substantially from person to person as observed by Wooding *et al.* (2010). They found that the magnitude of differences between the highest and lowest observed thresholds was 8196x for PTC, 256x for PROP and 64x for goitrin, whereas for salicin it was only 16x. This suggests that the variation in perceived bitterness of *Brassica* vegetables was to some extent attributable to goitrin but additionally to other factors as well (Wooding *et al.*, 2010). Furthermore, the levels and type of glucosinolates vary greatly in different *Brassica* vegetables and cultivars (Fenwick *et al.*, 1983a; Chin *et al.*, 1996; Rosa *et al.*, 1996; Hansen *et al.*, 1997; van Doorn *et al.*, 1998; Engel *et al.*., 2002; Schonhof *et al.*, 2004; Lee *et al.*, 2014; Mølmann *et al.*, 2015). Thus in some vegetables the content of glucosinolates may not reach bitter threshold levels. For example, sinigrin concentrations in white cabbage ranged, according to Drewnowski and Gomez-Carneros (2000), from 70 to 410 mg kg^{-1} and goitrin concentrations from 10 to 80 mg kg^{-1}. The bitterness thresholds for goitrin and sinigrin were reported by Fenwick *et al.* (1983a) to be 1.2 mg ml^{-1} and 10.6 mg ml^{-1}, respectively. The same authors reported that the multiple correlations between bitterness and the combined effects of progoitrin, gluconapin, sinigrin and glucobrassicin were 0.90, which explained 82% of the variation in bitterness score in Brussels sprouts. However, it is important to consider that masking effects of sucrose and other sugars in *Brassica* vegetables occur (Lawless, 1979; Calvino *et al.*, 1990; Keast & Breslin, 2002; Sharafi *et al.*, 2013). The lack of acceptance of vegetables due to bitterness may thus be reduced by the additional sweet taste, which could be a positive factor in increasing their acceptance (Beck *et al.*, 2014). Several studies showed that most consumers prefer *Brassica* vegetables with low content of bitter tasting glycosides and high sucrose content to bitter and strong tasting vegetables (van Doorn *et al.*, 1998; Drewnowski *et al.*, 2000; Schonhof *et al.*, 2004; Dinehart *et al.*, 2006).

Controversial aspects are linked to the nutritional effects of glucosinolates and their breakdown products. Goitrogenic and antinutritional effects were first been observed in rabbits in 1930 by Webster and Chesney which were designated 'cabbage' goiter. Overconsumption of glucosinolate-rich food may inhibit thyroid peroxidase and synthesis of thyroid hormone (Gaitan, 1990; Rider *et al.*, 1992; Fahey *et al.*, 2001; Hounsome *et al.*, 2008) and may lead to goiter. While iodine deficiency is the primary cause (Tepper, 2008), goitrogen-rich food may also promote goiter. Thus far there are no reports of deleterious health effects of glucosinolates in humans consuming normal amounts of *Brassica* vegetables (Vanderpas, 2006; Verkerk *et al.*, 2009; Wooding *et al.*, 2010). However, glucosinolates and their breakdown products exhibit cytotoxic and genotoxic effects (Musk *et al.*, 1995; Rouzaud *et al.*, 2004; Volden *et al.*, 2008). High intakes of cruciferous vegetables have also been associated with lower risks of different types of cancer (van Poppel *et al.*,1999; Stan *et al.*, 2008; Steevens *et al.*, 2011).

In some recent studies the relationship between glucosinolates and bitter taste has been questioned. Accordingly, other components such as phenolics may contribute substantially to bitterness in *Brassica* and many other vegetables (Drewnowski & Gomez-Carneros, 2000; Baik *et al.*, 2003; Padilla *et al.*, 2007; Kreutzmann *et al.*, 2008b; Doerr *et al.*, 2009; Wooding *et al.*, 2010; Lim & Padmanabhan, 2013; Zabaras *et al.*, 2013). Studies on broccoli and cauliflower reported feruloyl/sinapoyl gentibiosides and a large number of simple and complex flavanoid glycosides containing kaempferol and quercitin as the main agylcones (Cartea *et al.*, 2010; Zabaras *et al.*, 2013). Furthermore, several compounds acylated with sinapic, ferulic, caffeic and *p*-coumaric acids were found in *Brassica* vegetables (Llorach *et al.*, 2003; Vallejo *et al.*, 2004). Activation of human bitter taste receptors by different phenolic compounds were analysed by Soares *et al.* (2013) who found a significant contribution by phenolic compounds to the bitterness of fruits, vegetables and derived products even in very low concentrations.

4.3.2 Carrots: 6-Methoxymellein, polyacetylenes and phenolic acids

Carrots (*Daucus carota* L.) are among the best liked root vegetables and one of the predominant vegetables grown in many countries (Hoehn *et al.*, 2003; U.S. Census Bureau, 2012). A multitude of volatile and non-volatile compounds determines the flavor of carrots (Brückner, 2008; Jones, 2008). Sweetness is generally the most attractive requirement demanded by consumer for acceptable sensory quality. Whereas bitterness is considered an undesirable taste of carrots and is rejected by most consumers (Hoehn *et al.*, 2003; Kreutzmann *et al.*, 2007, 2008a; Kramer *et al.*, 2012b). A multiplicity of compounds may be causing bitter taste in carrots, including some amino acids (Chen *et al.*, 2014), phenolics, terpenoids, and polyacetylenes (Kreutzmann *et al.*, 2008a, 2008b; Schmiech *et al.*, 2008). 3-Methyl-6-methoxy-8-hydoxy-3,4-dihydroisocoumarin, named 6-methoxymellein (6-MM) (Fig. 4.5), was the first phenolic compound associated with the bitter taste of carrots (Sondheimer, 1957). 6-MM was identified as a phytoalexin by Condon and Kuc in 1960. Since then it has been demonstrated that carrot roots synthesize 6-MM under stress conditions caused by fungal pathogens but also by wounding and exposure to ethylene. Further studies showed that 6-MM is usually only detected in minor amounts in freshly harvested carrots which are much lower than sensory just noticeable levels (Carlton *et al.*, 1961; Mercier *et al.*, 1994; Chalutz *et al.*, 1969; Louarn *et al.*, 2012; Crespo *et al.*, 2012). In particular, exposure to ethylene during storage stimulates respiration and leads to formation of 6-MM formation via the acetate-malonic acid pathway and not the shikimic acid pathway as for other phenolic compounds (Sarkar & Phan, 1975, 1979; Lafuente *et al.*, 1996; Talcott & Howard 1999a; Fan *et al.*, 2000; Fan & Mattheis, 2000; Seljasen *et al.*, 2000; Heredia & Cisneros-Zevallos, 2009; Kramer *et al.*, 2012b). The production of 6-MM can be blocked by inhibiting ethylene action using 1-methylcyclopropene (1-MCP) (Fan *et al.*, 2000; Kramer *et al.*, 2012b). Sensory just noticeable levels of 6-MM ranged according to Talcott and Howard (1999b) from 48 to 71 mg kg^{-1} in strained carrots and 94 mg kg^{-1} exerted bitter taste in spiked carrot puree. A bitter taste detection threshold of 20 mg kg^{-1}

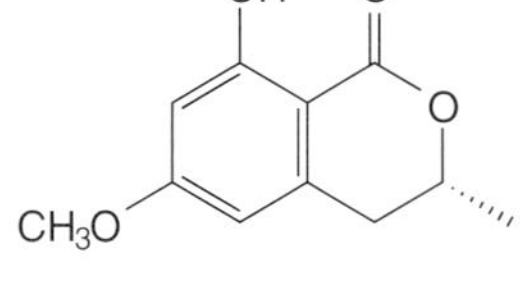

Figure 4.5 The phenolic bitter compound 3-Methyl-6-methoxy-8-hydoxy-3,4-dihydroisocoumarin named 6-Methoxymellein (6-MM) in carrots (Czepa and Hofmann 2003). Reproduced with permission of American Chemical Society.

in water was later found for 6-MM (Czepa & Hofmann, 2003). However, several authors pointed out that the high sugar content to some extent could mask the bitterness perception of 6-MM and other bitter compounds in carrots (Simon *et al.*, 1980, 1982; Seljasen *et al.*, 2000; Hoehn *et al.*, 2003; Kreutzmann *et al.*, 2008b) thus sensory noticeable levels in carrots may substantially exceed those found in water. Investigations on spatial distribution of 6-MM (Talcott & Howard, 1999b; Czepa & Hofmann, 2004) revealed that 70–80% of 6-MM was found in the peel (3 mm layer) and thus bitterness may substantially be reduced by peeling carrots.

Several other phenolics may also contribute to bitterness in carrots. The main phenolic acid, chlorogenic acid (5-caffeoylquinic acid) can constitute up to 60% of the total phenolics (Alasalvar *et al.*, 2001; Kreutzmann *et al.*, 2008b). According to Kreutzmann *et al.* (2008b), however, a di-caffeic derivative together with falcarindiol (FaDOH), a polyacetylene were highly related to bitterness in contrast to other potentially bitter compounds. Polyacetylenes such as Falcarinol (FaOH), falcarindiol (FaDOH) and falcarindiol 3-acetate (FaDOAc) are the main polyacetylenes found in carrots (Fig. 4.6). Their contribution to bitterness was first postulated by Czepa and Hofmann (2003, 2004). They found that the bitter detection threshold for FaDOH in water was 10 mg kg^{-1}, for FaOH 20 mg kg^{-1} and for FaDOAc 60 mg kg^{-1}. According to Czepa and Hofmann (2003) fresh carrots contained 41–45 mg kg^{-1} FaDOH. To estimate the contribution of FaDOH

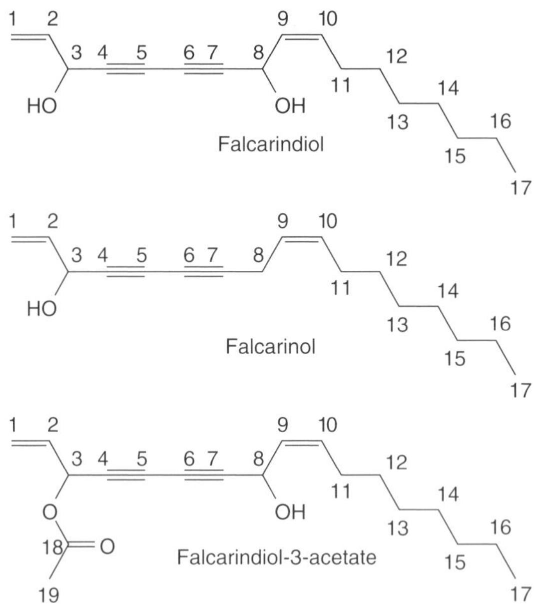

Figure 4.6 Polyacetylenes (Falcarinols) in carrot and other Apiaceae food plants (Czepa and Hofmann, 2003, Christenen and Brandt, 2006).

to the bitterness of carrots they calculated the bitter activity value (BAV) or the taste activity value (TAV) which is the ratio of compound content in carrots and its detection threshold. BAV above the value one indicate contribution towards bitterness. The same authors reported that the FaDOH concentration increased during storage to 87 mg kg^{-1}, corresponding to a BAV of 9. In a subsequent study, Czepa and Hofmann (2004) determined the falcarinols in eleven different carrot cultivars. FaOH concentrations ranged from 5.2 to 30.9 mg kg^{-1} corresponding to BAV's of 0.25–1.5, FaDOH concentrations ranged from 16.2 to 84.3 mg kg^{-1} corresponding to BAV's of 1.6–8.4 and FaDOAc concentrations ranged from 8.8 to 40.8 mg kg^{-1} corresponding to BAV's of .02–0.7. This study confirmed that FaDOH contributed to bitterness of all cultivars whereas FaOH and FaDOAc concentrations did not usually reach threshold values as indicated by BAV's below 1 indicating insignificant contribution to bitterness. In addition it was shown that some cultivars produced higher amounts of polyacetylenes than others (Metzger & Barnes, 2009). This was confirmed in other studies reporting on effects of cultivar, geographical location, growing conditions, state of development, storage as well as processing on polyacetylene concentrations in carrots (Kidmose *et al.*, 2004; Baranska *et al.*, 2005; Kjellenberg *et al.*, 2010; Koidis *et al.*, 2012; Kramer *et al.* 2012a,b; Aguiló-Aguayo *et al.*, 2014; Koidis *et al.*, 2015).

In light of processing and use of carrots it is important to consider the spatial distribution of polyacetylenes, as observed for 6-MM levels varies in the roots. FaDOH concentrations in the phloem were double to those found in the xylem. In addition the upper end contained higher concentrations of FaDOH than the lower end. Thus peeling and removing the upper green parts decreased FaDOH concentrations by about 50% and diminishes bitterness of carrots (Czepa & Hofmann, 2004; Kreutzmann *et al.*, 2008b). In carrots and other *Apiaceae* food plants such as celery, fennel, parsnip, parsley and others, the polyacetylenes are formed from oleic acid by dehydrogenations and β-oxidation leading to FaOH and further to FaDOH and FaDOAc (Hansen & Boll, 1986; Zidorn *et al.*, 2005; Christensen & Brandt, 2006; Rawson *et al.*, 2013a).

Polyacetylenes in plants provide resistance or act as defence systems against fungi, bacteria and insects. In addition, it is very important to note that these bitter compounds may promote some health benefits in humans such as the anti-cancer effects of carrots (Hansen & Boll, 1986; Brandt & Christensen, 2000; Christensen & Brandt, 2006; Tan *et al.*, 2014). However, polyacetylenes are toxic to humans if consumed in high amounts. Fortunately, the beneficial effects are only observed at low concentrations as found in carrots and parsnips (Kuklev *et al.*, 2013; Rawson *et al.*, 2013a). Thus, as Kreutzmann *et al.* (2008b) pointed out, in-depth knowledge is required to improve health promoting effects and sensory quality of carrots simultaneously. Investigations of bitterness suggest that high sugar content may mask the bitter perception of carrots (Roy, 1990; Hoehn *et al.*, 2003; Kreutzmann *et al.*, 2008b; Kramer *et al.*, 2012b). Interactions between other taste compounds and bitter compounds as well as interactions between bitter compounds determine sensory quality of carrots in a complex way and remain to be elucidated in future investigations (Soares *et al.*, 2013; Wilkie & Capaldi Phillips, 2014; Suess *et al.*, 2015).

4.3.3 Potatoes, tomatoes and other Solanum species: glycoalkaloids

Glycoalkaloids are nitrogen-containing compounds that are produced in vegetables of the *Solanaceae* family including potatoes, tomatoes, eggplants, and peppers (Cárdenas *et al.*, 2015). Bitterness in potatoes is mainly attributable to glycoalkaloids such as α-solanine, α-chaconine and β_2-chaconine (Fig. 4.7). The solanine contents of most commercial potato varieties range from 20 to 220 mg kg^{-1} but are usually less than 120 mg kg^{-1} (Zarzecka *et al.*, 2013). Glycoalkaloids concentrations are 3 to 10 times greater in the peel than in the flesh. Thus peeling reduces possible solanine contents substantially. Variations of solanine levels are mostly attributable to differences in variety of tubers. In addition growing conditions including use of pesticides (Hajslová *et al.*, 2005; Zarzecka *et al.*, 2013) and storage conditions may influence solanine concentrations. It is well documented that it is daylight exposure of potato tubers in the field, during storage, or on the store shelf or at home promoting greening as a consequence of chlorophyll synthesis but also synthesis of glycoalkaloids such as solanine. Solanine concentrations in light exposed tubers may reach 1800 mg kg^{-1} peel (Cantwell, 1996; Percival & Dixon, 1996; Percival *et al.*, 1996; Pavlista, 2001).

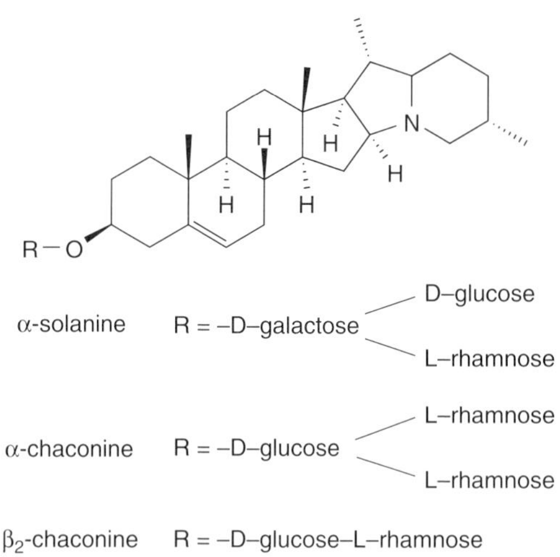

Figure 4.7 Glycoalkaloids in potato: α-solanine, α-chaconine and β_2-chaconine (Fenwick *et al.*, 1990).

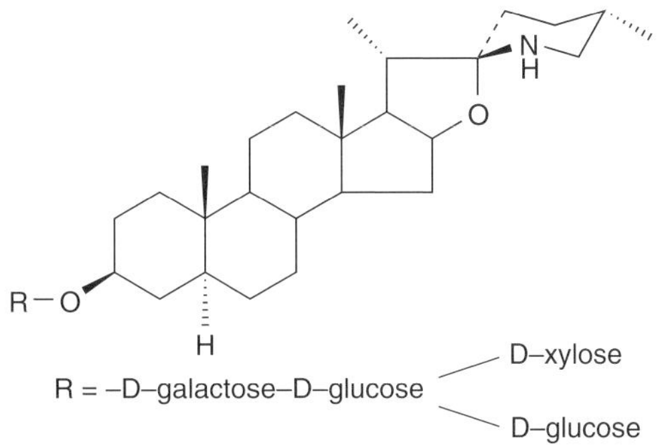

Figure 4.8 Glycoalkaloid Tomatin in tomato (Fenwick *et al.*, 1990).

This level exceeds the 200 mg kg^{-1} which is generally recognised as safe for human consumption (Health Canada, 2010). The toxicity of solanine is related to its inhibitory effects on cholinesterase preventing the breakdown of acetylcholine in the human body (Montario, 2015). In addition, potatoes with glycoalkaloid levels exceeding 100 mg kg^{-1} taste bitter and potatoes exceeding levels of 140 mg kg^{-1} exhibit a bitter and burning taste. This must be expected since the bitterness threshold levels in 0.02% lactic acid solution of α-solanine, α-chaconine and β_2-chaconine are 63 mg kg^{-1}, 16 mg $kg^{-1}$and 16 mg kg^{-1}, respectively (Zitnak & Filadelfi, 1985; Fenwick *et al.*, 1990). However, low concentrations of glycoalkaloids may improve the taste of potatoes (Valkonen *et al.*, 1996).

Tomatine (Fig. 4.8) is the main glycoalkaloid found in immature green tomatoes (Kozukue & Friedman, 2003; Kozukue *et al.*, 2004; Tohge *et al.*, 2014). Its bitterness threshold is 20 mg kg^{-1} (Fenwick *et al.*, 1990). During ripening tomatine is enzymatically degraded so that bitterness is absent in ripe red tomatoes. Apprehensions regarding genetically modified tomato lines such as the delayed ripening line 1345-5 that degradation of solanine was hampered by the genetic modification proved unfounded since genetic modification affects only ACC synthesis which plays a key role in ethylene formation (Health Canada, 2010).

Glycoalkaloids in *Solanacea* species have evolved to protect plants against predators and pathogens including bacteria, fungi, viruses, insects and animals (Valkonen *et al.*, 1996; Mulatu *et al.*, 2006; Ito *et al.*, 2007; Neilson *et al.*, 2013; Zarzecka *et al.*, 2013). Tomatidine the aglycone of tomatine has been shown to exhibit antibacterial properties against *Staphylococcus aureus* a highly antibiotic resistant human pathogen (Chagnon *et al.*, 2014) and inhibitory effects of human cancer cells (Friedman *et al.*, 2009). Thus breeding and biotechnological methodologies will probably be directed towards a reduction of glycoalkaloids in table potatoes but to increased contents of glycoalkaloids in disease resistant potato types and other *Solanacea* for other uses such as uses in pharmaceutical industry (Valkonen *et al.*, 1996).

4.4 FUTURE PROGRESS

The chemistry of bitter plant food compounds is a multifaceted subject area. Currently there is no common and reliable concept regarding the relationship between chemical structure and bitterness of compounds. Fruit and vegetable consumption is often insufficient because of an aversion to bitterness. Formerly, bitterness was generally linked with dietary danger and thus breeding and growing management of fruits and vegetables aimed at reducing bitter and often toxic compounds. Positive health effects of bitter components and secondary plant metabolites were only recognized in the last few decades. Progress in elucidating the genetics of bitter taste receptors lead to better understanding of bitter perception. Further studies concerning interactions between bitterness and other tastes may show new possibilities to mask bitterness and thereby increasing vegetable acceptance. Deeper insights generated in the last few decades encompassing the subject area bitterness lays the basis for new approaches and opportunities in the future to elucidate bitterness chemistry and its perception and possible reduction in fruits and vegetables.

REFERENCES

Acree, T. E. & Lindley, M. (2008). Structure-activity relationship and AH-B after 40 years. In *Sweetness and Sweeteners: Biology, chemistry and Psychophysics* Eds D. K. Weerasinghe & G. DuBois), pp. 96–108. Washington DC: A.C.S.

Aguiló-Aguayo, I. Brunton, N. Rai, D. K. Balagueró, E. Hossain, M. B. & Valverde, J. (2014). Polyacetylene levels in carrot juice, effect of pH and thermal processing. *Food Chem* **152**, 370–377.

Alasalvar, C. Grigor, J. M. Zhang, D. Quantick, P. C. & Shahidi, F. (2001). Comparison of volatiles, phenolics, sugars, antioxidant vitamins, and sensory quality of different colored carrot varieties. *J Agr Food Chem* **49**(3), 1410–1416.

Andreeva, V. A. Martin, C. Issanchou, S. Hercberg, S. Kesse-Guyot, E. & Méjean, C. (2013). Sociodemographic profiles regarding bitter food consumption. Cross-sectional evidence from a general French population. *Appetite* **67**, 53–60.

Arnold, R. A. Noble, A. C. & Singleton, V. L. (1980). Bitterness and astringency of phenolic fractions in wine. *J Agr Food Chem* **28**(3), 675–678.

Bai, Y. Suriano, L. & Wunderlich, S. M. (2014). Veggiecation: a novel approach to improve vegetable consumption among school-aged children. *J Nutr Educ Behav* **46**(4), 320–321.

Baik, H. Y. Juvik, J. A. Jeffery, E. H. Wallig, M. A. Kushad, M. & Klein, B. P. (2003). Relating glucosinolate content and flavor of broccoli cultivars. *J Food Sci* **68**(3), 1043–1050.

Bajec, M. R. & Pickering, G. J. (2008). Astringency: Mechanisms and Perception. *Crit Rev Food Sci* **48**(9), 858–875.

Bak, S. Paquette, S. Morant, M. Morant, A. Saito, S. Bjarnholt, N. Zagrobelny, M. Jørgensen, K. Osmani, S. Simonsen, H. Pérez, R. van Heeswijck, T. Jørgensen, B. & Møller, B. (2006). Cyanogenic glycosides: a case study for evolution and application of cytochromes P450. *Phytochemistry Reviews* **5**(2-3), 309–329.

Baldwin, E. Plotto, A. Manthey, J. McCollum, G. Bai, J. Irey, M. Cameron, R. & Luzio, G. (2010). Effect of *Liberibacter* infection (Huanglongbing disease) of citrus on orange fruit physiology and fruit/fruit juice quality: chemical and physical analyses. *J Agr Food Chem* **58**(2), 1247–1262.

Ballhorn, D. J. (2011). Cyanogenic glycosides in nuts and seeds. In *Nuts and Seeds in Health and Disease Prevention* Eds V. R. Preedy, R. R. Watson & V. B. Patel), pp. 129–136. San Diego: Academic Press.

Ballhorn, D. J. Kautz, S. & Rakotoarivelo, F. P. (2009). Quantitative variability of cyanogenesis in *Cathariostachys madagascariensis* - the main food plant of bamboo lemurs in Southeastern Madagascar. *Am J Primatol* **71**(4), 305–315.

Baranska, M. Schulz, H. Baranski, R. Nothnagel, T. & Christensen, L. P. (2005). In situ simultaneous analysis of polyacetylenes, carotenoids and polysaccharides in carrot roots. *J Agr Food Chem* **53**(17), 6565–6571.

Barceloux, D. G. (2009). Cyanogenic foods (cassava, fruit kernels, and cycad seeds). *Disease-a-Month* **55**(6), 336–352.

Barratt-Fornell, A. & Drewnowski, A. (2002). The taste of health: nature's bitter gifts. *Nutr Today* **37**(4), 144–150.

Beck, T. K. Jensen, S. Bjoern, G. K. & Kidmose, U. (2014). The masking effect of sucrose on perception of bitter compounds in *Brassica* vegetables. *J Sens Stud* **29**(3), 190–200.

Bedford, L. V. (1989). Sensory appraisal of Brussels sprouts for the freezing industry. *Acta Hortic* **244**, 57–64.

Beecher, G. R. (1999). Phytonutrients' role in metabolism: effects on resistance to degenerative processes. *Nutrition Reviews* **57**(9), 3–6.

Behrens, M. Gunn, H. C. Ramos, P. C. M. Meyerhof, W. & Wooding, S. P. (2013). Genetic, functional, and phenotypic diversity in TAS2R38-mediated bitter taste perception. *Chem Senses* **38**(6), 475–484.

Belitz, H. D. Grosch, W. & Schieberle, P. (2009). Aroma compounds. In *Food Chem* pp. 340–402. Springer Berlin Heidelberg.

Belitz, H. D. & Wieser, H. (1985). Bitter compounds: occurrence and structure-activity relationships. *Food Rev Int* **1**(2), 271–354.

Bolarinwa, I. F. Orfila, C. & Morgan, M. R. A. (2014). Amygdalin content of seeds, kernels and food products commercially-available in the UK. *Food Chem* **152**, 133–139.

Borràs, E. Amigo, J. M. van den Berg, F. Boqué, R. & Busto, O. (2014). Fast and robust discrimination of almonds (*Prunus amygdalus*) with respect to their bitterness by using near infrared and partial least squares-discriminant analysis. *Food Chem* **153**(0), 15–19.

Brandt, K. & Christensen, L. P. (2000). Vegetables as nutraceuticals – falcarinol in carrots and other root crops. In *Dietary Anticarcinogens and Antimutagens* Eds I. T. Johnson & G. R. Fenwick), pp. 386–391. Woodhead Publishing.

Bravo, L. (1998). Polyphenols: chemistry, dietary sources, metabolism, and nutritional significance. *Nutrition Reviews* **56**(11), 317–333.

Breslin, P. A. S. & Beauchamp, G. K. (1997). Salt enhances flavour by suppressing bitterness. *Nature* **387**(6633), 563–563.

Brückner, B. (2008). Consumer acceptance of fruit and vegetables: the role of flavour and other quality attributes. In *Fruit and Vegetable Flavour* Eds B. Brückner & S. G. Wyllie), pp. 11–17. Woodhead Publishing.

Brueckner, B. Schwarzbach, A. & Schrödter, R. (2010). Correlation between sugar and saponin contents and sensory attributes of white asparagus. *J Verbrauch Lebensm* **5**(3), 305–311.

Bufe, B. Breslin, P. A. S. Kuhn, C. Reed, D. R. Tharp, C. D. Slack, J. P. Kim, U.-K. Drayna, D. & Meyerhof, W. (2005). The molecular basis of individual differences in phenylthiocarbamide and propylthiouracil bitterness perception. *Curr Biol* **15**(4), 322–327.

Calvino, A. M. García-Medina, M. R. & Cometto-Muniz, J. E. (1990). Interactions in caffeine–sucrose and coffee–sucrose mixtures: evidence of taste and flavor suppression. *Chem Senses* **15**(5), 505–519.

Cantwell, M. (1996). A review of important facts about potato glycoalkaloids. *Perishables Handling Newsletter* (**87**), 26–27.

Capaldi, E. D. & Privitera, G. J. (2008). Decreasing dislike for sour and bitter in children and adults. *Appetite* **50**(1), 139–145.

Cárdenas, P. D. Sonawane, P. D. Heinig, U. Bocobza, S. E. Burdman, S. & Aharoni, A. (2015). The bitter side of the nightshades: genomics drives discovery in *Solanaceae* steroidal alkaloid metabolism. *Phytochemistry*, in press.

Carlton, B. C. Peterson, C. E. & Tolbert, N. E. (1961). Effects of ethylene & oxygen on production of a bitter compound by carrot roots. *Plant Physiol* **36**, 550–552.

Cartea, M. E. Francisco, M. Soengas, P. & Velasco, P. (2010). Phenolic compounds in *Brassica* vegetables. *Molecules* **16**(1), 251–280.

Carter, J. H. McLafferty, M. A. & Goldman, P. (1980). Role of the gastrointestinal microflora in amygdalin (laetrile)-induced cyanide toxicity. *Biochem Pharmacol* **29**(3), 301–304.

Cerf-Ducastel, B. Van de Moortele, P.-F. MacLeod, P. Le Bihan, D. & Faurion, A. (2001). Interaction of gustatory and lingual somatosensory perceptions at the cortical level in the human: a functional magnetic resonance imaging study. *Chem Senses* **26**(4), 371–383.

Cervellati, C. Paetz, C. Dondini, L. Tartarini, S. Bassi, D. Schneider, B. & Masia, A. (2012). A qNMR approach for bitterness phenotyping and QTL identification in an F1 apricot progeny. *J Biotechnol* **159**(4), 312–319.

Chagnon, F. Guay, I. Bonin, M.-A. Mitchell, G. Bouarab, K. Malouin, F. & Marsault, É. (2014). Unraveling the structure–activity relationship of tomatidine, a steroid alkaloid with unique antibiotic properties against persistent forms of *Staphylococcus aureus*. *European Journal of Medicinal Chemistry* **80**, 605–620.

Chalutz, E. DeVay, J. E. & Maxie, E. C. (1969). Ethylene-induced isocoumarin formation in carrot root tissue. *Plant Physiol* **44**, 235–241.

Chen, G. Li, J. Sun, Z. Zhang, S. Li, G. Song, C. Suo, Y. & You, J. (2014). Rapid and sensitive ultrasonic-assisted derivatisation microextraction (UDME) technique for bitter taste-free amino acids (FAA) study by HPLC–FLD. *Food Chem* **143**(0), 97–105.

Chin, H. W. Zeng, Q. & Lindsay, R. C. (1996). Occurrence and flavor properties of sinigrin hydrolysis products in fresh cabbage. *J Food Sci* **61**(1), 101–104.

Christensen, L. P. & Brandt, K. (2006). Bioactive polyacetylenes in food plants of the *Apiaceae* family: occurrence, bioactivity and analysis. *J Pharmaceut Biomed* **41**(3), 683–693.

Condon, P. & Kuc, J. (1960). Isolation of a fungitoxic compound from carrot root tissue inoculated with *Ceratocystis fimbriata*. *Phytopathology* **50**, 267–270.

Cox, D. N. Melo, L. Zabaras, D. & Delahunty, C. M. (2012). Acceptance of health-promoting *Brassica* vegetables: the influence of taste perception, information and attitudes. *Public Health Nutr* **15**(8), 1474–1482.

Crespo, P. Heller, W. Hoehn, E. & Baumgartner, D. (2012). Yield, quality traits and black root rot during storage and distribution of carrots grown in a *Chalara* infested field. *Europ. J. Hort. Sci.* **77**(2), 68.

Crozier, A. (2003). Classification and biosynthesis of secondary plant products: an overview. In *Plants: Diet and Health* pp. 27–48. Blackwell Science Ltd.

Czepa, A. & Hofmann, T. (2003). Structural and sensory characterization of compounds contributing to the bitter off-taste of carrots (*Daucus carota* L.) and carrot puree. *J Agr Food Chem* **51**(13), 3865–3873.

Czepa, A. & Hofmann, T. (2004). Quantitative studies and sensory analyses on the influence of cultivar, spatial tissue distribution, and industrial processing on the bitter off-taste of carrots (*Daucus carota* L.) and carrot products. *J Agr Food Chem* **52**(14), 4508–4514.

Dart, J. A. (2004). Bitter pit in apples. In *AGFACTS*.

Dawid, C. & Hofmann, T. (2012a). Identification of sensory-active phytochemicals in asparagus (*Asparagus officinalis* L.). *J Agr Food Chem* **60**(48), 11877–11888.

Dawid, C. & Hofmann, T. (2012b). Structural and sensory characterization of bitter tasting steroidal saponins from asparagus spears (*Asparagus officinalis* L.). *J Agr Food Chem* **60**(48), 11889–11900.

Dawid, C. & Hofmann, T. (2014). Quantitation and bitter taste contribution of saponins in fresh and cooked white asparagus (*Asparagus officinalis* L.). *Food Chem* **145**(0), 427–436.

Dea, S. Plotto, A. Manthey, J. A. Raithore, S. Irey, M. & Baldwin, E. (2013). Interactions and thresholds of limonin and nomilin in bitterness perception in orange juice and other matrices. *J Sens Stud* **28**(4), 311–323.

Delwiche, J. Buletic, Z. & Breslin, P. S. (2001). Covariation in individuals' sensitivities to bitter compounds: Evidence supporting multiple receptor/transduction mechanisms. *Percept Psychophys* **63**(5), 761–776.

Dias, J. S. & Ortiz, R. (2012). Transgenic vegetable breeding for nutritional quality and health benefits. *Food and Nutrition Sciences* **3**, 1209–1219.

Dinehart, M. E. Hayes, J. E. Bartoshuk, L. M. Lanier, S. L. & Duffy, V. B. (2006). Bitter taste markers explain variability in vegetable sweetness, bitterness, and intake. *Physiol Behav* **87**(2), 304–313.

Doerr, B. Wade, K. Stephenson, K. Reed, S. & Fahey, J. (2009). Cultivar effect on *Moringa oleifera* glucosinolate content and taste: a pilot study. *Ecology of Food and Nutrition* **48**(3), 199–211.

Drayna, D. (2005). Human taste genetics. *Annual Review of Genomics and Human Genetics* **6**(1), 217–235.

Drewnowski, A. (1997). Taste preferences and food intake. *Annu Rev Nutr* **17**(1), 237–253.

Drewnowski, A. & Gomez-Carneros, C. (2000). Bitter taste, phytonutrients, and the consumer: a review. *Am J Clin Nutr* **72**(6), 1424–1435.

Drewnowski, A. Henderson, S. A. & Barratt-Fornell, A. (2001). Genetic taste markers and food preferences. *Drug Metabolism and Disposition* **29**(4), 535–538.

Drewnowski, A. Henderson, S. A. Hann, C. S. Berg, W. A. & Ruffin, M. T. (2000). Genetic taste markers and preferences for vegetables and fruit of female breast care patients. *Journal of the American Dietetic Association* **100**(2), 191–197.

Duffy, V. B. Hayes, J. E. Davidson, A. C. Kidd, J. R. Kidd, K. K. & Bartoshuk, L. M. (2010). Vegetable intake in college-aged adults is explained by oral sensory phenotypes and TAS2R38 genotype. *Chemosens Percept* **3**(3-4), 137–148.

Engel, E. Baty, C. Le Corre, D. Souchon, I. & Martin, N. (2002). Flavor-active compounds potentially implicated in cooked cauliflower acceptance. *J Agr Food Chem* **50**(22), 6459–6467.

Erdogan-Orhan, I. & Kartal, M. (2011). Insights into research on phytochemistry and biological activities of *Prunus armeniaca* L. (apricot). *Food Research International* **44**(5), 1238–1243.

Fahey, J. W. Zalcmann, A. T. & Talalay, P. (2001). The chemical diversity and distribution of glucosinolates and isothiocyanates among plants. *Phytochemistry* **56**(1), 5–51.

Fan, X. & Mattheis, J. P. (2000). Reduction of ethylene-induced physiological disorders of carrots and iceberg lettuce by 1-methylcyclopropene. *HortScience* **35**(7), 1312–1314.

Fan, X. Mattheis, J. P. & Roberts, R. G. (2000). Biosynthesis of phytoalexin in carrot root requires ethylene action. *Physiologia Plantarum* **110**(4), 450–454.

Farnham, M. W. Simon, P. W. & Stommel, J. R. (1999). Improved phytonutrient content through plant genetic improvement. *Nutrition Reviews* **57**(9), 19–26.

Faus, I. (2000). Recent developments in the characterization and biotechnological production of sweet-tasting proteins. *Appl Microbiol Biot* **53**(2), 145–151.

Feeney, E. L. O'Brien, S. A. Scannell, A. G. M. Markey, A. & Gibney, E. R. (2014). Genetic and environmental influences on liking and reported intakes of vegetables in Irish children. *Food Qual Prefer* **32**, 253–263.

Femenia, A. Rossello, C. Mulet, A. & Canellas, J. (1995). Chemical composition of bitter and sweet apricot kernels. *J Agr Food Chem* **43**(2), 356–361.

Fenwick, G. R. Curl, C. L. Griffiths, N. M. Heaney, R. K. & Price, K. R. (1990). Bitter principles in food plants. In *Bitterness in Foods and Beverages* p. 46. Amsterdam, Pays-Bas: Elsevier.

Fenwick, G. R. Griffiths, N. M. & Heaney, R. K. (1983a). Bitterness in Brussels sprouts (*Brassica oleracea* L. var. *gemmifera*): the role of glucosinolates and their breakdown products. *J Sci Food Agr* **34**(1), 73–80.

Fenwick, G. R. Heaney, R. K. & Mullin, W. J. (1983b). Glucosinolates and their breakdown products in food and food plants. *Crit Rev Food Sci Nutr* **18**(2), 123–201.

Ferioli, F. Manco, M. A. & D'Antuono, L. F. (2015). Variation of sesquiterpene lactones and phenolics in chicory and endive germplasm. *J Food Compos Anal* **39**(0), 77–86.

Fox, A. L. (1932). The relationship between chemical constitution and taste. *P Natl Acad Sci USA* **18**(1), 115–120.

François, I. M. Wins, H. Buysens, S. Godts, C. Van Pee, E. Nicolaï, B. & De Proft, M. (2008). Predicting sensory attributes of different chicory hybrids using physico-chemical measurements and visible/near infrared spectroscopy. *Postharvest Biol Tec* **49**(3), 366–373.

Friedman, M. & Levin, C. E. (1995). α-Tomatine content in tomato and tomato products determined by HPLC with pulsed amperometric detection. *J Agr Food Chem* **43**(6), 1507–1511.

Friedman, M. Levin, C. E. Lee, S.-U. Kim, H.-J. Lee, I.-S. Byun, J.-O. & Kozukue, N. (2009). Tomatine-containing green tomato extracts inhibit growth of human breast, colon, liver, and stomach cancer cells. *J Agr Food Chem* **57**(13), 5727–5733.

Frydman, A. Liberman, R. Huhman, D. V. Carmeli-Weissberg, M. Sapir-Mir, M. Ophir, R. Sumner, L. W. & Eyal, Y. (2013). The molecular and enzymatic basis of bitter/non-bitter flavor of citrus fruit: Evolution of branch-forming rhamnosyltransferases under domestication. *Plant Journal* **73**(1), 166–178.

Gaitan, E. (1990). Goitrogens in food and water. *Annu Rev Nutr* **10**(1), 21–37.

Garcia-Burgos, D. & Zamora, M. C. (2015). Exploring the hedonic and incentive properties in preferences for bitter foods via self-reports, facial expressions and instrumental behaviours. *Food Qual Prefer* **39**(0), 73–81.

Geller, R. J. Barthold, C. Saiers, J. A. & Hall, A. H. (2006). Pediatric cyanide poisoning: causes, manifestations, management, and unmet needs. *Pediatrics* **118**(5), 2146–2158.

Gibbs, B. F. Alli, I. & Mulligan, C. (1996). Sweet and taste-modifying proteins: A review. *Nutrition Research* **16**(9), 1619–1630.

Glendinning, J. I. (1994). Is the bitter rejection response always adaptive? *Physiol Behav* **56**(6), 1217–1227.

Graziani, G. Ferracane, R. Sambo, P. Santagata, S. Nicoletto, C. & Fogliano, V. (2015). Profiling chicory sesquiterpene lactones by high resolution mass spectrometry. *Food Research International* **67**, 193–198.

Guadagni, D. G. Maier, V. P. & Turnbaugh, J. G. (1973). Effect of some citrus juice constituents on taste thresholds for limonin and naringin bitterness. *J Sci Food Agr* **24**(10), 1277–1288.

Guenther, P. M. Dodd, K. W. Reedy, J. & Krebs-Smith, S. M. (2006). Most Americans eat much less than recommended amounts of fruits and vegetables. *Journal of the American Dietetic Association* **106**(9), 1371–1379.

Hajslová, J. Schulzová, V. Slanina, P. Janné, K. Hellenäs, K. E. & Andersson, C. H. (2005). Quality of organically and conventionally grown potatoes: Four-year study of micronutrients, metals, secondary metabolites, enzymic browning and organoleptic properties. *Food Addit Contam* **22**(6), 514–534.

Hansen, J. L. Reed, D. R. Wright, M. J. Martin, N. G. & Breslin, P. A. S. (2006). Heritability and genetic covariation of sensitivity to PROP, SOA, quinine HCl, and caffeine. *Chem Senses* **31**(5), 403–413.

Hansen, L. & Boll, P. M. (1986). Polyacetylenes in *Araliaceae*: Their chemistry, biosynthesis and biological significance. *Phytochemistry* **25**(2), 285–293.
Hansen, M. Laustsen, A. M. Olsen, C. E. Poll, L. & Sørensen, H. (1997). Chemical and sensory quality of broccoli (*Brassica oleracea* L. var Italica). *Journal of Food Quality* **20**(5), 441–459.
Hasegawa, S. & Maier, V. P. (1990). Biochemistry of limonoid citrus juice bitter principles and biochemical debittering processes. In *Bitterness in Foods and Beverages* pp. 293–308. Amsterdam, PAYS-BAS: Elsevier.
Hayta, M. & Alpaslan, M. (2011). Apricot kernel flour and its use in maintaining health. In *Flour and Breads and their Fortification in Health and Disease Prevention* Eds V. R. Preedy, R. R. Watson & V. B. Patel), pp. 213–221. San Diego: Academic Press.
Health Canada (2010). Glycoalkaloids in foods.
Heredia, J. B. & Cisneros-Zevallos, L. (2009). The effect of exogenous ethylene and methyl jasmonate on pal activity, phenolic profiles and antioxidant capacity of carrots (Daucus carota) under different wounding intensities. *Postharvest Biol Tec* **51**(2), 242–249.
Hoehn, E. Schaerer, H. & Kuensch, U. (2003). Karottengeschmack - Beliebtheit, Süssigkeit und Bitterkeit. *AGRARForschung* **10**(4), 144–149.
Hounsome, N. Hounsome, B. Tomos, D. & Edwards-Jones, G. (2008). Plant metabolites and nutritional quality of vegetables. *J Food Sci* **73**(4), R48–R65.
Ito, S.-i. Ihara, T. Tamura, H. Tanaka, S. Ikeda, T. Kajihara, H. Dissanayake, C. Abdel-Motaal, F. F. & El-Sayed, M. A. (2007). α-Tomatine, the major saponin in tomato, induces programmed cell death mediated by reactive oxygen species in the fungal pathogen Fusarium oxysporum. *FEBS Letters* **581**(17), 3217–3222.
Jimenez-Garcia, S. N. Guevara-Gonzalez, R. G. Miranda-Lopez, R. Feregrino-Perez, A. A. Torres-Pacheco, I. & Vazquez-Cruz, M. A. (2013). Functional properties and quality characteristics of bioactive compounds in berries: Biochemistry, biotechnology, and genomics. *Food Research International* **54**(1), 1195–1207.
Jones, M. G. (2008). Formation of vegetable flavour. In *Fruit and Vegetable Flavour* Eds B. Brückner & S. G. Wyllie), pp. 71–102. Woodhead Publishing.
Joslyn, M. A. & Goldstein, J. L. (1964). Astringency of Fruits and Fruit Products in Relation to Phenolic Content. In *Advances in Food Research* Eds E. M. M. C.O. Chichester & G. F. Stewart), pp. 179–217. Academic Press.
Keast, R. S. J. & Breslin, P. A. S. (2002). An overview of binary taste-taste interactions. *Food Qual Prefer* **14**, 111–124.
Khan, M. K. Zill, E. H. & Dangles, O. (2014). A comprehensive review on flavanones, the major citrus polyphenols. *J Food Compos Anal* **33**(1), 85–104.
Kidmose, U. Hansen, S. L. Christensen, L. P. Edelenbos, M. Larsen, E. & Norbaek, R. (2004). Effects of genotype, root size, storage, and processing on bioactive compounds in organically grown carrots (*Daucus carota* L.). *J Food Sci* **69**(9), S388–S394.
King, J. C. Blumberg, J. Ingwersen, L. Jenab, M. & Tucker, K. L. (2008). Tree nuts and peanuts as components of a healthy diet. *The Journal of Nutrition* **138**(9), 1736S–1740S.
Kjellenberg, L. Johansson, E. Gustavsson, K.-E. & Olsson, M. E. (2010). Effects of harvesting date and storage on the amounts of polyacetylenes in carrots, *Daucus carota. J Agr Food Chem* **58**(22), 11703–11708.
Koidis, A. Rawson, A. & Brunton, N. (2015). Effect of Different Types of Processing and Storage on the Polyacetylene Profile of Carrots and Parsnips. In *Processing and Impact on Active Components in Food* (Ed V. Preedy), pp. 45–53. San Diego: Academic Press.
Koidis, A. Rawson, A. Tuohy, M. & Brunton, N. (2012). Influence of unit operations on the levels of polyacetylenes in minimally processed carrots and parsnips: An industrial trial. *Food Chem* **132**(3), 1406–1412.
Kozukue, N. & Friedman, M. (2003). Tomatine, chlorophyll, β-carotene and lycopene content in tomatoes during growth and maturation. *J Sci Food Agr* **83**(3), 195–200.

Kozukue, N. Han, J.-S. Lee, K.-R. & Friedman, M. (2004). Dehydrotomatine and α-tomatine content in tomato fruits and vegetative plant tissues. *J Agr Food Chem* **52**(7), 2079–2083.

Kramer, M. Bufler, G. Nothnagel, T. Carle, R. & Kammerer, D. R. (2012a). Effects of cultivation conditions and cold storage on the polyacetylene contents of carrot (*Daucus carota* L.) and parsnip (*Pastinaca sativa* L.). *J Hortic Sci Biotech* **87**(2), 101–106.

Kramer, M. Bufler, G. Ulrich, D. Leitenberger, M. Conrad, J. Carle, R. & Kammerer, D. R. (2012b). Effect of ethylene and 1-methylcyclopropene on bitter compounds in carrots (*Daucus carota* L.). *Postharvest Biol Tec* **73**, 28–36.

Kreutzmann, S. Christensen, L. P. & Edelenbos, M. (2008a). Investigation of bitterness in carrots (*Daucus carota* L.) based on quantitative chemical and sensory analyses. *LWT-Food Sci Technol* **41**(2), 193–205.

Kreutzmann, S. Svensson, V. T. Thybo, A. K. Bro, R. & Petersen, M. A. (2008b). Prediction of sensory quality in raw carrots (*Daucus carota* L.) using multi-block LS-ParPLS. *Food Qual Prefer* **19**(7), 609–617.

Kreutzmann, S. Thybo, A. K. & Bredie, W. L. P. (2007). Training of a sensory panel and profiling of winter hardy and coloured carrot genotypes. *Food Qual Prefer* **18**(3), 482–489.

Kubo, I. (1994). Structural basis for bitterness based on Rabdosia diterpenes. *Physiol Behav* **56**(6), 1203–1207.

Kuklev, D. V. Domb, A. J. & Dembitsky, V. M. (2013). Bioactive acetylenic metabolites. *Phytomedicine* **20**(13), 1145–1159.

Kurihara, K. & Beidler, L. M. (1968). Taste-modifying protein from miracle fruit. *Science* **161**(3847), 1241–1243.

Laaksonen, O. (2011). *Astringent food compounds and their interactions with taste properties.* University of Turku.

Laaksonen, O. Mäkilä, L. Tahvonen, R. Kallio, H. & Yang, B. (2013). Sensory quality and compositional characteristics of blackcurrant juices produced by different processes. *Food Chem* **138**(4), 2421–2429.

Laaksonen, O. Sandell, M. & Kallio, H. (2010). Chemical factors contributing to orosensory profiles of bilberry (*Vaccinium myrtillus*) fractions. *Eur Food Res Technol* **231**(2), 271–285.

Lafuente, M. T. López-Gálvez, G. Cantwell, M. & Yang, S. F. (1996). Factors influencing ethylene-induced isocoumarin formation and increased respiration in carrots. *J. Amer. Soc. Hort. Sci.* **121**(3), 537–542.

Lawless, H. T. (1979). Evidence for neural inhibition in bittersweet taste mixtures. *Journal of Comparative and Physiological Psychology* **93**(3), 538–547.

Lea, A. G. (1990). Bitterness and astringency - the procyanidins of fermented apple ciders. In *Bitterness in Foods and Beverages* pp. 123–143. Amsterdam, Pays-Bas: Elsevier.

Lea, A. G. H. & Arnold, G. M. (1978). The phenolics of ciders: Bitterness and astringency. *J Sci Food Agr* **29**(5), 478–483.

Lee, M.-K. Chun, J.-H. Byeon, D. H. Chung, S.-O. Park, S. U. Park, S. Arasu, M. V. Al-Dhabi, N. A. Lim, Y.-P. & Kim, S.-J. (2014). Variation of glucosinolates in 62 varieties of Chinese cabbage (*Brassica rapa* L. ssp. *pekinensis*) and their antioxidant activity. *LWT-Food Sci Technol* **58**(1), 93–101.

Legua, P. Forner, J. B. Hernández, F. & Forner-Giner, M. A. (2014). Total phenolics, organic acids, sugars and antioxidant activity of mandarin (*Citrus clementina* Hort. ex Tan.): Variation from rootstock. *Scientia Horticulturae* **174**, 60–64.

Lesschaeve, I. & Noble, A. C. (2005). Polyphenols: factors influencing their sensory properties and their effects on food and beverage preferences. *Am J Clin Nutr* **81**(1), 330S–335S.

Ley, J. (2008). Masking bitter taste by molecules. *Chemosens Percept* **1**(1), 58–77.

Lim, J. & Padmanabhan, A. (2013). Retronasal olfaction in vegetable liking and disliking. *Chem Senses* **38**(1), 45–55.

Liu, X. Maeda, S. Hu, Z. Aiuchi, T. Nakaya, K. & Kurihara, Y. (1993). Purification, complete amino acid sequence and structural characterization of the heat-stable sweet protein, mabinlin II. *European Journal of Biochemistry* **211**(1-2), 281–287.

Llorach, R. Gil-Izquierdo, A. Ferreres, F. & Tomás-Barberán, F. A. (2003). HPLC-DAD-MS/MS ESI characterization of unusual highly glycosylated acylated flavonoids from cauliflower (*Brassica oleracea* L. var. *botrytis*) agroindustrial byproducts. *J Agr Food Chem* **51**(13), 3895–3899.

Louarn, S. Nawrocki, A. Edelenbos, M. Jensen, D. F. Jensen, O. N. Collinge, D. B. & Jensen, B. (2012). The influence of the fungal pathogen *Mycocentrospora acerina* on the proteome and polyacetylenes and 6-methoxymellein in organic and conventionally cultivated carrots (*Daucus carota*) during post harvest storage. *Journal of Proteomics* **75**(3), 962–977.

Maga, J. A. (1990). Compound structure versus bitter taste. In *Bitterness in Food and Beverages* p. 14. Amsterdam, Pays-Bas: Elsevier.

Mazza, G. & Cottrell, T. (2008). Carotenoids and cyanogenic glucosides in saskatoon berries (*Amelanchier alnifolia* Nutt.). *J Food Compos Anal* **21**(3), 249–254.

Mennella, J. A. Pepino, M. Y. & Reed, D. R. (2005). Genetic and environmental determinants of bitter perception and sweet preferences. *Pediatrics* **115**(2), e216–e222.

Mercier, J. Arul, J. & Julien, C. (1994). Effect of food preparation on the iscoumarin, 6-methoxymellein, content of UV-treated carrots. *Food Research International* **27**, 401–404.

Metzger, B. T. & Barnes, D. M. (2009). Polyacetylene diversity and bioactivity in orange market and locally grown colored carrots (*Daucus carota* L.). *J Agr Food Chem* **57**(23), 11134–11139.

Meyerhof, W. (2005). Elucidation of mammalian bitter taste. In *Reviews of Physiology, Biochemistry and Pharmacology* pp. 37–72. Springer Berlin Heidelberg.

Milazzo, S. Ernst, E. Lejeune, S. Boehm, K. & Horneber, M. (2011). Laetrile treatment for cancer *Cochrane Database of Systematic Reviews* **11**, Art. No.: CD005476.

Miller, K. W. Anderson, J. L. & Stoewsand, G. S. (1981). Amygdalin metabolism and effect on reproduction of rats fed apricot kernels. *J Toxicol Env Health* **7**(3-4), 457–467.

Ming, D. & Hellekant, G. (1994). Brazzein, a new high-potency thermostable sweet protein from *Pentadiplandra brazzeana* B. *FEBS Letters* **355**(1), 106–108.

Mithen, R. F. (2001). Glucosinolates and their degradation products. In *Advances in Botanical Research* pp. 213–262. Academic Press.

Mølmann, J. A. B. Steindal, A. L. H. Bengtsson, G. B. Seljåsen, R. Lea, P. Skaret, J. & Johansen, T. J. (2015). Effects of temperature and photoperiod on sensory quality and contents of glucosinolates, flavonols and vitamin C in broccoli florets. *Food Chem* **172**, 47–55.

Montario, A. (2015). Potato glycoalkaloid toxicity: solanine.

Mulatu, B. Applebaum, S. W. Kerem, Z. & Coll, M. (2006). Tomato fruit size, maturity and α-tomatine content influence the performance of larvae of potato tuber moth Phthorimaea operculella (*Lepidoptera: Gelechiidae*). *Bulletin of Entomological Research* **96**(2), 173–178.

Musk, S. R. R. Smith, T. K. & Johnson, I. T. (1995). On the cytotoxicity and genotoxicity of allyl and phenethyl isothiocyanates and their parent glucosinolates sinigrin and gluconasturtiin. *Mutat Res Lett* **348**(1), 19–23.

Napoleone, G. Conliffe, C. Hayes, J. E. Kneeland, K. Sullivan, B. S. & Duffy, V. B. (2007). Modifying vegetable tastes to improve liking. *Journal of the American Dietetic Association* **107**(8, Supplement), A76.

Negri, P. Bassi, D. Magnanini, E. Rizzo, M. & Bartolozzi, F. (2008). Bitterness inheritance in apricot (*P. armeniaca* L.) seeds. *Tree Genetics & Genomes* **4**(4), 767–776.

Neilson, E. H. Goodger, J. Q. D. Woodrow, I. E. & Møller, B. L. (2013). Plant chemical defense: at what cost? *Trends Plant Sci* **18**(5), 250–258.

Noble, A. C. (1994). Bitterness in wine. *Physiol Behav* **56**(6), 1251–1255.

Padilla, G. Cartea, M. E. Velasco, P. de Haro, A. & Ordás, A. (2007). Variation of glucosinolates in vegetable crops of *Brassica rapa*. *Phytochemistry* **68**(4), 536–545.

Park, S. Valan Arasu, M. Lee, M.-K. Chun, J.-H. Seo, J. M. Lee, S.-W. Al-Dhabi, N. A. & Kim, S.-J. (2014). Quantification of glucosinolates, anthocyanins, free amino acids, and vitamin C in inbred lines of cabbage (*Brassica oleracea* L.). *Food Chem* **145**, 77–85.

Pavlista, A. D. (2001). Green potatoes: the problem and the solution. *NebGuide Horticulture*.

Peleg, H. Gacon, K. Schlich, P. & Noble, A. C. (1999). Bitterness and astringency of flavan-3-ol monomers, dimers and trimers. *J Sci Food Agr* **79**(8), 1123–1128.

Peltzer, K. & Pengpid, S. (2012). Fruits and vegetables consumption and associated factors among in-school adolescents in five Southeast Asian countries. *Int J Environ Res Public Health* **9**(10), 3575–3587.

Percival, G. & Dixon, G. R. (1996). Glycoalkaloid concentrations in aerial tubers of potato (*Solanum tuberosum* L.). *J Sci Food Agr* **70**(4), 439–448.

Percival, G. Dixon, G. R. & Sword, A. (1996). Glycoalkaloid concentration of potato tubers following exposure to daylight. *J Sci Food Agr* **71**(1), 59–63.

Peters, A. M. Haagsma, N. & van Amerongen, A. (1997). A pilot study on the effects of cultivation conditions of chicory (*Cichorium intybus* L.) roots on the levels of sesquiterpene lactones in chicons. *Zeitschrift für Lebensmitteluntersuchung und -Forschung A* **205**(2), 143–147.

Peters, A. M. & van Amerongen, A. (1996). Sesquiterpene lactones in chicory (*Cichorium intybus* L.): Distribution in chicons and effect of storage. *Food Research International* **29**(5–6), 439–444.

Peters, A. M. & van Amerongen, A. (1998). Relationship between levels of sesquiterpene lactones in chicory and sensory evaluation. *Journal of the American Society for Horticultural Science* **123**(2), 326–329.

Peterson, J. J. Beecher, G. R. Bhagwat, S. A. Dwyer, J. T. Gebhardt, S. E. Haytowitz, D. B. & Holden, J. M. (2006a). Flavanones in grapefruit, lemons, and limes: a compilation and review of the data from the analytical literature. *J Food Compos Anal* **19**, S74–S80.

Peterson, J. J. Dwyer, J. T. Beecher, G. R. Bhagwat, S. A. Gebhardt, S. E. Haytowitz, D. B. & Holden, J. M. (2006b). Flavanones in oranges, tangerines (mandarins), tangors, and tangelos: a compilation and review of the data from the analytical literature. *J Food Compos Anal* **19**, S66–S73.

Plazas, M. Andujar, I. Vilanova, S. Hurtado, M. Gramazio, P. Herraiz, F. J. & Prohens, J. (2013). Breeding for chlorogenic acid content in eggplant: interest and prospects. *Not Bot Horti Agrobo* **41**(1), 26–35.

Plotto, A. Baldwin, E. McCollum, G. Manthey, J. Narciso, J. & Irey, M. (2010). Effect of *Liberibacter* infection (huanglongbing or "greening" disease) of citrus on orange juice flavor quality by sensory evaluation. *J Food Sci* **75**(4), S220–S230.

Price, K. R. Dupont, M. S. Shepherd, R. Chan, H. W. S. & Fenwick, G. R. (1990). Relationship between the chemical and sensory properties of exotic salad crops—coloured lettuce (*lactuca sativa*) and chicory (*cichorium intybus*). *J Sci Food Agr* **53**(2), 185–192.

Puri, A. (1990). Removal of bitter compounds from citrus products by adsorption techniques. In *Bitterness in Foods and Beverages* pp. 325–336. Amsterdam, Pays-Bas: Elsevier.

Qian, H. Sun, B. Miao, H. Cai, C. Xu, C. & Wang, Q. (2015). Variation of glucosinolates and quinone reductase activity among different varieties of Chinese kale and improvement of glucoraphanin by metabolic engineering. *Food Chem* **168**, 321–326.

Rawson, A. Brunton, N. P. Rai, D. K. McLoughlin, P. Tiwari, B. K. & Tuohy, M. G. (2013a). Stability of falcarinol type polyacetylenes during processing of *Apiaceae* vegetables. *Trends in Food Science & Technology* **30**(2), 133–141.

Rawson, A. Hossain, M. B. Patras, A. Tuohy, M. & Brunton, N. (2013b). Effect of boiling and roasting on the polyacetylene and polyphenol content of fennel (*Foeniculum vulgare*) bulb. *Food Research International* **50**(2), 513–518.

Reed, D. R. Knaapila, A. & Claude, B. (2010). Genetics of taste and smell: poisons and pleasures. In *Progress in Molecular Biology and Translational Science* pp. 213–240. Academic Press.

Rider, E. D. Ikegami, M. & Jobe, A. H. (1992). Localization of alveolar surfactant clearance in rabbit lung cells. *Am J Physiol-Lung C* **263**(2), L201–L209.

Robichaud, J. L. & Noble, A. C. (1990). Astringency and bitterness of selected phenolics in wine. *J Sci Food Agr* **53**(3), 343–353.

Rosa, E. A. S. Heaney, R. K. Fenwick, G. R. & Portas, C. A. M. (1996). Glucosinolates in crop plants. In *Horticultural Reviews* pp. 99–215. John Wiley & Sons, Inc.

Rouphael, Y. Cardarelli, M. Bassal, A. Leonardi, C. Giuffrida, F. & Colla, G. (2012). Vegetable quality as affected by genetic, agronomic and environmental factors. *Food, Agriculture and Environment* **10**(3&4), 680–688.

Rouseff, R. L. (1990). *Bitterness in foods and beverages*. Amsterdam; New York; New York, NY, U.S.A.: Elsevier; Distributors for the United States and Canada, Elsevier Science Pub. Co.

Rouseff, R. L. & Matthews, R. F. (1984). Nomilin, taste threshold and relative bitterness. *J Food Sci* **49**(3), 777–779.

Rouzaud, G. Young, S. A. & Duncan, A. J. (2004). Hydrolysis of glucosinolates to isothiocyanates after ingestion of raw or microwaved cabbage by human volunteers. *Cancer Epidemiology Biomarkers & Prevention* **13**(1), 125–131.

Roy, G. (1992). Bitterness: reduction and inhibition. *Trends in Food Science & Technology* **3**, 85–91.

Roy, G. M. (1990). The applications and future implications of bitterness reduction and inhibition in food products. *Crit Rev Food Sci* **29**(2), 59–71.

Sabolu, S. Kathiria, K. B. Mistry, C. R. & Kumar, S. (2014). Generation mean analysis of fruit quality traits in eggplant (*Solanum melongena* L.). *Australian Journal of Crop Science* **8**(2), 243–250.

Saltmarsh, M. Crozier, A. & Ratcliffe, B. (2003). Fruit and Vegetables. In *Plants: Diet and Health* pp. 107–133. Blackwell Science Ltd.

Samek, Z. Holub, M. Drożdż, B. Iommi, G. Corbella, A. & Gariboldi, P. (1971). Sesquiterpenic lactones of the *Cynara scolymus* L. species. *Tetrahedron Letters* **12**(50), 4775–4778.

Sánchez-Pérez, R. Arrázola, G. Martín, M. L. Grané, N. & Dicenta, F. (2012). Influence of the pollinizer in the amygdalin content of almonds. *Scientia Horticulturae* **139**, 62–65.

Sánchez-Pérez, R. Howad, W. Garcia-Mas, J. Arús, P. Martínez-Gómez, P. & Dicenta, F. (2010). Molecular markers for kernel bitterness in almond. *Tree Genetics and Genomes* **6**(2), 237–245.

Sánchez-Pérez, R. Jørgensen, K. Olsen, C. E. Dicenta, F. & Møller, B. L. (2008). Bitterness in almonds. *Plant Physiology* **146**(3), 1040–1052.

Sandell, M. A. & Breslin, P. A. S. (2006). Variability in a taste-receptor gene determines whether we taste toxins in food. *Curr Biol* **16**(18), R792–R794.

Sarkar, S. K. & Phan, C. T. (1975). The biosynthesis of 8-hydroxy-6-methoxy-3-methyl-3,4-dihydroisocoumarin and 5-hydroxy-7-methoxy-2-methylchromone in carrot root tissues treated with ethylene. *Physiologia Plantarum* **33**(1), 108–112.

Sarkar, S. K. & Phan, C. T. (1979). Naturally-occurring and ethylene-induced phenolic compounds in the carrot root. *Journal of Food Protection* **42**(6), 526–534.

Schmiech, L. Uemura, D. & Hofmann, T. (2008). Reinvestigation of the bitter compounds in carrots (*Daucus carota* L.) by using a molecular sensory science approach. *J. Agric Food Chem* **56**, 10252–10260.

Schonhof, I. Krumbein, A. & Brückner, B. (2004). Genotypic effects on glucosinolates and sensory properties of broccoli and cauliflower. *Food / Nahrung* **48**(1), 25–33.

Schwarzbach, A. Schreiner, M. & Knorr, D. (2006). Effect of cultivars and deep freeze storage on saponin content of white asparagus spears (*Asparagus officinalis* L.). *Eur Food Res Technol* **222**(1-2), 32–35.

Seljasen, R. Bengtsson, G. B. Hoftun, H. & Vogt, G. (2001). Sensory and chemical changes in five varieties of carrot (*Daucus carota* L) in response to mechanical stress at harvest and post-harvest. *J. Sci. Food Agric* **81**, 436–447.

Seljasen, R. Bengtsson, G. B. Skrede, G. & Vogt, G. (2000). Rapid analysis of 6-methoxymellein in carrots by boiling water extraction, solid phase extraction and HPLC. *Food Chem* **70**(3), 397–401.

Shallenberger, R. S. & Acree, T. E. (1971). Chemical structure of compounds and their sweet and bitter taste. In *Taste* (Ed *L. M. Beidler)*, pp. 221–277. Springer Berlin Heidelberg.

Sharafi, M. Hayes, J. E. & Duffy, V. B. (2013). Masking vegetable bitterness to improve palatability depends on vegetable type and taste phenotype. *Chemosens Percept* **6**(1), 8–19.

Shragg, T. A. Albertson, T. E. & Fisher Jr, C. J. (1982). Cyanide poisoning after bitter almond ingestion. *West J Med* **136**(1), 65–69.

Simon, P. W. Peterson, C. E. & Lindsay, R. C. (1980). Genetic and environmental influences on carrot flavor. *Journal of the American Society for Horticultural Science* **105**, 416–420.

Simon, P. W. Peterson, C. E. & Lindsay, R. C. (1982). Genotype, soil, and climate effects on sensory and objective components of carrot flavor. *J. Amer. Soc. Hort. Sci* **107**(4), 644–648.

Singleton, V. L. & Esau, P. (1969). Phenolic substances in grapes and wine, and their significance. *Advances in Food Research Supplement*, 1–261.

Singleton, V. L. & Noble, A. C. (1976). Wine flavor and phenolic substances. In *Phenolic, Sulfur, and Nitrogen Compounds in Food Flavors* pp. 47–70. American Chemical Society.

Soares, S. Kohl, S. Thalmann, S. Mateus, N. Meyerhof, W. & De Freitas, V. (2013). Different phenolic compounds activate distinct human bitter taste receptors. *J Agr Food Chem* **61**(7), 1525–1533.

Sondheimer, E. (1957). The isolation and identification of 3-methyl-6-methoxy-8-hydroxy-3,4-dihydroisocoumarin from carrots. *J. Am. Chem. Soc.* **79**, 5036–5039.

Stan, S. D. Kar, S. Stoner, G. D. & Singh, S. V. (2008). Bioactive food components and cancer risk reduction. *Journal of Cellular Biochemistry* **104**(1), 339–356.

Steevens, J. Schouten, L. J. Goldbohm, R. A. & van den Brandt, P. A. (2011). Vegetables and fruits consumption and risk of esophageal and gastric cancer subtypes in the Netherlands cohort study. *International Journal of Cancer* **129**(11), 2681–2693.

Steiner, J. E. Glaser, D. Hawilo, M. E. & Berridge, K. C. (2001). Comparative expression of hedonic impact: affective reactions to taste by human infants and other primates. *Neuroscience & Biobehavioral Reviews* **25**(1), 53–74.

Suess, B. Festring, D. & Hofmann, T. (2015). Umami compounds and taste enhancers. In *Flavour Development, Analysis and Perception in Food and Beverages* Eds J. K. Parker, J. S. Elmore & L. Methven), pp. 331–351. Woodhead Publishing.

Sun, H.-J. Cui, M.-l. Ma, B. & Ezura, H. (2006). Functional expression of the taste-modifying protein, miraculin, in transgenic lettuce. *FEBS Letters* **580**(2), 620–626.

Talcott, S. T. & Howard, L. R. (1999a). Chemical and sensory quality of processed carrot puree as influenced by stress-induced phenolic compounds. *J Agric Food Chem* **47**, 1362–1366.

Talcott, S. T. & Howard, L. R. (1999b). Determination and distribution of 6-methoxymellein in fresh and processed carrot puree by a rapid spectrophotmetric assay. *J Agric Food Chem* **47**, 3237–3242.

Tan, K. W. Killeen, D. P. Li, Y. Paxton, J. W. Birch, N. P. & Scheepens, A. (2014). Dietary polyacetylenes of the falcarinol type are inhibitors of breast cancer resistance protein (BCRP/ABCG2). *European Journal of Pharmacology* **723**, 346–352.

Tepper, B. J. (2008). Nutritional implications of genetic taste variation: the role of PROP sensitivity and other taste phenotypes. *Annu Rev Nutr* **28**(1), 367–388.

Theerasilp, S. & Kurihara, Y. (1988). Complete purification and characterization of the taste-modifying protein, miraculin, from miracle fruit. *J Biol Chem* **263**(23), 11536–11539.
Tiwari, U. & Cummins, E. (2013). Factors influencing levels of phytochemicals in selected fruit and vegetables during pre- and post-harvest food processing operations. *Food Research International* **50**(2), 497–506.
Tohge, T. Alseekh, S. & Fernie, A. R. (2014). On the regulation and function of secondary metabolism during fruit development and ripening. *Journal of Experimental Botany* **65**(16), 4599–4611.
Toomey, V. M. Nickum, E. A. & Flurer, C. L. (2012). Cyanide and amygdalin as indicators of the presence of bitter almonds in imported raw almonds. *Journal of Forensic Sciences* **57**(5), 1313–1317.
U.S. Census Bureau (2012). Table 862. Commercial vegetable and other specified crops. p. 552. Statistical Abstract of the United States.
Valkonen, J. P. T. Keskitalo, M. Vasara, T. Pietilä, L. & Raman, K. V. (1996). Potato glycoalkaloids: A burden or a blessing? *Critical Reviews in Plant Sciences* **15**(1), 1–20.
Vallejo, F. Tomás-Barberán, F. A. & Ferreres, F. (2004). Characterisation of flavonols in broccoli (*Brassica oleracea* L. var. *italica*) by liquid chromatography–UV diode-array detection–electrospray ionisation mass spectrometry. *Journal of Chromatography A* **1054**(1–2), 181–193.
Van der Wel, H. Larson, G. Hladik, A. Hladik, C. M. Hellekant, G. & Glaser, D. (1989). Isolation and characterization of pentadin, the sweet principle of *Pentadiplandra brazzeana* Baillon. *Chem Senses* **14**(1), 75–79.
van Doorn, H. E. van der Kruk, G. C. van Holst, G.-J. Raaijmakers-Ruijs, N. C. M. E. Postma, E. Groeneweg, B. & Jongen, W. H. F. (1998). The glucosinolates sinigrin and progoitrin are important determinants for taste preference and bitterness of Brussels sprouts. *J Sci Food Agr* **78**(1), 30–38.
van Poppel, G. Verhoeven, D. H. Verhagen, H. & Goldbohm, R. A. (1999). *Brassica* vegetables and cancer prevention. In *Advances in Nutrition and Cancer 2* Eds V. Zappia, F. Della Ragione, A. Barbarisi, G. Russo & R. Iacovo), pp. 159–168. Springer.
Vanderpas, J. (2006). Nutritional epidemiology and thyroid hormone metabolism. *Annu Rev Nutr* **26**(1), 293–322.
VanEtten, C. H. Daxenbichler, M. E. Kwolek, W. F. & Williams, P. H. (1979). Distribution of glucosinolates in the pith, cambial-cortex, and leaves of the head in cabbage, *Brassica oleracea* L. *J Agr Food Chem* **27**(3), 648–650.
Verkerk, R. Schreiner, M. Krumbein, A. Ciska, E. Holst, B. Rowland, I. De Schrijver, R. Hansen, M. Gerhäuser, C. Mithen, R. & Dekker, M. (2009). Glucosinolates in *Brassica* vegetables: the influence of the food supply chain on intake, bioavailability and human health. *Molecular Nutrition & Food Research* **53**(S2), S219–S265.
Vetter, J. (2000). Plant cyanogenic glycosides. *Toxicon* **38**(1), 11–36.
Vincken, J.-P. Heng, L. de Groot, A. & Gruppen, H. (2007). Saponins, classification and occurrence in the plant kingdom. *Phytochemistry* **68**(3), 275–297.
Volden, J. Wicklund, T. Verkerk, R. & Dekker, M. (2008). Kinetics of changes in glucosinolate concentrations during long-term cooking of white cabbage (*Brassica oleracea* L. ssp. *capitata f. alba*). *J Agr Food Chem* **56**(6), 2068–2073.
Walters, D. E. (1996). How are bitter and sweet tastes related? *Trends in Food Science & Technology* **7**(12), 399–403.
Webster, B. & Chesney, A. M. (1930). Studies in the etiology of simple goiter. *American Journal of Pathology* **6**, 275–284.
Wilkie, L. M. & Capaldi Phillips, E. D. (2014). Heterogeneous binary interactions of taste primaries: Perceptual outcomes, physiology, and future directions. *Neuroscience & Biobehavioral Reviews* **47**, 70–86.

Wilkie, L. M. Capaldi Phillips, E. D. & Wadhera, D. (2013). Sucrose and non-nutritive sweeteners can suppress the bitterness of vegetables independent of PTC taster phenotype. *Chemosens Percept* **6**(3), 127–139.

Wilkie, L. M. Capaldi Phillips, E. D. & Wadhera, D. (2014). Sodium chloride suppresses vegetable bitterness only when plain vegetables are perceived as highly bitter. *Chemosens Percept* **7**(1), 10–22.

Wirthensohn, M. G. Chin, W. L. Franks, T. K. Baldock, G. Ford, C. M. & Sedgley, M. (2008). Characterising the flavour phenotypes of almond (*Prunus dulcis* Mill.) kernels. *J Hortic Sci Biotech* **83**(4), 462–468.

Wooding, S. Gunn, H. Ramos, P. Thalmann, S. Xing, C. & Meyerhof, W. (2010). Genetics and bitter taste responses to goitrin, a plant toxin found in vegetables. *Chem Senses* **35**(8), 685–692.

Wulfkuehler, S. Gras, C. & Carle, R. (2013a). Sesquiterpene lactone content and overall quality of fresh-cut witloof chicory (*Cichorium intybus* L. var. *foliosum* Hegi) as affected by different washing procedures. *J Agr Food Chem* **61**(32), 7705–7714.

Wulfkuehler, S. Kurfiss, L. Kammerer, D. Weiss, A. Schmidt, H. & Carle, R. (2013b). Impact of different washing procedures on quality of fresh-cut iceberg lettuce (*Lactuca sativa* var. *capitata* L.) and endive (*Cichorium endivia* L.). *Eur Food Res Technol* **236**(2), 229–241.

Wulfkuehler, S. Gras, C. & Carle, R. (2014). Influence of light exposure during storage on the content of sesquiterpene lactones and photosynthetic pigments in witloof chicory (*Cichorium intybus* L. var. *foliosum* Hegi). *LWT-Food Sci Technol* **58**(2), 417–426.

Yada, S. Lapsley, K. & Huang, G. (2011). A review of composition studies of cultivated almonds: Macronutrients and micronutrients. *J Food Compos Anal* **24**(4–5), 469–480.

Yamashita, H. Theerasilp, S. Aiuchi, T. Nakaya, K. Nakamura, Y. & Kurihara, Y. (1990). Purification and complete amino acid sequence of a new type of sweet protein taste-modifying activity, curculin. *J Biol Chem* **265**(26), 15770–15775.

Yusof, S. Ghazali, H. M. & King, G. S. (1990). Naringin content in local citrus fruits. *Food Chem* **37**(2), 113–121.

Zabaras, D. Roohani, M. Krishnamurthy, R. Cochet, M. & Delahunty, C. M. (2013). Characterisation of taste-active extracts from raw *Brassica oleracea* vegetables. *Food & function* **4**(4), 592–601.

Zarzecka, K. Gugala, M. & Mystkowska, I. (2013). Glycoalkaloid contents in potato leaves and tubers as influenced by insecticide application. *Plant Soil Environ* **59**(4), 183–188.

Zeinstra, G. G. Koelen, M. A. Kok, F. J. & de Graaf, C. (2010). The influence of preparation method on children's liking for vegetables. *Food Qual Prefer* **21**(8), 906–914.

Zhang, S. Miao, H. Sun, R. Wang, X. Huang, S. Wehner, T. C. & Gu, X. (2013). Localization of a new gene for bitterness in cucumber. *J Hered* **104**(1), 134–139.

Zidorn, C. Johrer, K. Ganzera, M. Schubert, B. Sigmund, E. M. Mader, J. Greil, R. Ellmerer, E. P. & Stuppner, H. (2005). Polyacetylenes from the *Apiaceae* vegetables carrot, celery, fennel, parsley, and parsnip and their cytotoxic activities. *J Agr Food Chem* **53**(7), 2518–2523.

Zitnak, A. & Filadelfi, M. A. (1985). Estimation of taste thresholds of three potato glycoalkaloids. *Can I Food Sc Tech J* **18**(4), 337–339.

5 Bitterness in Beverages

Ayyappan A. Aachary and Michael N. A. Eskin

5.1 INTRODUCTION

The world consumption of beverages continues to increase annually. A number of these beverages contain bitter components that make them unique yet still quite acceptable by many consumers. Hot tea still remains the most popular beverage worldwide and accounts for around 21% of all beverages consumed. Other popular beverages with bitter flavors include hot coffee, hot cocoa or chocolate, beer, wine and cider. This chapter will discuss those compounds responsible for bitterness in these beverages.

5.2 BITTERNESS IN TEA

Next to water, tea remains the second most popular beverage in the world with China and India accounting for close to 60% of world production. The three major categories of tea, based on their manufacture, are unfermented green tea, partially fermented Oolong tea, and completely fermented black tea (Mihara *et al.*, 2004; Bhattacharyya *et al.*, 2007). About 78% of total tea production is focused on black tea that is consumed primarily in Western countries (Kraujalytė *et al.*, 2016). Oolong and green teas are very popular in India and China, although they are starting to make inroads in Western countries.

The presence of high levels of catechins, particularly in green tea, is responsible for bitterness while polymerized catechins, theaflavins and thearubigins, contribute to the quality of fermented black tea (Hilton & Ellis, 1972). HPLC detection of catechins and caffeine successfully differentiated between green and black teas (Fernández *et al.*, 2000). Around 3% of catechins in some green teas is contributed by (-)-epigallocatechin gallate (EGCG) and (-)-epigallocatechin (EGC) compared to much lower levels of (-)- epicatechingallate (ECG), the main catechin in black tea samples. Caffeine ranged from 1-3.5% but was higher in the instant teas. A later

Bitterness: Perception, Chemistry and Food Processing, First Edition. Edited by Michel Aliani & Michael N. A. Eskin.

study by Lee and Lee (2008) reported a much higher content of EGCG and caffeine in green tea extracts which ranged from 0.30-2.04 and 0.85-10.22 mg/g, respectively. This compared to the corresponding levels of 0.24-0.32 and 1.01-5.26 mg/g for EGCG and caffeine in black tea extracts, respectively.

The four major catechins in green tea, (−)–epicatechin (EC), ECG, EGC and EGCG all contribute to its bitterness, and astringency (Fig. 5.1) (Narukawa *et al.*, 2011). These researchers evaluated bitterness of green tea catechins using a cell-based assay with the human taste receptor hTAS2R39. The strongest response to bitter taste htAS2R39 was observed for ECG, followed by ECGC. Increasing the content of catechins in green tea, especially ECG, directly correlated with the taste intensity that ultimately reduced its palatability. These observations were later confirmed with mouse behavioral assays and taste sensor analysis with the greater bitter intensity of ECG or EGCG in humans attributed to the presence of galloyl groups. This could explain why both EC and EGC elicited weaker

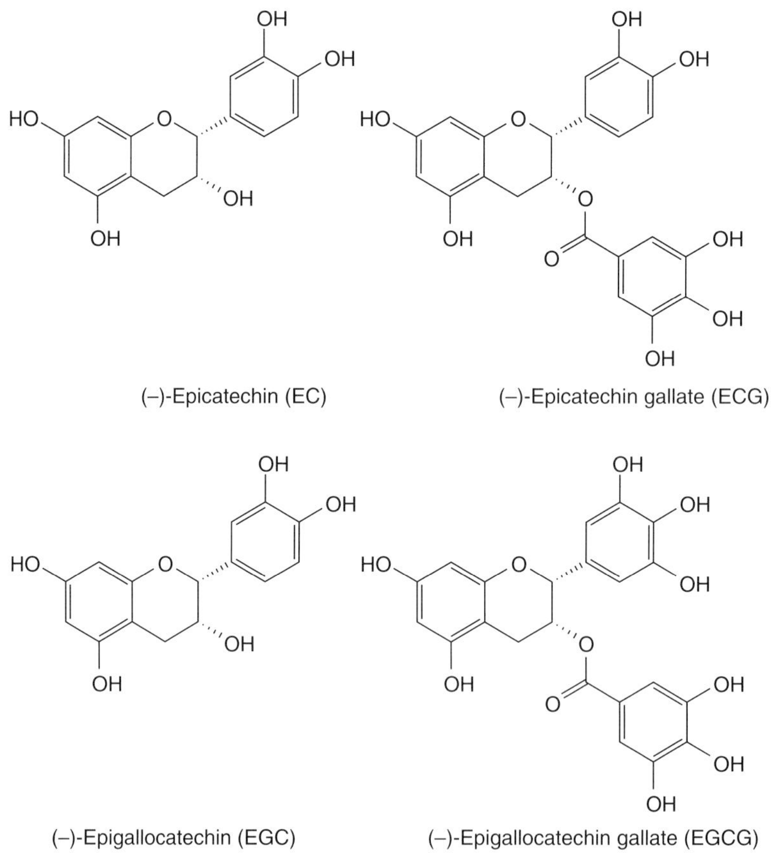

Figure 5.1 Major catechins of green tea (Narukawa *et al.*, 2011) Reproduced with permission of Elsevier.

bitterness intensity. A recent study by Yamazaki *et al.* (2013) examined the effect of green tea catechins on 25 human bitter-taste receptors (TAS2Rs). They found that, in addition to the bitter receptor hTAS2R39, hTA2SR14 also responded to green tea catechins. Of the catechins studied, however, only ECG and EGCG activated htA2SR14 as it was not activated by either EC or EGC.

A spectrum of compounds including phenolics, purine alkaloids, nucleotides, amino acids, organic acids, ions, and sugars determine the taste attributes of green tea (Yu *et al.*, 2014; Wang & Ruan, 2009; Liang *et al.*, 2008). Next to tea polyphenols, caffeine, a purine alkaloid, is the main contributor to tea bitterness. Generally the non-volatile compounds of green tea are the major source of bitterness. However these compounds have different bitter taste transduction pathways.

A bitter-taste-receptor-independent activation of bitter taste was observed previously with caffeine (Rosenzweig *et al.*, 1999). Green tea also contains a few compounds with umami qualities such as guanylic acid (GMP) and inosinic acid (IMP). Approximately one-half of the total amino acids in green tea are contributed by a non-proteinogenic unique amino acid namely L-theanine (5-*N*-ethyl-l-glutamine), which imparts sweet, brothy and umami characteristics to green tea (Juneja *et al.*, 1999). A recent sensory study pointed to the contribution of EGCG, caffeine and L-glutamic acid to the taste of ready-to-drink green tea. While the first two compounds elicited a bitter taste, L-glutamic acid contributed its characteristic umami taste (Yu *et al.*, 2014).

5.3 BITTERNESS IN COFFEE

Coffee consumption is extremely popular in developed countries where it accounts for over 70% of global consumption. Recent research has shown that caffeine, 5-hydroxymethyl-2-furanaldehyde, furfuryl alcohol, diketopiperazines, pyrazines, and trigonelline are the potential candidates for bitterness in coffee. The roasting of coffee beans results in a spectrum of volatile key odorants and non-volatile bitter compounds in coffee (Semmelroch & Grosch, 1995, 1996, Mayer *et al.*, 2000; Frank *et al.*, 2006, 2007, 2008; Blumberg *et al.*, 2010) (Scheme 5.1). Information available on the latter group of molecules, however, is still incomplete. The trans-esterification, epimerization, and lactonisation of non-bitter caffeoylquinic acids (3-*O*-, 4-*O*-, and 5-*O*-) during coffee roasting generate caffeoyl quinides (Clifford, 1979; Frank *et al.*, 2006, 2008; Blumberg *et al.*, 2010). These are highly bitter tasting compounds which are also generated from dicaffeoylquinic acids, during toasting. The formation of 4-vinylcatechol from corresponding lactones of caffeoylquinic acids have also been reported. Interestingly, oligomerization of 4-vinylcatechol generated a group of polyhydroxylated phenylindans with harsh and lingering bitter taste attributes (Blumberg *et al.*, 2010; Frank *et al.*, 2007). Blumberg *et al.* (2010) also studied the importance of roasting time and temperature

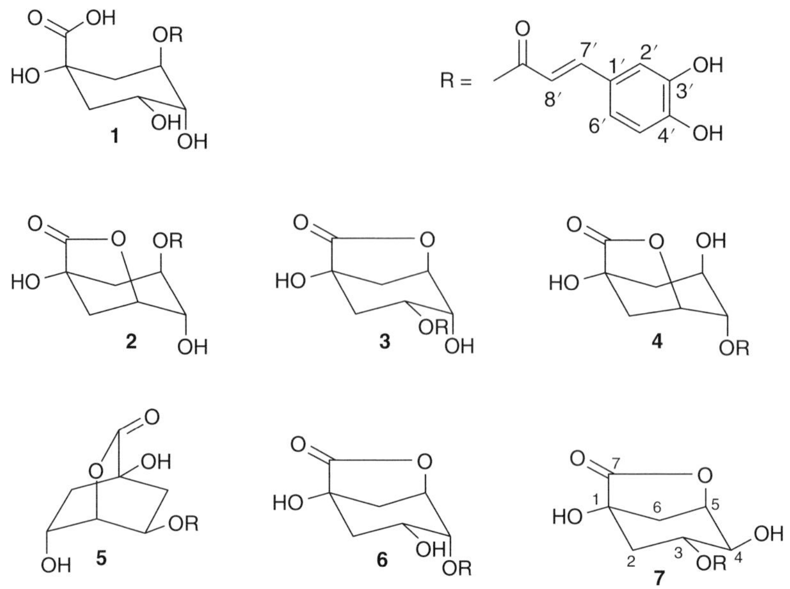

Scheme 5.1 Bitter lactones identified in coffee. 5-*O*-caffeoyl-muco-γ-quinide (2), 3-*O*-caffeoyl-γ-quinide (3), 4-0-caffeoyl-muco-γ-quinide (4), 5-*O*-caffeoyl-epi-δ-quinide (5), and 4-*O*-caffeoyl-γ-quinide (6), as well as the novel 3-0-caffeoyl-epi-γ-quinide (7) formed upon thermal treatment (30 min., 230°C) of 5-*O*-caffeoylquinic acid (1) (Frank *et al.*, 2006).

on the stability of bitter precursors; 4-*O*-caffeoyl quinic acid, 5-*O*-caffeoyl quinic acid and 3-*O*-caffeoyl quinic acid and on the formation of bitter compounds.

Degradation of *O*-caffeoylquinic acids also generates a family of hydroxybenzenes (di/tri), such as hydroxyhydroquinone, pyrogallol, 4-methylcatechol, catechol, and 4-ethylcatechol (Tressl *et al.*, 1978; Clifford, 1979, Haffenden & Yaylayan, 2005; Lang *et al.*, 2006). Bitterness in coffee is also provided by certain products of Maillard-type and caramelization reactions of carbohydrates such as 5-(hydroxymethyl)furan-2-aldehyde (Richards, 1956; Belitz, 1977; Antal *et al.*, 1990; Lewkowski, 2001; Moon & Shibamoto, 2009). Furfuryl alcohol, another furan derivative, also adds to coffee's bitterness (Shibamoto *et al.*, 1981). There is no data, however, on whether these compounds act as transient intermediates for the production of bitter taste compounds in coffee during roasting.

Most of the bitter taste compounds in coffee are prone to oxidation and are highly unstable, which makes it difficult to identify them based on fractionation. A synthetic-constructive strategy followed by Kreppenhofer *et al.* (2011), however, provided new information. Thermal treatment of binary mixtures of a furan derivative and di/ trihydroxybenzene generated (furan-2-yl) methylated benzene diols and triols with their bitter threshold assessed using a sensory panel. Kreppenhofer *et al.* (2011) was able to identify new categories of bitter compounds from roasted coffee including 4-(furan-2-ylmethyl)benzene-1,2,3-triol, 4-(furan-2-

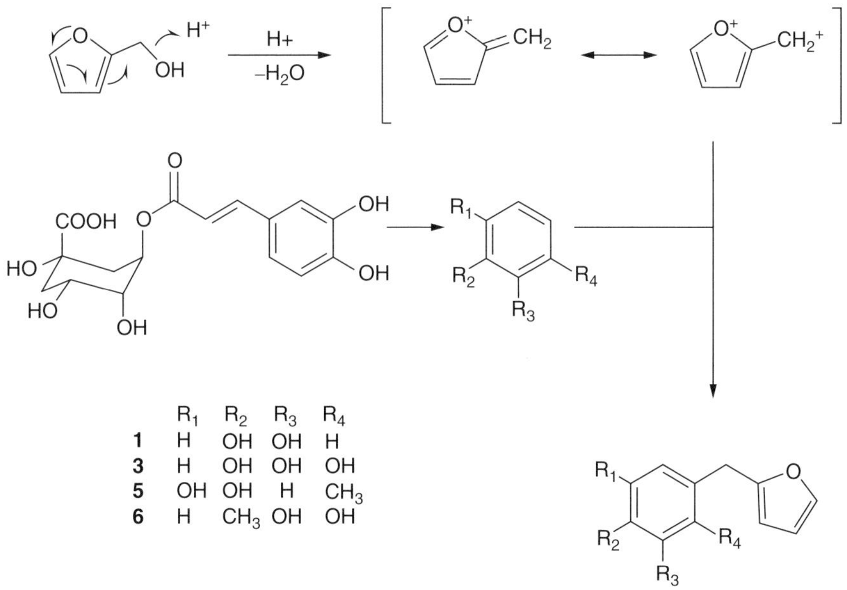

Scheme 5.2 Reaction mechanism leading to the formation of (furan-2-yl) methylated benzene diols and triols **1,3,5,6** from furfuryl alcohol and 5-0-chlorogenic acid upon coffee roasting (Kreppenhofer *et al.*, 2011).

ylmethyl)benzene-1,2-diol, 3-(furan-2-ylmethyl)-6-methylbenzene-1,2-diol and 4-(furan-2-ylmethyl)-5-methylbenzene-1,2-diol (Scheme 5.2).

Coffee alkaloids (caffeine and trigonelline) are not the primary elicitors of bitter taste in coffee as roasting-generated compounds such as furfurylalcohol, 5-hydroxymethyl-2-furancarboxaldehyde, pyrazines and 2,5-diketopiperazines were more responsible for this attribute (Belitz, 1977; Chen, 1979; Shibamoto *et al.*, 1981; Ginz & Engelhardt, 2000). Quinic acid, produced by thermal degradation of chlorogenic acids, exhibited an aspirin-like bitter taste with a threshold level of 10 ppm. Its concentration is far higher than this level in roasted coffee, indicating its contribution to bitterness (Maga & Katz, 1978; McCamey *et al.*, 1990). Rizzi *et al.* (2004) provided evidence for the degradation of *O*-caffeoylquinic acids which might also contribute to the bitter taste of roasted coffee.

As discussed earlier, roasting converts the non-bitter 5-*O*-caffeoylquinic acid into various bitter-tasting chlorogenic acid lactones. These compounds include 5-*O*-caffeoyl-*epi*-δ-quinide, 3-*O*-caffeoyl-γ-quinide, 5-*O*-caffeoyl-*muco*-γ-quinide, 4-*O*-caffeoyl-γ-quinide, and 4-*O*-caffeoyl-*muco*-γ-quinide (Frank *et al.*, 2006). Other bitter taste molecules formed during roasting of coffee include 4-*O*-feruloyl-γ-quinide, 3-*O*-feruloyl-γ-quinide, 3,5-*O*-dicaffeoyl-*epi*-δ-quinide, 3,4-*O*-dicaffeoyl-γ-quinide, and 4,5-*O*-dicaffeoyl-*muco*-γ-quinide (Scheme 5.1). The bitter threshold levels of these compounds vary depending upon on their structure. Frank *et al.* (2008) later identified a previously unknown bitter compound in coffee, namely 3-*O*-caffeoyl-*epi*-γ-quinide, which exhibited a low bitter recognition threshold of 19.5 ppm. The data reported so far in the literature is rather contradictory, as the molecular nature of the key bitter tasting compounds in coffees is still unclear. Lang *et al.* (2015) recently isolated and identified furokaurane glucoside mozambioside, a highly polar, bitter-tasting subfraction during sensory-guided fractionation of a roasted coffee beverage. The bitter taste threshold of furokaurane glucoside mozambioside was 60(±10)µmol/L. Arabica coffee beans proved a rich source ranging from 396-1188 nmol/g mozambioside compared to only trace amounts (<5 nmol/g) detected in the corresponding Robusta coffee beans. The impact of roasting on the level of mozambioside, however, still remains to be examined.

5.4 BITTERNESS IN COCOA/HOT CHOCOLATE

The two major steps in the primary processing of the cocoa beans, fermentation and drying, are responsible for the development of most of the flavor precursors. Microbial fermentation of cocoa pulp as well as enzymatic degradation of carbohydrates, proteins and polyphenols of cocoa beans contribute to flavor development. The role of endogenous enzymes in flavor development is also very significant in the case of cocoa. (Ziegleder, 1991; Jinap *et al.*, 1995; Wood & Lass, 2008). Polyphenol compounds and alkaloids both contribute to the astringency and bitterness of cocoa. The major alkaloids of cocoa are caffeine, methylxanthines, and theobromine, while the important polyphenols include proanthocyanidins and flavan-3-ols (epicatechin and catechin) (Jinap *et al.*, 1995; Ziegleder, 1991; Misnawi *et al.*, 2003; Wollgast & Anklam, 2000). A 30% and 20% reduction of alkaloids and polyphenols during the microbial fermentation of cocoa beans results in a significant reduction in bitterness and astringency.

The formation of flavor precursors during roasting of dried, deshelled cocoa beans or nibs, impart characteristics chocolate flavors (Ziegleder, 1991; Ziegleder & Biehl, 1988). However, the type of cocoa beans, processing time and temperature, fermentation method and its parameters (pH, temperature, etc.) significantly affect the flavor characteristics (Meyer & Biehl, 1989; Biehl *et al.*, 1985, 1990; Baker *et al.*, 1994; Hansen *et al.*, 1998; Hashim *et al.*, 1998a,b; de Brito *et al.*, 2000, Wollgast & Anklam, 2000).

The major compounds involved in the astringent taste of cocoa are flavan-3-ol group, anthocyanins, hydrolyzable tannins and condensed tannins (Bate-Smith, 1973; Haslam & Lilley, 1988). During fermentation the astringency is reduced due to oxidation of flavan-3-ol and tannins, followed by their increased polymerization and complexation with proteins (Bonvehi & Coll, 2000). To better understand and evaluate the bitterness during the roasting of cocoa beans, the typical aroma of cocoa develops through Maillard-type reactions (Mohr *et al.*, 1976; Barel *et al.*, 1985; Hoskin & Dimick, 1994; Jinap *et al.*, 1998; Puziah *et al.*, 1998).

If the fermentation of cocoa beans is incomplete, roasting will not develop the chocolate flavor, but increase astringency and bitterness (Biehl & Voigt, 1996; Puziah *et al*., 1998). The (-)-epicatechin content (3,3',4',5,7-pentahydroxyflavan) of unfermented Forastero cocoa beans was reported to account for approximately 35% of the total polyphenols (Forsyth, 1952; Forsyth & Quesnel, 1963). Using HPLC, Kim and Keeny (1984) showed that the (-) epicatechin content of different varieties of cocoa beans ranged from 21.89-43.27 mg/g of the defatted cocoa bean samples. Experimental microwave and sun drying of beans after harvest was attributed for the lower (-) epicatechin content (21.89 mg/g in defatted cocoa beans) of Trinidad-Jamaican beans. Kim and Keeny (1984) also reported a much lower concentration of (-)-epicatechin, ranging from 2-10 mg/g in the corresponding fermented defatted cocoa beans. As discussed previously, polyphenols contribute more to the development of astringency and bitterness than any other endogenous compounds (Bonvehi & Coll, 2000; Luna *et al*., 2002).

5.5 BITTERNESS IN BEER

Beer, a fermented beverage brewed from malt, is produced by brewing and fermentation of starches and sugar (primarily sucrose). Most beer is flavored with hops in which bitterness is an important factor contributing to its acceptance. The hop plant (*Humulus lupulus*) is the most important herb used in the brewing industry (Mudura *et al*., 2008, Mudura & Coldea, 2015). Its contribution to the bitterness of beer results in a more balanced and satiating palate to the final product (Malowicki & Shellhammer, 2005). The golden colored resinous lupulins-granules of hop plant flowers have bitter taste and preservative attributes, which help to produce a fine quality beer (Sakamoto, 2003). Hop acids can be α-acids or β-acids, which make the characteristics of hop resins more complex (de Keukeleire, 1999). These acids do not exhibit any bitter taste and are poorly soluble in water. Depending up on the growing conditions and the variety of hop plant, the concentration of α-acids may vary between 2-15% (Bamforth, 2000).

Three important analogues of hop α-acids, humulone, cohumulone, and adhumulone were characterized, along with the minor α-acids posthumulone and prehumulone (Jaskula *et al*., 2007). They vary structurally due mainly to differences in the side chains (Bamforth, 2000). Kolpin (2010) reported that the amount of the α-acid portion of resins from various hop varieties with cohumulone, and humulone contributing the major portion (20-50%) followed by adhumulone (15%). While a higher level of α-acids was reported in wort, beer exhibited only lower levels of α-acids (Haseleu *et al*., 2010), possibly because of the thermal isomerization of the α-acids to the iso-α-acids during the wort boiling process (de Keukeleire, 1999). Generally the percentage conversion of α-acids to iso-α-acids never goes beyond 50% and at the same time only about 25% of the original bittering potential survives in the beer (Bamforth, 2000) (Fig. 5.2).

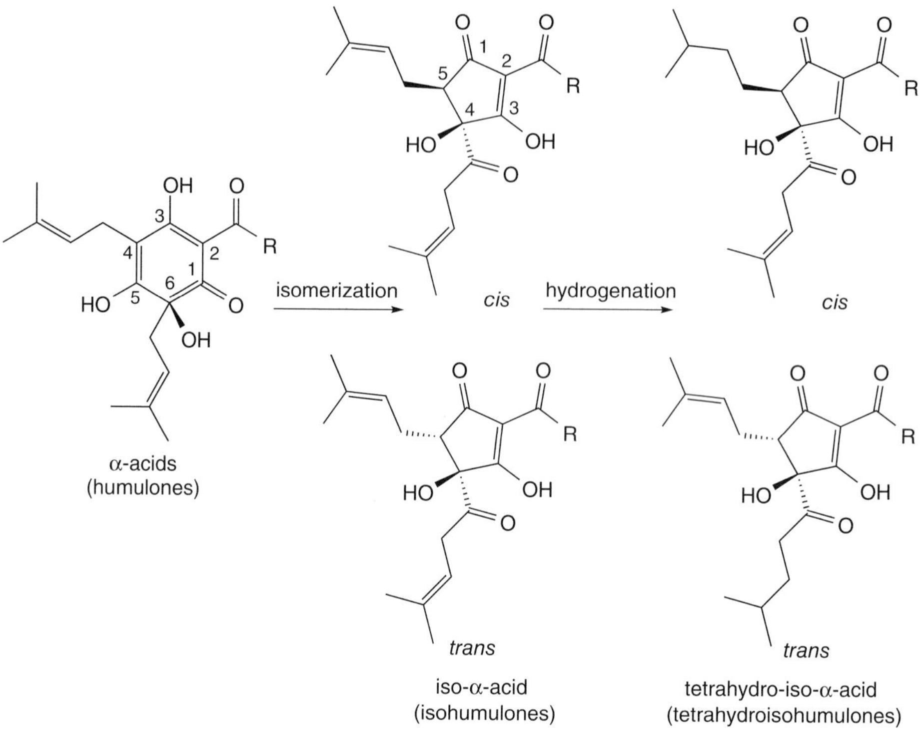

Figure 5.2 Scheme showing conversion of α-acids to iso-α-acids and tetra-iso-α-acids (Humulones: R = isobutyl, cohumulones: R = isopropyl, adhumulones: R = *sec* butyl) (Urban *et al.*, 2013).

In terms of pKa values, the iso-α-acids are more acidic (pH~3) than hop acids (pH~5.5) and therefore iso-α-acids exhibit a better solubility in lager beers (pH 4.2-4.4) which results in intense bitterness (de Keukeleire *et al.*, 1992; Huvaere *et al.*, 2004a). In addition, iso-α-acids contribute to beer stability (Bamforth & Kanauchi, 2003; Ferreira *et al.*, 2005; Blanco *et al.*, 2006) owing to their stereoisomeric and hydrophobic properties. The stereoisomerism of iso-α-acid resulted from the spatial arrangement of the prenyl side chain at C5 and the tertiary alcohol function at C4 (de Cooman *et al.*, 2000). The *trans*-isomer of isohumulones is the major isohumulones in the beer foam compared to its *cis*-isomer because of its higher hydrophobic potential (Kappler *et al.*, 2010a,b).

Spoilage due to microbial action and as a result haze formation were the two important factors that defined the quality of beer. It was reported that iso-α-acids result in significant microbiological stability of beer (Sakamoto, 2003; Vaughan *et al.*, 2005; Suzuki *et al.*, 2006; Blanco *et al.*, 2007, Van Cleemput *et al.*, 2009; Hazelwood *et al.*, 2010), as well as beneficial health effects (Yajima *et al.*, 2004; Shimuraa *et al.*, 2005; Obara *et al.*, 2009). The reduction in aroma and bitterness from storage-induced reactions has a great consequence on the shelf-stability of beers by the formation of off-flavor compounds (Pozdrik *et al.*, 2006; Vanderhaegen *et al.*, 2006). Those compounds are mainly the products of two important reactions such as photodegradation and radical-assisted oxidation of iso-α-acids (Burns *et al.*, 2001; Huvaere *et al.*, 2004b). Research on nonvolatile off-taste development in beer, however, is scarce. The major consequences of such reactions are long-lasting, lingering, and harsh bitterness of beer (King & Duineveld, 1999; de Cooman *et al.*, 2000). Data on such off-flavour developments at the molecular level is still not available.

Even though the iso-α-acids formation best fit to a first-order reaction, the rate constants for the *trans*-isomers are much lower than *cis*-isomers. This difference is more obvious at higher temperatures (~100 °C) and is possibly attributed to relatively higher activation energy for *cis*-isomer formation ((Malowicki & Shellhammer, 2006; Jaskula *et al.*, 2008). Thermodynamic stability of *cis*-isomer is also explained with respect to its lower Gibbs' free energy (Hughes, 2006) and minimal electrostatic interactions due to its special arrangement of side chains at C4 and C5 on opposite sides of the five-membered ring (Jaskula *et al.*, 2007).

In beer, various iso-α-acids are present in the following order: *cis/trans*-isohumulone (43%) > *cis/trans*-isocohumulone (39%) > *cis/trans*-isoadhumulone (16.5%) (Intelmann *et al.*, 2009). The initial *trans/cis* ratio of freshly brewed beer is ~0.4, however this ratio changes during storage, possibly due to the higher stability of *cis*-isomers, which in turn results in changes in flavor and taste (de Keukeleire, 1999; Intelmann *et al.*, 2009; Nimubona, 2010).

About 80% of the bitter taste of beers is attributed to the high concentrations of iso-α-acids (15-100 mg/L) as their sensory threshold detection is much lower than this concentration (5mg/L) (de Keukeleire *et al.*, 1992; Baxter & Hughes, 2001; Heyerick *et al.*, 2003). A comparison of the bitterness of *cis* and *trans* isomers of isohumolones found that *cis*-isohumolones had a significantly stronger bitter taste than the *trans*-isomers. In addition, the iscohumolones were significantly less bitter than the corresponding isohumolones (Hughes & Simpson, 1996; Hughes, 2000; Kappler *et al.*, 2010 a,b).

Even though, the spectrophotometric detection of beer bitterness at 275 nm results in an approximate idea of beer bitterness (Analytica EBC, 1997), the most accurate methods involve chromatographic methods such as HPLC, which helps to identify and quantify specific bittering agents (Kappler *et al.*, 2010a). A polypyrrole sensor array based electronic tongue was used to estimate the bitter intensity of beer with the results in line with the iso-α-acid content of various beer samples examined by HPLC (Arrieta *et al.*, 2010).

A gradual reduction in the intensity of bitterness and increase in sweetness during storage was observed in all beer types (Dalgliesh, 1977). Pangborn *et al.* (1977) compared the bitterness and degree of liking of commercial lagers with the chemical analysis of bitterness units (BU) when stored at different temperatures up to 132 days. BU units represent the amount of iso-alpha acids in beer and are now referred to as International Bitter Units (IBU) or European Bitter Units (EBU).

Using a trained panel, BU values decreased over time and temperature of storage while temperature did not affect sensory bitterness which only decreased with storage time (Pangborn *et al.*, 1977; McMurrough & Byrne, 1992). Oxidation of isohumulones during beer storage at 37 °C for 10 days was reported by Kaneda *et al.* (1992) where the reduction was accelerated by hydrogen peroxide and iron ions. When the beer was stored at 40 °C for 156 days, Walters *et al.* (1997) reported a 71% reduction of total iso-α-acids, however in the control beer samples stored at 25°C such changes in iso-α-acids content were not observed. Iso-α-acids at a range of 20-40 mg/L is generally considered to be bitter (Crombecq, 1995) while a bitterness range of 17.5-25 EBU was found to be preferred by a sensory panel when lager beer was prepared without hops, but added with mixtures of isohumulone/isoadhumulone/isocohumulone (Collin *et al.*, 1994). Cepicka *et al.* (1992) made time–intensity measurements to understand temporal aspects of flavor perception that would be lost in traditional scaling of bitterness, but without much success. Previously, Pangborn *et al.* (1983) used a chart-recorder TI method and Hughes and Simpson (1994) used computerized-TI measurements to study the bitterness of beer.

To minimize degradation of iso-α-acids during the storage of beer requires good brewing practices that ensure oxygen levels in beer are as low as possible (<100 mg/L). This can be achieved by minimizing the accumulation of oxygen during filling (Stewart, 2004). In addition, packaging in brown bottles is essential so that visible light (300-500 nm) cannot penetrate and trigger the photodegradation of iso-α-acids by riboflavin (vitamin B2) (Caballero *et al.*, 2012).

5.6 BITTERNESS IN WINE

A recent paper by Atero *et al.* (2015) indicated the unanimity among health professionals regarding the beneficial effects of moderate wine consumption by individuals suffering from diabetes, osteoporosis, cardiovascular disease, neurological diseases and longevity. Flavanol polymers such as proanthocyanidins or condensed tannins greatly influence bitterness and astringency of wine (Lea & Arnold, 1978; Fischer & Noble, 1994). The fermentation process leaches out these compounds from the grape seeds and skins into red wine (Gawel *et al.*, 2001).

Assessment of wine sensory properties is made more difficult by the confusion between the sour and bitter taste and the sensation of astringency (Lee & Lawless, 1991). In addition, the presence of other basic substances also affects the perception of astringency (Brannan *et al.*, 2001). Many physical and chemical properties are involved in the complex mechanisms that elicit astringency and bitterness. In addition to the content of proanthocyanidin, the degree of polymerization, and extent of galloylation also contribute to bitterness and astringency (Noble, 1994; Vidal *et al.*,

2003, 2004). Other factors which contribute to bitterness and astringency perception in wine include the pH, level of ethanol and viscosity (Ishikawa & Noble, 1995; Kallithraka *et al.*, 1997; DeMiglio *et al.*, 2002).

Ethanol content in wine is one of the major contributing factors of bitterness (Mattes & DiMeglio, 2001; Fontoin *et al.*, 2008). However, noting that few parameters of model wine solutions influence bitterness, it cannot be excluded that some other molecules are involved in the perception of the bitterness of red wines (Brannan *et al.*, 2001). In red wine, tannin composition and concentration are of course important, but are insufficient to totally explain the variation of astringency and bitterness perception which are also influenced by the external physiochemical factors and other compounds (polysaccharides, peptides, ions, volatile compounds, etc.). Ethanol content and pH have a greater role in modifying astringency and bitterness of wine (Fontoin *et al.*, 2008). However, more research is needed in this direction to confirm the effects of ethanol in the astringency and bitterness of wine.

Irrespective of being bitter, some of these compounds do not contribute to bitter taste of wine. For example, the amount of tyramine and tyrosol in wine is not present at sufficient levels needed to elicit bitterness (Singleton & Esau, 1969). A high threshold value of many of these compounds incuding esculin (a bitter glycoside) is the reason for their inability to affect bitterness. This is also the case for the addition of caffeic acid derivatives, caffeoyl tartaric acid and 2-S-glutathionyl caffeoyltartaric acid, when added at levels normally present in wine they did not affect the overall bitterness (Verette *et al.*, 1988). Addition of caffeic acid and its gallic ester, chlorogenic acid, to wine also failed to produce a detectable bitter taste (Verette *et al.*, 1988; Nagel *et al.*, 1987).

While flavonols impart bitterness to wine, the higher molecular weight flavonoids contribute to astringency, which is a tactile sensation rather than a taste recognized by taste receptors. Other constituents affecting the taste perception of flavonols, such as lowering of pH and subsequent increase in astringency or increased level of ethanol and subsequent reduction in bitterness. In grapes and wines, catechin and epicatechin monomers and their polymers contribute to bitterness. These flavonoids are generally found as polymers with eight or more monomeric untis in older wines, while their presence in the younger wines are restricted to dimeric or trimeric form (Ribereau-Gayon, 1972). Kallithraka *et al.* (1997) reported that high concentrations of (-)-epicatechin were more bitter than (+)-catechin. These results were in agreement with that reported by Thorngate and Noble (1995) who attributed this to the higher lipophilicity of (-)-epicatechin. In addition, hydroxybenzoic and hydroxycinnamic acids provide a puckering and velvety astringency (Hufnagel & Hofmann, 2008). Previously it was reported that the monomeric flavan-3-ols, procyanidin dimers and trimers contribute largely to astringency, however, this is now being contradicted (Hufnagel & Hofmann, 2008).

Aroma, flavor and mouthfeel are all interlinked. The aroma substance retention by catechin showed that changes in polyphenol content caused a reduction in aroma compounds (Dufour & Bayonove, 1999). This was attributed to the molecular complexation, which in turn related to the structure of the phenolics. Consequently fruity/floral aromas tend to decrease with increase in polyphenol content (Goldner *et al.*, 2011). Changes in astringency and flavour characteristics were also observed following the addition of a grape seed extract to wines (Cliff *et al.*, 2012).

5.7 BITTERNESS IN CIDER

Cider or apple wine is a popular alcoholic beverage in Europe, North America and Australia. Because apples are a predominant crop in China, cider is also becoming a promising segment in the fruit industry. The four major groups of phenolic compounds present in cider, a fermented product of apple juice, are chlorogenic and *p*-coumaric acids, phenolic acids, dihydrochalcones, catechins and procyanidins (Fig. 5.3). Dihydrochalcones mainly include phloretin xyloglucoside and phloridizin. Both catechins and procyanidins together form the flavan-3-ols of cider. In the context of preparing cider, the content of polyphenols is important as it influences (1) the color of final product and (2) the balance between bitterness to astringency. One of the factors determining the colloidal stability of cider is the content of polyphenols. They are part of alcoholic and malolactic fermentations and exhibit antimicrobial activity as well as in the development of cider aroma.

The important contribution by polyphenols to the color, bitterness and astringency of cider led to an examination of the impact that different apple varieties have on phenolic compounds and sensory properties of cider. Rickstina-Dolge and coworkers (2014) fermented 12 different apple varieties with *Saccharomyces cerrevisiae* yeast '71B-1122' for the production cider in the Faculty of Food Technology at Latvia University of Agriculture. Significant differences ($p<0.05$) in bitterness and astringency were observed among apple samples particularly in ciders made from crab apple varieties *Riku* and *Hyslop*. Crab apples were highest in total phenol content with chlorogenic and caffeic acids the dominating polyphenols in all cider samples. The sensory properties of the finished cider product depended on the physiological composition of the apples used. They concluded that to produce a quality cider product required selecting those apple varieties with an intensive aroma such as the desert apple *Suksis* and moderate astringency from crab apple varieties as *Riku* and *Hyslop*.

The polymeric procyanidins are the main contributors for bitterness and astringency of cider. However, another bitter compound present in cider, phloridzin, is present at too low concentrations to elicit bitterness or astringency characteristics (Lea & Timberlake, 1974; Lea & Arnold, 1978). The bitterness of cider is also contributed by heterofermentative lactic microbiota such as *Lactobacillus* species. The lactic acid bacteria in cider utilize the residual fructose, glycerol, and lactic acid as carbon sources, resulting in changes in ropiness, acetification and bitterness (Dueñas *et al.*, 1994, 1995).

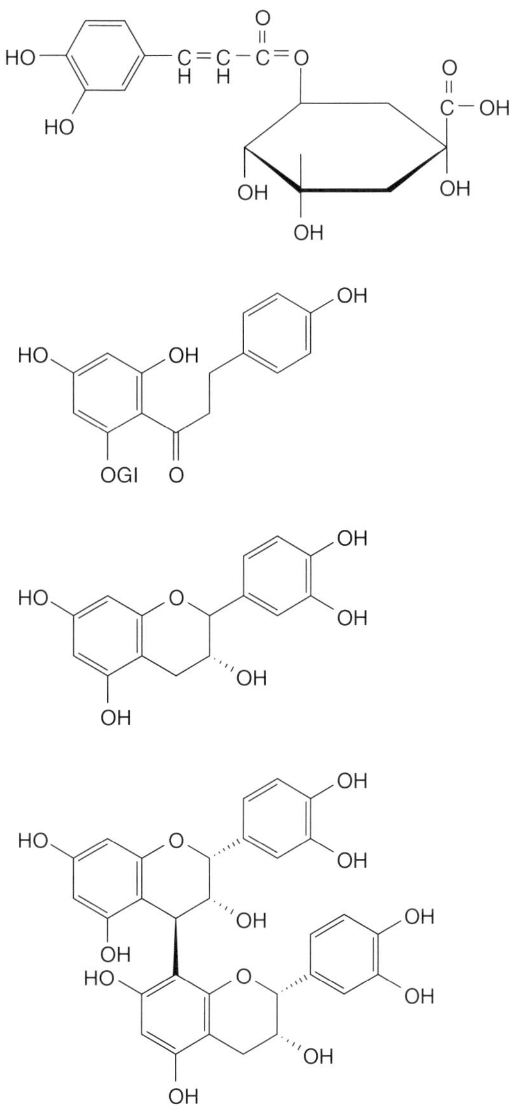

Chlorogenic Acid

Phloridzin

(-)Epicatechin

Procianidin B2

Figure 5.3 Phenolics in cider.

Ye and coworkers (2014) found significant changes in the content of polyphenols and organic acids, which increased during fermentation, many of the other polyphenols, including chlorogenic acid, (-)-epicatechin, and phloridzin decreased by different degrees. Such changes would impact the final flavor of the cider including bitterness. With the exception of protocatechuic acid, which increased during fermentation, many of the other polyphenols, including chlorogenic acid, (-)-epicatechin, and phloridzin decreased by different degrees. Such changes would impact the final flavor of the cider including bitterness. Because of the popularity of French cider, Symoneaux *et al.* (2014a) investigated the impact of degree of polymerization and concentration of procyanidins (or tannins) on the sensory properties of a cider model (a water solution of ethanol, fructose and malic acid). Polyphenols contributed to both the astringency and bitterness of both cider and wine but in cider the phenolic compounds are less polymerized compared to wine. Examination of 90 ciders by Le Quere and coworkers (2006) confirmed that the procyanidins were smaller with an average degree of polymerization ranging between 1.61 to 3.69. This contrasted with wine procyanidins where the average degree of polymerization ranged from 1.8 to 13 (Monogas *et al.*, 2003; Chira *et al.*, 2011; Kassara & Kennedy, 2011). Using 15 trained panelists, Symoneaux and coworkers (2014a) found that different concentrations of four purified procyanidin fractions modified the bitterness, astringency, sweetness and sourness characteristics of the cider model. In the presence of a high concentration of medium procyanidins (a pentamer, DP5) the cider model was much more bitter compared to the model containing a high degree of polymerization (DP) procyanidins. This was consistent with early work by Lea and Arnold (1978) who also reported greater bitterness in the presence of shorter (tetrameric) procyanidins. Symoneaux *et al.* (2014b) then used a fractional factorial design to study the impact on the sweetness, sourness, bitterness and astringency of the cider model by the degree of procyanidin polymerization and its interactions with fructose, ethanol and acidity. The largest increase in bitterness and astringency was observed in the presence of 750 mg/L of procyanidin tetramers and pentamers with no effect on either sweetness or sourness. These researchers suggested that further work is needed to examine the behavior of different polymerization (DP) products of procyanidins in real cider.

REFERENCES

Analytica EBC. (1997). *Brauerei und Getränke Rundschau* (5th ed.) CH-8047, Zürich, Switzerland.

Antal, M.J. Mok, W.S.L. & Richards, G.N. (1990). Mechanism of formation of 5-hydroxymethyl-2-furaldehyde from d-fructose and sucrose. *Carbohyd. Res.* **199**, 91–109.

Arrieta, A.A., Rodriguez-Mendez, M.L., de Saja, J.A., Blanco, C.A. & Nimubona, D. (2010). Prediction of bitterness and alcoholic strength in beer using an electronic tongue. *Food Chem.* **123**, 642–646.

Artero, A., Artero, A., Tarín, J.J. & Cano, A. (2015). The impact of moderate wine consumption on health. *Maturitas.* **80**(1), 3–13.

Baker, D.M., Tomlins, K.I. & Gray, C. (1994). Survey of Ghanaian cocoa farmer fermentation practices and their influence on cocoa flavour. *Food Chem.* **51**, 425–431.

Bamforth, C.W. (2000). Beer: an ancient yet modern biotechnology. *Chem. Ed.* **5**, 102–112.

Bamforth, C.W. & Kanauchi, M. (2003). Interactions between polypeptides derived from barley and other beer components in model foam systems. *J. Sci. Food Agric.* **83**, 1045–1050.

Barel, M., Leon, D. & Vincent, J.C. (1985). Café Cacao. *Thé*, **4**, 277–286.

Bate-Smith, E.C. (1973). Haem analysis of tannins: the concept of relative astringency. *Phytochem.* **12**, 907–912.

Baxter, E.D. & Hughes, P.S. (2001). *Beer: Quality, Safety and Nutritional Aspects.* Royal Society of Chemistry, Cambridge, U.K.

Belitz, H.D. (1977). Flavor-active substances in coffee. *ASIC 7th Colloquium, Hamburg*, ASIC, Bremen, pp. 243–252.

Bhattacharyya, N., Seth, S., Tudu, B., Pradip, T., Arun, J., Devdulal, G., Rajib, B. & Manabendra, B. (2007). Monitoring of black tea fermentation process using electronic nose. *J. Food Eng.* **80**, 1146–1156.

Biehl, B., Brunner, E., Passern, D., Quesnel, V.C. & Adomako, D. (1985). Acidification, proteolysis and flavour potential in fermenting cocoa beans. *J. Sci. Food Agric.* **36**, 583–598.

Biehl, B., Meyer, B., Said, M.B. & Samarakoddy, R.J. (1990). Bean spreading: a method of pulp preconditioning to impair strong nib acidification during cocoa fermentation in *Malaysian. J. Food Agric.* **51**, 35–45.

Biehl B. and Voigt J. (1996). *Biochemistry of chocolate flavour precursors.* International Cocoa Conference, Salvador De Bahia, Brazil.

Blanco, C.A., Rojas, A., Caballero, P., Ronda, F., Gomez, M. & Caballero, I. (2006). A better control of beer properties by predicting acidity of hop iso-α-acids. *Trends Food Sci. Technol.* **17**, 373–377.

Blanco, C.A., Rojas, A. & Nimubona, D. (2007). Effects of acidity and molecular size on bacteriostatic properties of beer hop derivates. *Trends Food Sci. Technol.* **18**, 144–149.

Blumberg, S., Frank, O. & Hofmann, T. (2010). Quantitative studies on the influence of bean roasting parameters and hot water percolation on the concentrations of bitter compounds in coffee brew. *J. Agric. Food Chem.* **58**, 3720–3728.

Bonvehi, J.S. & Coll, F.V. (2000). Evaluation of purine alkaloids and diketopiperazines contents in processed cocoa powder. *Eur. Food Res. Technol.* **210**, 189–195.

Brannan, G.D., Setser, C.S., & Kemp, K.E. (2001). Interaction of astringency and taste characteristics. *J. Sen. Stud.* **16**, 179–197.

Burns, C.S., Heyerick, A., de Keukeleire, D. & Forbes, M.D.E. (2001). Mechanism for formation of the lightstruck flavor in beer revealed by time-resolved electron paramagnetic resonance. *Chemistry: a European Journal.* **7**, 4553–4561.

Caballero, I., Blanco, C. A. & Porras, M. (2012). Iso-α-acids, bitterness and loss of beer quality during storage. *Trends. Food Sci. Technol.* **26**(1), 21–20.

Cepicka, J., Strejcek, F., & Pokorny, J. (1992). Development of the sensory bitterness during the tasting of beer. *Monatsschrift fuÈr Brauwissenschaft.* **10**, 329–331.

Chen, W.C. (1979). Studies on the bitter taste of roasted coffee. Relationship between structure and bitter taste of some organic compounds (in German). Ph.D. Thesis, University of Munich, Munich.

Chira, K., Lorrain, B., Ky, I. & Teissedre, P.L. (2011). TYannin composition of cabernet-sauvignon and merlot grapes from the Bordeaux area for different vintages (2006-2009) and comparison totannin profile of five 2009 vintage Mediterranean grapes varieties. *Mol.* **16**(2), 1519–1532.

Cliff, M.A., Stanich, K., Edwards, J.E. & Saucier, C.T. (2012). Adding grape seed extract to wine affects astringency and other sensory attributes. *J. Food Qual.* **35**(4), 263–271.

Clifford, M.N. (1979). Chlorogenic acids – Their complex nature and routine determinations in coffee beans. *Food Chem.* **4**, 63–71.

Collin, S., Derdelinck, G. & Dufour, J-P. (1994). Relationships between the chemical composition and sensory evaluation of lager beers. *Food Qual. Pref.* **5**, 145–149.

Crombecq, P. (1995). *Bier Jaarboek 1995-1996, Kosmos Z & K:Utrecht/Antwerpen.*

Dalgliesh, C.E. (1977). Flavour stability. *Proc. Congr. Eur. Brew. Conv.* **9**, 623–659.

de Brito, E.S., Pezoa García, N.H., Gallão, M.I., Cortelazzo, A.L., Fevereiro, P.S. & Braga, M.R. (2000). Structural and chemical changes in cocoa (*Theobroma cacao* L) during fermentation, drying and roasting. *J. Sci. Food Agric.* **81**, 281–288.

de Cooman, L., Aerts, G. & Overmeire, H. (2000). Alterations of the profiles of iso-alpha-acids during beer ageing, marked instability of *trans*-iso-alpha-acids and implications for beer bitterness consistency in relation to tetrahydro-iso-alpha-acids. *J. Inst. Brew.* **106**(3), 169–178.

de Keukeleire, D. (1999). Fundamentals of beer and hop chemistry. *Química Nova.* **23**(1), 108–112.

de Keukeleire, D., Vindevogel, J., Sziics, R. & Pat, S. (1992). The history and analytical chemistry of beer bitter acids. *Trends Anal. Chem.* **77**(8), 275–280.

De Miglio, P., Pickering, G.J. & Reynolds, A.G. (2002). Astringent subqualities elicited by red wine: the role of ethanol and pH. In: Cullen, C.W., Pickering, G. J., and Phillips, R., Eds. *Proceedings of the International Bacchus to The Future Conference, St Catharines, ON, May 23–25, 2002*. St. Catharines, ON: Brock University Press.

Dueñas, M., Irastorza, A., Fernandez, C., Bilbao, A. & Huerta, A. (1994). Microbial populations and malolactic fermentation of apple cider using traditional and modified methods. *J. Food Sci.* **59**, 1060–1064.

Dueñas, M., Irastorza, A., Fernandez, C. & Bilbao, A. (1995). Heterofermentative *Lactobacilli* causing ropiness in Basque country ciders. *J. Food Prot.* **58**, 76–80.

Dufour, C. & Bayonov, C.L. (1999). Interactions between wine polyphenols and aroma substances : An insight a the molecular level. *J. Agric. Food Chem.* **47**, 678–684.

Fernández, P. L., Martín, M. J., González, A. G. & Pablos, F. (2000). HPLC determination of catechins and caffeine in tea. Differentiation of green, black and instant teas. *The Analyst.* **125**, 421–425.

Ferreira, I., Jorge, K., Nogueria, L., Silva, F. & Trugo, L. (2005). Effects of of the combination of hydrophobic amino acids, iso-α-acids, and malto-oligosaccharides on beer foam stability. *J. Agric. Food Chem.* **53**, 4976–4981.

Fischer, U. & Noble, A.C. (1994). The effect of ethanol, catechin concentration, and pH on sourness and bitterness of wine. *Am. J. Enol. Vitic.* **45**, 6–10.

Fontoin, H., Saucier, C., Teissedre, P-L. & Gloroes, Y. (2008). Effect of pH, ethanol and acidity on astringency and bitterness of grape seed tannin oligomers in model wine solution. *Food Qual. Pref.* **19**, 286–291.

Forsyth, W.G.C. (1952). Cocoa polyphenolic substances: Changes during fermentation. *Biochem. J.* **51**, 516–520.

Forsyth, W.G.C. & Quesnel, V.C. (1963). Mechanism of cocoa curing. *Adv. Enzymol.* **25**, 457–492.

Frank, O., Blumberg, S., Krümpel, G. & Hofmann, T. (2008). Structure determination of 3-*O*-caffeoyl-*epi*-γ-quinide, an orphan bitter lactone in roasted coffee. *J. Agric. Food Chem.* **56**, 9581–9585.

Frank, O., Blumberg, S., Kunert, C., Zehentbauer, G. & Hofmann, T. (2007). Structure determination and sensory analysis of bitter-tasting 4-vinylcatechol oligomers and their identification in roasted coffee by means of LC-MS/MS. *J. Agric. Food Chem.* **55**, 1945–1954.

Frank, O., Zehentbauer, G. & Hofmann, T. (2006). Bioresponse-guided decomposition of roast coffee beverage and identification of key bitter taste compounds. *Eur. Food Res. Technology*, **222**, 492–508.

Gawel, R., Iland, P.G. & Francis, I.L. (2001). Characterizing the astringency of red wine: a case study. *Food Qual. Pref.* **12**, 83–94.

Ginz, M. & Engelhardt, U.H. (2000). Identification of proline-based diketopiperazines in roasted coffee. *J. Agric. Food Chem.* **48**, 3528–3532.

Goldner, M.C., Leo Lira, di., van Baren, C. & Bandoni, A. (2011). Influence of polyphenol levels on the perception of arma in *Vitis vinifera* cv. Malbec wines. *S. Afr. J. Enol. Vitic.* **32**, 21–27.

Haffenden, L.J.W. & Yaylayan, V.A. (2005). Mechanism of formation of redox-active hydroxylated benzenes and pyrazine in ^{13}C-labeled glycin/D-glucose model systems. *J. Agric. Food Chem.* **53**, 9742–9746.

Hansen, C.E., del Olmo, M. & Burri, C. (1998). Enzyme activities in cocoa beans during fermentation. *J. Sci. Food Agric.* **77**, 273–281.

Haseleu, G., Lagemann, A., Stepha, A., Intelmann, D., Dunkel, A. & Hofmann, T. (2010). Quantitative sensomics profiling of hop-derived bitter compounds throughout a full-scale beer manufacturing process. *J. Agric. Food Chem.* **58**, 7930–7939.

Hashim, P., Selamat, J., Muhammad, S.K.S. & Ali, A. (1998a). Changes in free amino acid, peptide-N, sugar and pyrazine concentration during cocoa fermentation. *J. Sci. Food Agric.* **78**, 535–542.

Hashim, P., Selamat, J., Muhammad, S.K.S. & Ali, A. (1998b). Effect of mass and turning time on free amino acid, peptide-N, sugar and pyrazine concentration during cocoa fermentation. *J. Sci. Food Agric.* **78**, 543–550.

Haslam, E., & Lilley, T.H. (1988). Natural astringency in foodstuffs - A molecular interpretation. *Crit. Rev. Food Sci. Nutr.* **27**, 1–40.

Hazelwood, L.A., Walsh, M.C., Pronk, J.T. & Daran J.C. (2010). Involvement of vacuolar sequestration and active transport in tolerance of *Saccharomyces cerevisiae* to hop iso-α-acids. *Appl. Environ. Microbiol.* **76**, 318–328.

Heyerick, A., Zhao, Y.N., Sandra, P., Huvaere, K., Roelens, F. & de Keukeleire, D. (2003). Photolysis of hop-derived *trans*-iso-alpha-acids and *trans*-tetrahydroiso-alpha-acids: product identification in relation to the lightstruck flavour of beer. *Photochem. Photobiol. Sci.* **2**, 306–314.

Hilton, P.J. & Ellis, R.T. (1972). Estimation of market value of Central African tea by theaflavin analysis. *J. Sci. Food. Agric.* **23**, 227–232.

Hoskin, J.C. & Dimick, P.S. (1994). Chemistry of flavour development in chocolate. In Beckett, S.T. *Industrial Chocolate Manufacture and Use*. 2nd ed. Glasgow: Chapman & Hall.

Hufnagel, J.C. & Hofmann, T. (2008). Orosensory-directed identification of astringent mouthfeel and bitter-tasting compounds in red wine. *J. Agric. Food Chem.* **56**, 1378–1386.

Hughes, P.S. (2000). The significance of iso-alpha-acids for beer quality–Cambridge prize paper. *J. Inst. Brew.* **106**(5), 271–276.

Hughes, P.S. (2006). Differences in the UV-spectra of the hop-derived *cis*- and *trans*-iso-α-acids. *J. Amer. Soc. Brew. Chem.* **64**(1), 47–51.

Hughes, P.S. & Simpson, W.J. (1994). Sensory impact of hop-derived compounds. *Monograph-European Brewing Convention*, **22**, 128–140.

Hughes, P.S. & Simpson, W.J. (1996). Bitterness of congeners and stereoisomers of hop-derived bitter acids found in beer. *J. Amer. Soc. Brew. Chem.* **54**, 234–237.

Huvaere, K., Olsen, K., Andersen, M.L., Skibsted, L.H., Heyerick, A. & de Keukeleire, D. (2004a). Riboflavin-sensitized photooxidation of isohumulones and derivatives. *Photochem. Photobiol. Sci.* **3**, 337–340.

Huvaere, K., Sinnaev, B., Van Bocxlaer J. & de Keukeleire, D. (2004b). Photooxidative degradation of beer bittering principles: product analysis with respect to light struck flavour formation. *Photochem. Photobiol. Sci.* **3**, 854–858.

Intelmann, D., Haseleu, G. & Hofmann, T. (2009). LC-MS/MS quantization of hop-derived bitter compounds in beer using the ECHO technique. *J. Agric. Food Chem.* **57**(4), 1172–1182.

Ishikawa, T. & Noble, A.C. (1995). Temporal perception of astringency and sweetness in red wine. *Food Qual. Pref.* **6**, 27–33.

Jaskula, B., Goiris, K., de. Rouck, G., Aerts, G. & de Cooman, L. (2007). Enhanced quantitative extraction and HPLC determination of hop and beer bitter acids. *Inst. Brew. Dist.* **113**(4), 381–390.

Jaskula, B., Kafarski, P., Aerts, G. & de Cooman L. (2008). A kinetic study on the isomerization of hop α-acids. *J. Agric. Food Chem.* **56**(15), 6408–6415.

Jinap, S., Dimick, P.S. & Hollender, R. (1995). Flavour evaluation of chocolate formulated from cocoa beans from different countries. *Food Cont.* **6**, 105–110.

Jinap, S., Wan Rosli, W.I., Russly, A.R. & Nurdin, L.M. (1998). Effect of roasting time and temperature on volatile components profile during nib roasting of cocoa beans (*Theobroma cacao*). *J. Sci. Food Agric.* **77**, 441–448.

Juneja, L.R., Chu, D.-C., Okubo, T., Nagato, Y. & Yokogoshi, H. (1999). L-Theanine- A unique amino acid of green tea and its relaxation effect in humans. *Trends Food Sci. Technol.* **10**(6–7), 199–204.

Kallithraka, S., Bakker, J. & Clifford, M.N. (1997). Effect of pH on astringency in model solutions and wines. *J. Agric. Food Chem.* **45**, 2211–2216.

Kaneda, H., Kano, Y., Koshino, S. & Ohya-Nishiguchi, H. (1992). Behavior and role of iron ions in beer deterioration. *J. Agric. Food Chem.* **40**, 2102–2107.

Kappler, S., Krahl, M., Gessinger, C., Becker, T., & Krottenhaler, M. (2010a). Degradation of iso-α-acids during wort boiling. *J. Inst. Brew.* **116**(4), 332–338.

Kappler, S., Schönberger, C., Krottenthaler, M. & Beckers, T. (2010b). Isohumulones–a review. *Brew. Sci.* **63**, 105–111.

Kassara, S. & Kennedy, J.A. (2011). Relationship between red wine grade and phenolics. 2: Tannin composition and size. *J. Agric. Food Chem.* **59**, 8409–8412.

Kim, H. & Keeney, P.G. (1984). (-)-Epicatechin content in fermented and unfermented cocoa beans. *J. Food Sci.* **49**, 1090–1092.

King, B.M. & Duineveld, C.A.A. (1999). Changes in bitterness as beer ages naturally. *Food Qual. Pref.* **10**, 315–324.

Kolpin, K.M. (2010). The human bitterness detection threshold of iso-alpha-acids andtetrahydro-alpha-acids in lager beer. Doctoral thesis, University of Oregon.

Kraujalytė, V., Pelvan, E. & Alasalvar, C. (2016). Volatile compounds and sensory characteristics of various instant teas produced from black tea. *Food Chem.* **194**, 864–872.

Kreppenhofer, S., Frank, O. & Hofmann, T. 2011. Identification of (furan-2-yl) methylated benzene diols and triols as a novel class of bitter compounds in roasted coffee. *Food Chem.* **126**(2), 441–449.

Lang, R., Müller, C. & Hofmann, T. (2006). Development of a stable isotope dilution analysis with liquid chromatography–tandem mass spectrometry detection for the quantitative analysis of di and trihydroxybenzenes in foods and model systems. *J. Agric. Food Chem.* **54**, 5755–5762.

Lang, R., Klade, S., Dunkel, A. & Hofmann, T. (2015). Moxambioside is an aribca-specific bitter-tasting furokaurane glucoside in coffee beans. *J. Agric. Food Chem.* **63**, 10492–10499.

Lea, A.G. & Arnold, G.M. (1978). The phenolics of cider. Bitterness and astringency. *J. Sci. Food Agric.* **29**, 478–483.

Lea, A.G. & Timberlake, F. (1974). The phenolics of ciders: 1. Procyanidins. *J. Sci. Food Agric.* **25**(12), 1537–1545.

Lee, C.B. & Lawless, H.T. (1991). Time-course of astringent sensations. *Chem. Senses*, **16**, 225–238.

Lee, K. J. & Lee, S.H. (2008). Extraction behavior of caffeine and EGCG from green and black tea. *Biotechnol. Bioprocess Eng.* **13**, 646–649.

Le Quere. J.-M., Husson, F., Rennard, C.M.G.C. & Primault, J. (2006). French cider characterization by sensory, technological and chemical evaluation. *LWT Food Sci. Technol.* **39**(9), 1033–1044.

Lewkowski, J. (2001). Synthesis, chemistry and application of 5-hydroxymethyl-furfural and its derivatives. *ARKIVOC.* **17**, 54.

Liang, Y.R., Ye, Q., Jin, J., Liang, H., Lu, J.L. & Du, Y.Y. (2008). Chemical and instrumental assessment of green tea sensory preference. *Int. J. Food Prop.* **11**(2), 258–272.

Luna, F., Crouzillat, D., Cirou, L. & Bucheli, P. (2002). Chemical composition and flavour of Ecuadorian cocoa liquor. *J. Agric. Food Chem.* **50**, 3527–3532.

Maga, J.A. & Katz. 1978. Simple phenol and phenolic compounds in food flavor. *CRC Crit. Rev. Food Sci.Nutr.* **10**, 323–372.

Malowicki, M.G.; Shellhammer, T.H. (2005). Isomerization and degradation kinetics of hop (*Humulus lupulus*) acids in a model wort-boiling system. *J. Agric. Food Chem.* **53**, 4434–4439.

Malowicki, M.G. & Shellhammer, T.H. (2006). Factors affecting hop bitter acid isomerization kinetics in a model wort boiling system. *J. Amer. Soc. Brew. Chem.* **64**(1), 29–32.

Mattes, R.D. & DiMeglio, D. (2001). Ethanol perception and ingestion. *Physiol. Behav.* **72**(1–2), 217–229.

Mayer, F., Czerny, M., & Grosch, W. (2000). Sensory study of the character impact aroma compounds of a coffee beverage. *Eur. Food Res. Technol.* **211**, 272–276.

Meyer, B. & Biehl, B. (1989). Post-harvest pod storage: a method for pulp preconditioning to impair strong nib acidification during cocoa fermentation in Malaysia. *J. Sci. Food Agric.* **48**, 285–304.

McCamey, D.A., Thorpe, T.M. & McCarthy, J.P. (1990). In: Rouseff RL (ed), *Bitterness in Foods and Beverages*, Elsevier, London, pp 169–182.

McMurrough, I. & Byrne, J.R. (1992). HPLC analysis of bittering substances, phenolic compounds, and various compounds of alcoholic bevarages. In *Food analysis by HPLC*. Nollet, L.M.L. (Ed.), New York: Marcel Dekker, (Chapter 16).

Mihara, R., Mitsunaga, T., Fukui, Y., Nakai, M., Yamaji, N., & Shibata, H. (2004). A novel acylated quercetin tetraglycoside from oolong tea (*Camelia sinensis*) extracts. *Tetrahedron Letters*, **45**(26), 5077–5080.

Misnawi, S.J., Jamilah, B. & Nazamid, S. (2003). Effects of incubation and polyphenol oxidase enrichment on colour, fermentation index, procyanidins and astringency of unfermented and partly fermented cocoa beans. *Int. J. Food Sci. Tech.* **38**, 285–295.

Mohr, W., Landschreiber, E. & Severin, T. (1976). Zur spezifitat des kakaoaromas. *Fette Seifen Anstrichmittel*, **78**, 88–95.

Monogas, M., Gomez-Cordoves, C., Bartolome, B., Laureano, O. & Ricardo, J. M. (2003). Mopnopmeric, oligomeric, and polymeric flavonol-3-ol composition of wines aqnd grapes from Vitis vinifera L. cv. Gracioano, Tempranillo, and Cabernet sauvignon. *J. Agric. Food Chem.* **51**(22), 6475–6481.

Moon, J.-K. & Shibamoto, T. (2009). Role of roasting conditions in the profile of volatile flavor chemicals formed from coffee beans. *J. Agric. Food Chem.* **57**, 5823–5831.

Mudura, E. & Coldea, T. (2015). Hop-derived prenylflavonoids and their importance in brewing technology – A review. *Bulletin of University of Agricultural Sciences and Veterinary Medicine Cluj-Napoca: Food Sci. Technol.* **72**(1), 1–10.

Mudura, E., Muste, S., Tofana, M., Socaci, S.A. & Goina, V. (2008). Improving the hop utilization in the beer biotechnology. *Bulletin of University of Agricultural Sciences and Veterinary Medicine of Cluj-Napoca: Food Sci. Technol.* **65**(2), 281–286.

Nagel, C.W., Herrick, I.W. & Graber, W.R. (1987). Is chlorogenic acid bitter? *J. Food Sci.* **52**(1), 213

Narukawa, M., Noga, C., Ueno, Y., Sato, T., Misaka, T. & Watanabe, T. (2011). Evaluation of the bitterness of green tea catechins by a cell-based assay with the human bitter taste receptor hTAS2R39. *Biochem. Biophys. Res. Comm.* **405**(4), 620–625.

Nimubona, D. (2010). Doctoral thesis, Spain: University of Valladolid.

Noble, A.C. (1994). Bitterness in wine. *Physiol. Behav.* **6**, 1251–1255.

Obara, K., Mizutani, M., Hitomi, Y., Yajima, H. & Kondo, K. (2009). Isohumulones, the bitter component of beer, improve hyperglycemia and decrease body fat in Japanese subjects with prediabetes. *Clin. Nutr.* **28**, 278–284.

Pangborn, R.M., Lewis, M.L. & Tanno, L.A.S. (1977). Sensory quantification of bitterness and flavour of beer during storage. *J. Inst. Brew.* **83**, 244–250.

Pangborn, R.M., Lewis, M.L. & Yamashita, J.F. (1983). Comparison of time-intensity with category scaling of bitterness of iso-a-acids in model systems and in beer. *J. Inst. Brew.* **89**, 349–355.

Pozdrik, R., Roddick, F.A., Rogers, P.J. & Nguyen, T. (2006). Spectrophotometric method for exploring 3-methyl-2-butene-1-thiol (MBT) formation in lager. *J. Agric. Food Chem.* **54**, 6123–6129.

Puziah, H., Jinap, S., Sharifah, K.S.M. & Asbi, A. (1998). Effect of mass and turning time on free amino acid, peptide-n, sugar and pyrazine concentration during cocoa fermentation. *J. Sci. Food Agric.* **78**, 543–550.

Ribereau-Gayon, P. (1972). The anthocyanins of grapes and wines. In *Anthocyanins as Food Colors*. P. Markakis, ed. Academic Press, NY (London). pp 209–247.

Richards, E.L. (1956). Nonenzymatic browning: The reaction between glucose and glycerine the dry state. *Biochem. J.* **64**, 639–644.

Rickstina-Dolge, R., Kruma, Z., Dimins, F., Straumite, E. & Karklina, D. (2014). Phenolic composwition and sensory properties of ciders produced from Latvian applies. *Proc. Latv. Univ. Agr.* **31**(326), 39–45.

Rizzi, G.P., Boekley, L.J. & Ekanayake, A. 2004. In: Shahidi F, Weerasinghe DK (eds) *Nutraceutical beverages—chemistry, nutrition, and health effects*, ACS Symposium Series 871, Washington, DC, pp 229–236.

Rosenzweig, S., Yan, W., Dasso, M. & Spielman, A.I. (1999). Possible novel mechanism for bitter taste mediated through cGMP. *J Neurophysiol*, **81**, 1661–1665.

Sakamoto, K. (2003). Beer spoilage bacteria and hop resistance. *Int. J. Food Microbiol.* **89**, 105–124.

Semmelroch, P. & Grosch, W. (1995). Analysis of roasted coffee powders and brews by gas chromatography-olfactometry of headspace samples. *LWT Food Sci Technol.* **28**, 310–313.

Semmelroch, P. & Grosch, W. (1996). Studies on character impact odorants of coffee brews. *J. Agric. Food Chem.* **44**, 537–543.

Shibamoto, T., Harada, K., Mihara, J., Nishimura, K., Yamaguchi, A., Aitoku, A. & Fukada, T. (1981). Application of HPLC for evaluation of coffee flavor quality. In *The Quality of Foods and Beverages*, Charambous, G.; Inglett, G., Eds.; Academic Press: New York, Vol. **2**, p 311.

Shimuraa, M., Hasumia, A., Minatoa, T., Hosonoa, M., Miuraa, Y. & Mizutania, S. (2005). Isohumulones modulate blood lipid status through the activation of PPARα. *Biochim. Biophys. Acta.* **1736**, 51–60.

Singleton, V.L. and Esau, P. 1969. Phenolic sunbstances in grapes and wine and their significance. *Adv. Food Res.* **1**, 1–261.

Stewart, G.G. (2004). The chemistry of beer instability. *J. Chem. Ed.* **81**(7), 963–968.

Suzuki, K., Iijima, K., Sakamoto, K., Sami, M. & Yamashita, H. 2006. A review of hop resistance in beer spoilage lactic acid bacteria. *J. Inst. Brew.* **112**(2), 173–191.

Symoneaux, R., Baron, A., Marnet, N., Bauduin, R. & Chollet, S. (2014a). Impact of apple procyanidins on sensory perception in model cider (part 1): Polymerization degree and concentration. *LWT-Food Sci. Technol.* **57**(1), 22–27.

Symoneaux, R., Chollet, S., Bauduin, R., Le Quere, J.M. & Baron, A. (2014b). Impact of apple procyanidins on sensory perception in model cider (part 2): Degree of polymerization and interactions with the matrix components. *LWT-Food Sci. Technol.* **57**(1), 28–34.

Thorngate, J.H. & Noble, A.C. (1995). Sensory evaluation of bitterness and astringency of 3R(−)-epicatechin and 3S(+)-catechin. *J. Sci. Food Agric.* **67**, 531–535.
Tressl, R., Bahri, D., Koeppler, H. & Jensen, A. (1978). Diphenols and caramel compounds in roasted coffees of different varieties- II. *Zeitschrift fur Lebensmittel-untersuchung und -forschung*, **167**, 111–114.
Urban, J.U., Clinton J.D., Brian J.C. & Werner, K. (2013). Absolute Configuration of Beer's Bitter Compounds. *Angewandte Chemie International Edition*, **52**(5), 1553–1555.
Van Cleemput, M., Cattoor, K., De Bosscher, K., Haegeman, G., de Keukeleire, D. & Heyerick, A. (2009). Hop (*Humulus lupulus*)-derived bitter acids as multipotent bioactive compounds. *J. Nat. Prod.* **72**(6), 1220–1230.
Vanderhaegen, B., Delvaux, F., Daenen, L., Verachtert, H. & Delvaux, F.R. (2006). Aging characteristics of different beer types. *Food Chem.* **103**, 404–412.
Vaughan, A., O'Sullivan, T. & Van Sinderen, D. (2005). Enhancing the microbiological stability of malt and beer–a review. *J. Inst. Brew.* **111**(4), 355–371.
Verette, E., Noble, A.C. & Somers, C. (1988). Hydroxycinnamates of *Vitis vinifera* : Sensory assessment in relation to bitterness in white wine. *J. Sci. Food Agric.* **45**, 267–272.
Vidal, S., Francis, L., Guyot, S., Marnet, N., Kwiatkowski, M., Gawel, R., Cheynier, V. & Waters, E.J. (2003). The mouth-feel properties of grape and 1745 apple proanthocyanidins in a wine-like medium. *J. Sci. Food Agric.* **83**, 564–573.
Vidal, S., Francis, L., Noble, A., Kwiatkowski, M., Cheynier, V. & Waters, E. (2004). Taste and mouthfeel properties of different types of tannin-like polyphenolic compounds and anthocyanins in wine. *Anal. Chim. Acta.* **513**, 57–65.
Walters, M.T., Heasman, A.P. & Hughes, P.S. (1997). Comparison of (+)-catechin and ferulic acid as natural antioxidants and their impact on beer flavor stability. Part 2: extended storage trials. *J. Amer. Soc. Brew. Chem.* **55**(3), 91–98.
Wang, K. & Ruan, J. (2009). Analysis of chemical components in green tea in relation with perceived quality, a case study with Longjing teas. *Int. J. Food Sci. Technol.* **44**(12), 2476–2484.
Wollgast, J. & Anklam, E. (2000). Review on polyphenols in Theobroma cacao: changes in composition during the manufacture of chocolate and methodology for identification and quantification. *Food Res Int.* **33**, 423–447.
Wood, G.A.R. & Lass, R.A. (2008). *Cocoa*, 4th edition. Longman Group, London.
Yajima, H., Ikeshima, E., Shiraki, M., Kanaya, T., Fujiwara, D. & Odai, T. (2004). Isohumulones, bitter acids derived from hops, activate both peroxisome proliferator-activated receptor and reduce insulin resistance. *J. Biol. Chem.* **279**(32), 33456–33462.
Yamazaki, T., Narukawa, M., Mochizuki, M., Misaka, T. & Watanabe, T. (2013). Activation of the hTAS2R14 human bitter-taste receptor by (-)- epigallocatechin gallate and (-)-epicatechin gallate. *Biosci. Biotechnol. Biochem.* **77**(9), 1981–1983.
Ye, M., Yue, T. & Yuan, Y. (2014). Evaluation of polyphenols and organic acids during the fermentation of apple cide. *J. Sci. Food Agric.* **94**, 2951–2957.
Yu, P., Yeo, A.S-L., Low, M-Y. & Zhou, W. (2014). Identifying key non-volatile compounds in ready-to-drink green tea and their impact on taste profile. *Food Chem.* **155**, 9–16.
Ziegleder, G. (1991). Composition of flavor extracts of raw and roasted beans. *Z Lebensmitt Unters-Forsch*, **192**, 521–525.
Ziegleder, G. & Biehl, B. (1988). Analysis of cocoa flavour compounds and precursors. In *Modern Methods of Plant Analysis*, vol. **8**, ed. by Linkskens HF and Jackson JF. Springer, Berlin, pp 361–393.

6 Structural Characteristics of Food Protein-Derived Bitter Peptides

Rotimi E. Aluko

6.1 INTRODUCTION

The abundance of food proteins, especially from plant sources has made them a good source of substrates for enzymatic conversion into more functional or nutritive products. Protein hydrolysis is carried out for several purposes but one of the most important reasons is the need to provide amino acids in short peptide chains (usually <20 residues) as a means of increasing absorption from the gastrointestinal tract. In this form, the short-chain peptides can be used to formulate nutritious foods for children and elderly people with impaired digestion. Enzymatic hydrolysis is also a preferred method to reduce allergenicity of food proteins because the process cleaves and destroys epitopes. Recently, the rapid growth of the functional foods and nutraceuticals industry has involved development of enzymatic food protein digests (protein hydrolysates) that contain bioactive short-chain peptides. These bioactive peptides have been proposed as natural therapeutic alternatives to drugs for the effective management of metabolic disorders such as hypertension, diabetes, obesity and even cancer. While protein hydrolysis can be tailored to yield peptides with desired functional or bioactive properties, the taste properties are difficult to control. The most encountered taste attribute of enzymatic protein hydrolysates is bitterness, which can lead to reduced consumer acceptability of formulated food products. But natural products can also develop bitter peptides as a result of food processing or during cheese ripening. For example, cheese taste can be influenced by proteolysis that produces certain peptides and amino acids. It has been established that cheese bitterness, which could be considered as a principal defect is mainly as a result of the release of hydrophobic peptides during manufacture or aging (McSweeney, 1997; Fallico *et al.*, 2005; Taborda *et al.*, 2008). Previous reports have confirmed the linear relationship between amount of β-casein-derived peptides from f193-209 and bitterness intensity of aged cheddar cheese

Bitterness: Perception, Chemistry and Food Processing, First Edition. Edited by Michel Aliani & Michael N. A. Eskin.

(Broadbent *et al.*, 2002; Singh *et al.*, 2010; Karametsi *et al.*, 2014). Soeryapranata *et al.* (2002) showed that this relationship was dependent on length of aging; a longer period gave stronger bitterness intensity (Fig. 6.1). Cheese is very prone to bitter taste development because of various factors such as the production of extracellular proteinases by indigenous milk microflora (e.g., *Pseudomonas*), which can lead to extensive proteolysis during storage or ripening (Hicks *et al.*, 1986). Moreover during cheese manufacture, bacteriophage proliferation or antibiotics in the milk can reduce starter cell numbers, which lowers lactococcal and associated bitter peptide-degrading peptidases levels (Sullivan *et al.*, 1973; McSweeney, 1997). Other factors involved in cheese curd bitter flavor development include the use of certain bitterness-promoting starter cultures, chymosin retention, pH, and low salt content (McSweeney, 1997). For example, a starter culture consisting of *Lactococcus lactis* ssp. *lactis* S3 produced bitter ripened cheese, whereas *Lactococcus lactis* ssp. *cremoris* S1 and *Lactococcus lactis* ssp. *cremoris* S2 produced non-bitter cheeses even after 6 months of ripening (Broadbent *et al.*, 1998). A starter culture that contained *Lactococcus lactis* ssp. *lactis* and *Lactococcus lactis* ssp. *cremoris* produced a Cheddar cheese that had higher bitterness intensity than a similar cheese made with the same microorganism but with added *Lactobacillus* (Borsting *et al.*, 2012). However, *Lactococcus lactis* ssp. *lactis* 527 may be another alternative starter culture microbe since it was very effective in degrading the bitter heptapeptide (Gly-Pro-Phe-Pro-Ile-Ile-Val) released from β-casein during cheese making or ripening (Shimamura *et al.*, 2009). Reduced peptide bitterness intensity has been associated with two endopeptidases present in *Lactobacillus helveticus* CNRZ32, a bacterium strain used commercially as an adjunct to reduce bitterness in Cheddar and Gouda cheeses (Sridhar *et al.*, 2005; Borsting *et al.*, 2012). The genes that encode these *L. helveticus* proteases have been cloned and could be incorporated into the genome of *Lactococcus lactis* as a means of providing a more effective means of producing cheeses with lower levels of bitter peptides. The effect of chymosin is due to its ability to hydrolyze β-casein, which contains several bitter peptide sequences while pH affects protease activity (Exterkate *et al.*, 1995). Plasmin-dependent proteolysis of β-casein does not occur at pH <5.4 but increases in the pH 5.4-6.0 range while for chymosin, the release of bitter peptides from β-casein is highest at pH 6.0 (Larsson *et al.*, 2006). But the source of chymosin is also important; for example, bovine chymosin produced peptides with stronger bitterness intensity than camel chymosin (Borsting *et al.*, 2012). The lactococcal cell envelope associated proteinase (CEP) also acts on β-casein to generate bitter peptides in an environment of low salt content. Moreover, the peptides generated are resistant to further proteolysis by chymosin and CEP (Exterkate *et al.*, 1995), which leads to accumulation of these bitterness factors. High salt content is therefore, desirable not only to inhibit CEP activity but to enhance protein aggregation, which prevents their hydrolysis to bitter peptides (McSweeney, 1997). Broadbent *et al.* (2002) showed that lactococcal strains that produced group 'h' CEP was associated with most intense bitterness of Cheddar

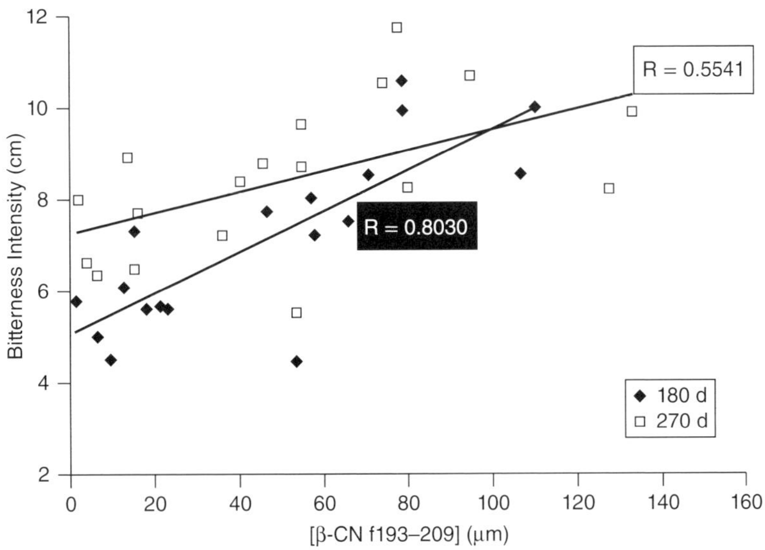

Figure 6.1 Correlation between bitterness intensity of cheese and β-casein f193-209 concentration in the aqueous extract of cheese at 180 days (■) and 270 days (□). Reprinted with permission from Soeryapranata *et al.* (2002). Copyright 2015 American Chemical Society.

cheese. Therefore, CEP negative or those that produce group 'a' or 'e' proteases could be developed as starter cultures for producing cheeses with low levels of bitter peptides. However, high salt content can lead to low water activity that limits enzymatic activity of bitter peptide-degrading peptidases, which promotes bitter taste intensity in the cheese (Fallico *et al.*, 2005). It has also been shown that partial replacement of whole milk with milk protein concentrate can reduce residual levels of chymosin and plasmin in the cheese curd, which limits secondary proteolysis (responsible for bitter peptide production) during storage and favours production of cheese with low bitterness intensity (Shakeel-Ur-Rehman *et al.*, 2003). Bitterness attribute of cheese is also dependent on storage period because of the potential for continuous proteolysis, which leads to increased bitterness intensity as ripening or shelf life increases (Gomez *et al.*, 1997).

Generally, most proteins have no bitter taste but upon proteolysis by appropriate enzymes, the resultant protein hydrolysates usually develop various degrees of bitterness intensity. Rye and wheat flours were hydrolyzed with a protease (corolase) or with several carbohydrases; only the protein digest had bitter taste, which confirms the role of peptides in determining bitterness intensity of hydrolysates (Heinio *et al.*, 2012). Bitterness of protein hydrolysates is dependent on several factors such as enzyme type, degree of hydrolysis, substrate protein and duration of hydrolysis. For example, extensive protein hydrolysis will produce protein hydrolysates with smaller size and potentially peptides with higher bitter intensity than limited hydrolysis. This is because taste is a contact attribute and small size peptides are more likely to be able to interact with bitter receptors when compared to bigger peptides. Casein hydrolysate prepared with *Bacillus* protease had higher contents of low molecular weight (<10 kDa) and hydrophobic peptides and will likely possess stronger bitterness intensity than the hydrolysate from bromelain hydrolysis with lower contents of these peptides (Gallagher *et al.*, 1994). Protein hydrolysates are also prepared through microbial fermentation of protein-rich foods, especially milk and soybeans. These fermented foods may contain peptides that impart some degree of bitterness, which depends on the fermenting microorganism. Milk was fermented with various yeast isolates and then subjected to bitterness testing. *Clavispora lusitaniae* KL4 produced the bitterest fermented milk while *Pichia kudriavzevii* KL84A and *Torulaspora delbrueckii* KL66A produced the least bitter product (Chaves-Lopez *et al.*, 2012).

There are two main approaches at reducing the bitterness intensity of food protein hydrolysates. First, the choice of protease is important because certain enzymes produce highly bitter protein hydrolysates while others produce less bitter hydrolysates. For example, Humiski and Aluko (2007) showed that papain, α-chymotrypsin, and trypsin produced protein hydrolysates with reduced bitterness when compared to flavourzyme and alcalase. Kodera *et al.* (2006) showed that protease D3 obtained from germinating soybean cotyledons could be used to produce soybean and casein hydrolysates with reduced bitterness intensity when compared to other proteases such as trypsin, subtilisin and thermolysin. Second, once the protein hydrolysate has been produced, bitterness can be reduced by passing the product through a hydrophobic column or by using an exopeptidase to perform additional hydrolysis (Cheung *et al.*, 2015). The principle involved for the column separation is that most bitter peptides contain highly hydrophobic amino acids (Ishibashi *et al.*, 1987a,b; Matoba & Hata, 1972; Lee & Warthesen 1996; Gomez *et al.*, 1997; Wu & Aluko, 2007) and will bind to a non-polar matrix, while the unbound less hydrophobic peptides will flow through and be collected as a less bitter product (Helbig *et al.*, 1980; Ma *et al.*, 1983). For the exopeptidase treatment, the enzyme action leads to sequential removal of hydrophobic amino acids that may be present at the terminal ends (Arai *et al.*, 1970a,b; Fujimaki *et al.*, 1970). Carboxypeptidase A seems to be most commonly used since it removes amino acids from the C-terminal end that determines the bitterness intensity of some peptides (Arai *et al.*, 1970a). A reduction in the amount of hydrophobic amino acids present in the peptide, especially at the C-terminal will lead to the production of a less bitter hydrolysate. However, application of these debittering methods may lead to inactivation or reduced peptide potency since the removed amino acids may be critical for imparting bioactive effects. In the following sections, structural characteristics and detailed methods for bitter peptide preparation, characterization and taste modification are discussed.

6.2 BITTER PEPTIDES PREPARATION AND TASTE EVALUATION

Dairy products have been the most characterized for bitter peptides and cheese is a common starting material for bitter peptides isolation. This is because during aging, casein is slowly hydrolyzed by active proteases that are present in the cheese to form various peptides some of which have bitter taste attributes. Due to their small sizes (usually <15 amino acid residues) several bitter peptides are soluble in aqueous solvents and can be extracted from cheese by simply blending with water (Lee & Warthesen, 1996; Toelstede & Hofmann, 2008a). The water extract can then be subjected to liquid chromatography-tandem mass spectrometry quantification and identification of peptides (Toelstede & Hofmann, 2008a). Or the water extract can be centrifuged, the top lipid-containing layer removed and the aqueous supernatant filtered through various devices to remove large sized peptides. For example, after filtering through glass wool, an aqueous cheese extract was separated into different molecular weight peptides using ultrafiltration membranes with 500 or 3000 Da molecular weight cut-offs (Lee & Warthesen, 1996). To determine bitterness intensity, a trained human sensory evaluation panel can be used during which the peptide taste is compared with that of a standard bitter compound. Typical standard solutions used for determining peptide bitterness intensity or taste attributes include MgSO4, salicin, caffeine, and quinine sulfate (Humiski & Aluko, 2007; Singh *et al.*, 2005; Toelstede & Hofmann, 2008a; Karametsi *et al.*, 2014). For example, a human sensory panel was used to evaluate the three cheese extract peptide fractions; the smallest size peptide fraction (<500 Da) had the least bitterness intensity when compared to the 500-3000 Da and >3000 Da fractions (Lee & Warthesen, 1996). Similarly, early eluting fractions (big size peptides) from Sephadex G25 column had a linear relationship with bitterness score (Sorensen *et al.*, 1996). The results suggest that longer peptide chains with higher numbers of amino acids provided the basis for higher bitterness intensity of peptides. The 500-3000 and >3000 Da fractions were each subsequently fractionated on a C18 column using the principle of reverse-phase HPLC; collected fractions were evaluated for bitterness intensity by the human sensory panel. In general, the fractions that eluted early, especially within the first 5 min had the highest bitterness intensity when compared to fractions that eluted at later time periods (Lee & Warthesen, 1996). Since a hydrophobic column was used, the early eluting fraction will be highly hydrophilic and this was confirmed by showing that main amino acid was glutamic/glutamine. The HPLC fractions also contained low levels of hydrophobic amino acids such as leucine, valine and isoleucine but were virtually devoid of the aromatic amino acids, phenylalanine and tyrosine. The results suggest that contrary to some reports, hydrophilic amino acids can also impart bitter taste; however, it is possible that they are present in peptide sequences as combinations with hydrophobic amino acids. For example, hydrophilic amino acid-containing peptides such as Ala-Asp-Glu, Glu-Glu-Asn, and Glu-Pro-Ala-Asp have been shown to be present in bitter fractions of food protein hydrolysates (Maehashi *et al.*, 1999).

6.3 ROLE OF AMINO ACID COMPOSITION AND POSITION ARRANGEMENT IN DETERMINING PEPTIDE BITTERNESS INTENSITY

6.3.1 Relationship between peptide hydrophobicity and bitterness intensity

Peptide bitterness has been attributed mainly to the net hydrophobicity rather than any particular amino acid sequence (Ney, 1979). But as will be discussed below, bitterness intensity of a peptide is also dependent on the type(s) of amino acids present at the N- and C-terminals. The mean hydrophobicity of peptide is represented as ‘Q’ and can be calculated using the following equation (McSweeney, 1997): $\sum \Delta f_t/n$ where Δf_t is side chain hydrophobicity (free energy of transfer, Δf_t) and n = number of amino acid residues on the peptide chain. According to Ney (1979), peptides with molecular weight less than 10 kDa and Q <1300 calories per residue should not have bitter taste. However, peptides with sizes up to 6 kDa and Q >1400 calories per residue will most likely have a bitter taste. This principle was aptly demonstrated for rapeseed protein hydrolysates fractions; the less bitter RP55 had a Q of 1466 while the bitterest RP85 had 1673, which indicates some direct relationship between hydrophobicity and peptide bitterness (Zhang *et al.*, 2007).

But Q value alone does not determine peptide bitterness because in a bitter Camembert cheese for example, most of the bitter peptides were concentrated in the 0.4-1.0 kDa size range though peptides within the 1.0-2.8 kDa range were also detected but in minor concentrations (Engel *et al.*, 2001). Aubes-Dufau *et al.* (1995) also showed that bitterness intensity of peptic hemoglobin hydrolysates was influenced by molecular size; the >10 kDa fraction had no bitter taste, 5-10 kDa was slightly bitter while the 0.5-5 kDa had a very bitter taste. A similar result was obtained by Cho *et al.* (2004) who showed that bitterness intensity increased with decrease in peptide size to 2-3 kDa but smaller peptides (<1 kDa) had weak intensity. Most peptides that are bigger than 6 kDa will have no bitter taste even if the Q value >1400 calories per residue. Apart from the Q value, bitterness intensity of a peptide is also dependent on location of hydrophobic amino acids on the peptide chain (amino acid sequence). For example, three bitter peptides that were isolated from cheddar cheese had Q values <1400, which shows that amino acid sequence also plays an important role in determining peptide bitterness intensity (Lee *et al.*, 1996). The three peptides contained Pro residues within the chain in addition to

Arg, Val, or Leu at the N- or C-terminals; these features have been shown to be common structural components of bitter peptides (Fujimaki *et al.*, 1968; Otagiri *et al.*, 1985; Shinoda *et al.*, 1985; Ishibashi *et al.*, 1987a; Kukman *et al.*, 1995). Kim and Li-Chan (2006) suggested that the presence of bulky hydrophobic amino acids at the C-terminal and bulky positively charged residues at the N-terminal enhance bitterness intensity of peptides. Other workers proposed that for di- and tripeptides, bitterness intensity is potentiated by the presence of hydrophobic amino acids at the C-terminal and the presence of an adjacent amino acid with a bulky group (Wu & Aluko, 2007). Several peptides isolated from Gouda cheese also had no strict relationship between Q value and bitterness intensity. This is evident in the data which showed that Ile-Pro-Pro-Leu with highest Q value of 2658 had a threshold value of >6 mM, whereas Tyr-Pro-Phe-Pro-Gly-Pro-Ile-His-Asn-Ser with a Q value of 1688 had the lowest threshold value of 0.05 mM (Toelstede & Hofman, 2008b). There was also no direct relationship between Q values and bitterness intensity of soy protein hydrolysate peptide fractions (Cho *et al.*, 2004).

6.3.2 Influence of peptide chain length and N- or C-terminal amino acid residue

One of the earliest reports on bitter peptides identified the amino acid sequence of seven pepsin hydrolyzed fragments from soybean protein as follows: Gly-Leu, Leu-Phe, Ser-Lys-Gly-Leu, Leu-Lys, Phe-Ile/Leu-Gln-Gly-Val, Arg-Leu-Leu, and Arg-Leu (Fujimaki *et al.*, 1968). A common structural feature of these peptides is the presence of Leu at the C- or N-terminal of six of them, which indicated the important contribution of hydrophobic amino acids, especially Leu to peptide bitterness. Arai *et al.* (1970b) also showed that for the Tyr-Phe-Leu tripeptide, removal of Tyr resulted in a dipeptide (Phe-Leu) that has similar bitterness intensity. In contrast Leu removal gave a dipeptide (Tyr-Phe) that was less bitter than Tyr-Phe-Leu, which supports the role of Leu in potentiating peptide bitterness. Even though Phe-Ile/Leu-Gln-Gly-Val does not have Leu at the C- or N-terminal, the presence of hydrophobic residues in the form of Phe and Val at these positions may have contributed to giving the peptide a bitter taste. Another bitter peptide with amino acid sequence pyrrolidone carboxyl-Gly-Ser-Ala-Ile-Phe-Val-Leu and containing leucine at the C-terminal was also identified from the peptic digest of soybean hydrolysate (Yamashita *et al.*, 1969). These authors showed that bitterness intensity of a mixture of the component amino acids was lower than the measured bitterness intensity for pyrrolidone carboxyl-Gly-Ser-Ala-Ile-Phe-Val-Leu. The results implied that amino acid polymerization can lead to increased bitterness of the resultant peptide probably as a result of the synergistic cooperative effect of the component bitter amino acids. Therefore, Yamashita *et al.* (1969) demonstrated the principle that extensive proteolysis to reduce peptide chain length can be an effective method to reduce peptide bitterness. In addition to potential synergistic effects, other researchers showed that the presence of amino or carboxyl groups in

Table 6.1 Bitterness intensity of peptides (X-X) and equivalent free amino acid mixtures (X+X).

Sample	Threshold value (mM)
Leu + Leu	20.0
Leu-Leu	3.7
Leu + Phe	15
Leu-Phe	1.3
Phe-Leu	1.3
Phe + Phe	7.0
Phe-Phe	0.6

Adapted from Matoba and Hata (1972).

free amino acids contributes to weakening of bitterness intensity. It was reported that free tyrosine or phenylalanine has less bitterness intensity when compared to the phenylalanine where amino and carboxyl groups have been modified by acetylation and esterification, respectively (Matoba *et al.*, 1970). A previous report also indicated that esterification of the C-terminal carboxylic group was more effective in reducing peptide bitterness intensity when compare to acetylation of the N-terminal amino group (Arai *et al.*, 1970b). Based on these early works, it would seem that the C-terminal amino acids are major determinants of peptide bitterness intensity. Therefore, polymerization of the amino acids to form peptide chains leads to elimination of several of these amino or carboxyl groups and is a major contributing factor for the higher bitterness intensity of peptides when compared to a mixture of amino acids (Matoba *et al.*, 1970; Matoba & Hata, 1972). Table 6.1 compares the bitterness intensity of peptides and their equivalent free amino acid mixtures. This principle was well demonstrated by work of Otagiri *et al.* (1985), which showed that the dipeptides Arg-Arg and/or Pro-Pro had about three times the bitterness intensity of a mixtures of the free amino acid forms Arg and Pro. Similarly Phe-Phe and Phe-Phe-Phe had 15 and 100 times, respectively the bitterness intensity of the free amino acid form of Phe.

Three bitter peptides have been reported from a tryptic hydrolysate of casein with amino acid sequences of Gly-Pro-Phe-Pro-Val-Ile, Phe-Phe-Val-Ala-Pro-Phe-Pro-Glu-Val-Phe-Gly-Lys, and Phe-Ala-Leu-Pro-Gln-Tyr-Leu-Lys (Matoba *et al.*, 1970). These peptides also contain hydrophobic amino acid residues either at the N- or C-terminal, which supports the theory that amino acid position contributes to peptide bitterness intensity. A subsequent work used *Bacillus subtilis* alkaline protease to hydrolyze dairy casein followed by isolation of a bitter peptide, which was identified as Arg-Gly-Pro-Pro-Phe-Ile-Val (Minamiura *et al.*, 1972). Structure-function experiments showed that the bitterness intensity of Arg-Gly-Pro-Pro-Phe-Ile-Val was due to the core Gly-Pro-Pro-Phe sequence.

This is because enzymatic removal of N-terminal Arg residue or the C-terminal Ile-Val residues did not affect peptide bitterness intensity. However, cleavage of Arg-Gly by treatment with an amino peptidase resulted in a loss of peptide bitterness intensity (Minamiura *et al.*, 1972). The dipeptide Arg-Gly had higher bitterness intensity than Gly-Arg, which indicates that presence of Arg at the N-terminal was more important than when present at the C-terminal (Otagiri *et al.*, 1985). Same pattern of results were reported for the dipeptide Arg-Pro, which has stronger bitterness intensity than peptides that contain only Arg or Pro. Reversing the amino acid sequence to form Pro-Arg led to reduced bitterness intensity, which confirms that the N-terminal Arg residue is an important determinant of bitterness intensity of dipeptides that contain Arg. Replacing the N-terminal Arg with another basic amino acid such as Lys still produced a bitter peptide but with a reduced intensity when compared to dipeptides with Arg at the N-terminal. Interesting, the tripeptide Arg-Gly-Pro, which contains an inserted Gly residue has less bitterness intensity than Arg-Pro but Gly-Pro-Arg or Pro-Arg-Gly maintained similar intensity as Arg-Pro. Thus it was suggested that in addition to Arg being present at the N-terminal, close proximity of Arg and Pro is required for potentiating bitterness intensity. Increasing the distance between these two amino acids by insertion of another amino acid produces unfavorable alignment that reduces bitterness intensity. In contrast, the dipeptide Pro-Gly has sweet taste attributes, whereas Gly-Pro is bitter, which indicates that the presence of Pro at the C-terminal contributes to bitterness intensity of this peptide; similarly, Phe-Gly had higher bitterness intensity than Gly-Phe. For peptides with three or more amino acid residues, bitterness intensity is influenced by synergistic effects, whereby higher numbers of certain residues and location at the C-terminal enhances bitterness (Otagiri *et al.*, 1985). This synergistic effect seems to be relevant when the amino acids are Phe, Pro, Arg, Leu, Ile, and Tyr whose multiplicity on a peptide chain and presence at the C-terminal increases bitterness intensity. This was illustrated with synthetic peptides where Arg-Pro-Phe-Phe, Arg-Arg-Pro-Phe-Phe, Arg-Arg-Pro-Pro-Phe-Phe, and Arg-Arg-Pro-Pro-Pro-Phe-Phe had 30, 50, 143, and 500 times the bitterness of caffeine (Otagiri *et al.*, 1985). A recent work also showed that stimulation of human taste receptors was directly related to the number of amino acid residues on the peptide chain (Kohl *et al.*, 2013).

6.3.3 Amino acid type and position on peptide chain

The importance of amino acid sequence in determining peptide bitterness was further studied using various synthetic equivalents of β-casein fragments. The synthetic octapeptide Arg-Gly-Pro-Phe-Phe-Ile-Ile-Val has an extremely bitter taste that is 250 times that of caffeine and a threshold value of 0.004 mM (Shinoda *et al.*, 1985). Another β-casein peptide sequence, the highly hydrophobic hexapeptide Pro-Val-Leu-Gly-Pro-Val-Thr had twice the bitterness intensity of caffeine. The decapeptide Pro-Val-Arg-Gly-Pro-Phe-Pro-Ile-Ile-Val had similar bitterness intensity as Arg-Gly-Pro-Phe-Pro-Ile-Ile-Val (Shinoda *et al.*, 1985), which suggests that the N-terminal Pro-Val dipeptide sequence has minimal contribution to bitterness intensity of the decapeptide. The tetradecapeptide Pro-Val-Leu-Gly-Pro-Val-Arg-Gly-Pro-Phe-Pro-Ile-Ile-Val essentially consists of two separate bitter peptide sequences (Pro-Val-Leu-Gly-Pro-Val and Arg-Gly-Pro-Phe-Pro-Ile-Ile-Val) but its bitterness intensity (0.015 mM threshold value) is lower than that of Arg-Gly-Pro-Phe-Pro-Ile-Ile-Val (Shinoda *et al.*, 1985). Moreover, when the amino acid sequence was reversed to form Val-Ile-Ile-Pro-Phe-Pro-Gly-Arg, the bitterness intensity was reduced with a threshold value of 0.14 mM when compared to 0.004 mM for Arg-Gly-Pro-Phe-Pro-Ile-Ile-Val, which confirms the importance of amino acid sequence for this peptide (Shinoda *et al.*, 1985). In contrast, the heptapeptide Arg-Gly-Pro-Phe-Pro-Ile-Val had a threshold value of 0.11 mM, which is similar to the value obtained for the reversed sequence peptide Val-Ile-Pro-Phe-Pro-Gly-Arg (0.07 mM). Thus, the bitterness intensity depends not only on the sequence but for some peptides, the type of amino acids is an important determinant. Removal of the C-terminal Arg residue produced a heptapeptide Val-Ile-Ile-Pro-Phe-Pro-Gly with half the bitterness intensity (0.26 mM threshold value) of Val-Ile-Ile-Pro-Phe-Pro-Gly-Arg. The importance of the N-terminal Arg residue was confirmed by showing that the hexapeptide Pro-Phe-Pro-Ile-Ile-Val with a threshold value of 0.13 mM has less bitterness intensity than Arg-Gly-Pro-Phe-Pro-Ile-Ile-Val (0.004 mM). Since the roles of amino acid position and sequence seem important in determining bitterness intensity of peptides, circular dichroism was performed to elucidate effects of spatial structure. It was shown that peptides with similar bitterness intensity (Arg-Gly-Pro-Phe-Pro-Ile-Val and Val-Ile-Pro-Phe-Pro-Gly-Arg) tended to have similar secondary structure. In contrast, peptides that differ in bitterness intensity (Val-Ile-Ile-Pro-Phe-Pro-Gly-Arg and Arg-Gly-Pro-Phe-Pro-Ile-Ile-Val) showed differences in secondary structure (Shinoda *et al.*, 1985). The authors concluded that the bitterness intensity of these β-casein peptide fragments was positively related to the presence of positively charged as well as hydrophobic amino acids. A recent work showed that all the bitter peptides isolated from aged Cheddar cheese contained a hydrophobic amino acid (Val, Pro, Ile, or Phe) at the C-terminal, including Gly-Pro-Val-Arg-Gly-Pro-Phe-Pro-Ile-Ile-Val, which had the strongest bitterness intensity (Karametsi *et al.*, 2014).

Additional information on the structure-bitterness relationship of peptides was provided through the use of synthetic Pro-Phe-Pro-Gly-Pro-Ile-Pro and Tyr-Pro-Phe-Pro-Gly-Pro-Ile bitter peptides (β-casein fragments commonly found in cheese bitter fractions) as well as their amino acid substituted equivalents. Reductions in peptide chain length led to decreased bitterness with Pro-Phe-Pro-Gly-Pro-Ile, Pro-Gly-Pro-Ile-Pro and Pro-Ile-Pro having threshold values of 0.44, 0.80 and 1.40 mM, respectively when compared to 0.25 mM for Pro-Phe-Pro-Gly-Pro-Ile (Shinoda *et al.*, 1986c). Thus, synergistic effects of multiple hydrophobic amino acids seem to be responsible for the intense bitterness of Pro-Phe-Pro-Gly-Pro-Ile. Substitution of Gly for Phe and Ile to form a Pro-Gly-Pro-Gly-Pro-Gly peptide resulted in substantial increase in threshold value to 2.5 mM, which suggests that amino acid hydrophobic chains are critical determinant of bitterness intensity for this peptide. Tyr-Pro-Phe-Pro-Gly-Pro-Ile had a bitterness threshold value of 0.16 but removal of the N-terminal Tyr residue led to an increase to 0.44 mM and production of a less bitter Pro-Phe-Pro-Gly-Pro-Ile (Shinoda *et al.*, 1986c). But the N-terminal Pro-Phe-Pro fragment had similar (0.4 mM threshold value) bitterness intensity as Pro-Phe-Pro-Gly-Pro-Ile. In contrast Phe-Pro, Pro-Phe and Tyr-Pro peptides had threshold values of 1.5, 38.0, and 19.0 mM, respectively when compared to Tyr-Pro-Phe (0.4 mM), which provides evidence that the more hydrophobic Tyr residue has stronger contributions than Pro to peptide bitterness intensity. From the results of Shinoda *et al.* (1986c) it seems that peptides containing Gly and Pro residues may have practical use in formulating food products with decreased bitterness intensity.

Another β-casein peptide fragment (Val-Val-Val-Pro-Pro-Phe-Leu) was also used for structure-bitterness studies and the results showed that Arg residue at the N-terminal potentiates bitterness intensity more than bulky or hydrophobic amino acids (Shinoda *et al.*, 1986b). The heptapeptide Val-Val-Val-Pro-Pro-Phe-Leu had a bitterness threshold value of 0.14 mM, which increased to 0.38, 3.75, 4.5, and 25 mM for peptides Val-Val-Val-Pro-Pro-Phe-Leu-Gln-Pro, Val-Val-Val-Pro-Pro, Val-Val-Val, and Val-Val, respectively. The results confirm earlier reports that longer peptide chains of amino acids with hydrophobic side groups have stronger bitterness intensities than shorter chains of similar amino acid composition. The importance of amino acid sequence was further demonstrated when it was shown that peptide Val-Gly had no detectable bitterness threshold but Gly-Val had a value of 4.5 mM (Shinoda *et al.*, 1986b). Also important is the position of an amino acid with a hydrophobic side group because Val-Gly-Gly and Gly-Val-Gly had no detectable bitterness threshold but Gly-Gly-Val still had a weak value of 38.0 mM. To demonstrate role of amino acid side group in potentiating peptide bitterness, the Val-Val-Val-Pro-Pro-Phe-Leu peptide was compared to Arg-Gly-Pro-Pro-Phe-Ile, which replaced the bulky hydrophobic Val groups with Arg (positively charged) and Gly (no side group). Peptide Arg-Gly-Pro-Pro-Phe-Ile had a bitterness threshold value of 0.025 mM when compared to the 0.14 mM for Val-Val-Val-Pro-Pro-Phe-Leu and 1.5 mM for Phe-Ile; thus the positively charged amino acid potentiates bitterness intensity better than the bulky group. However, a peptide chain length of hydrophobic amino acids was again confirmed as critical structural feature for stronger bitterness intensity because Arg-Gly had a 10.0 mM threshold value when compared to the 0.025 mM for Arg-Gly-Pro-Pro-Phe-Ile (Shinoda *et al.*, 1986b).

The role for Leu in potentiating bitterness intensity was further explored using a series of synthetic peptides that contained only Leu or in combination with Gly.

L-Leu alone has a bitterness threshold of 20 mM, which is 20 times less than that of caffeine; interestingly, D-Leu has a sweet taste with a threshold value of 6 mM (Ishibashi *et al.*, 1987a). The methyl ester form of L-Leu has a bitterness intensity that is six times greater than that of free Leu, which is consistent with other reports that presence of free carboxylic groups reduces bitterness intensity. This principle is exemplified by the observation that Leu-Leu has a bitterness threshold of 2.5 mM when compared to the 20 mM for L-Leu. Increasing the number of Leu residues to 3 (tripeptide) and 4 (tetrapeptide) led to bitterness threshold values of 1.2 and 0.6, respectively, which also suggests synergistic effects (Ishibashi *et al.*, 1987a). Since free Gly has a sweet taste but not a bitter taste and Gly dipeptide is tasteless, this amino acid was used as a spacer to study the effect of Leu position on bitterness intensity of various peptides. As expected, incorporation of Gly into Leu dipeptides did not have any substantial effect on bitterness threshold because Leu-Gly and Gly-Leu had similar values as Leu only (Ishibashi *et al.*, 1987a). Combination of Leu with other amino acids showed that the dipeptides with Phe or N-terminal Ile had bitterness threshold values of <2 mM. When the Ile was at the C-terminal or when Leu was combined with Val in either position, bitterness was reduced and threshold value increased to ≥4.0 mM. The HCl form of dipeptides that contain Leu and Asp acid at the N-terminal were very bitter with bitterness threshold values of <2 mM. The HCl salts of Leu and Glu dipeptides also had bitterness threshold values of <2 mM, irrespective of amino acid position. In contrast, dipeptides of Leu with Gly had weak bitterness intensities with threshold values 20 or 25 mM, irrespective of the amino acid position or isomeric form of Gly and Leu. However, for tripeptides, the presence of Leu residues and their position on the peptide chain had substantial effects on bitterness intensity. A tripeptide containing only Gly (Gly-Gly-Gly) had no detectable taste but replacement of the N-terminal Gly with Leu to form Leu-Gly-Gly produced a peptide with a weak bitterness threshold value of 75 mM. When the middle or C-terminal Gly was replaced with Leu to form Gly-Leu-Gly or Gly-Gly-Leu, the bitterness threshold decreased substantially to 10 mM. Replacement of two Gly residues with Leu to form Leu-Leu-Gly, Leu-Gly-Leu and Gly-Leu-Leu produced peptides with bitterness threshold values of 5, 5, and 1.5 mM, respectively (Ishibashi *et al.*, 1987a). For tetra- and pentapeptides, all the amino acid combinations produced peptides with bitterness threshold values of ≥13 mM, with the exception of when the Leu was at the C-terminal end. Thus the tetrapeptide Gly-Gly-Gly-Leu and pentapeptide Gly-Gly-Gly-Gly-Leu had bitterness threshold values of 4.5 and 2.2 mM, respectively, which are very close to the 1.2 mM value for Leu tripeptide (Leu-Leu-Leu). Therefore, the C-terminal position seems to be the major determinant of the bitterness intensity of Leu-containing peptides. This hypothesis is supported by the work of Spellman *et al.* (2005) who isolated a bitter peptide (Val-Glu-Glu-Leu-Lys-Pro-Thr-Pro-Glu-Gly-Asp-Leu-Glu-Ile-Leu) from whey protein hydrolysate. The peptide, which is characterized by the presence of Leu at the C-terminal end, had a linear positive relationship with bitterness intensity of the whey protein hydrolysate.

Bitterness intensity of Phe and Tyr containing peptides has also been studied to demonstrate role of amino acid type and position. This is because of previous reports that have implicated aromatic amino acids as one of the main primary determinants of peptide bitterness intensity. For example, the bitter taste fraction from Manchego cheese extract was due to the presence of Phe, Tyr and Trp (Taborda *et al.*, 2008). L-Phe has a bitter taste with threshold value of 20 mM, whereas D-Phe is very sweet with a threshold value of 2.2 mM (Ishibashi *et al.*, 1987b). Just as observed for leucine, the amino acid isomeric form did not affect bitterness intensity of dipeptides that consist of only Phe. Bitterness intensity of Phe-containing dipeptides was shown to be independent of amino acid position with Gly-Phe, Phe-Leu, and Ile-Phe having threshold values of 1.2, 1.5, and 1.5 mM, respectively (Ono *et al.*, 1988). But bitterness intensity increased substantially for a Phe only tripeptide (Phe-Phe-Phe), which had a threshold value of 0.2 mM. Ligand binding tests showed Phe-Phe-Phe as the strongest T2R activator (Fig. 6.2) with an EC_{50} (effective concentration that caused 50% activation) of 0.37 mM when compared to 7.2-7.4 mM for the Phe-containing dipeptides (Upadhyaya *et al.*, 2010). The results suggest that a higher hydrophobicity of the Phe tripeptide may have enhanced interaction with the T2R receptor, which led to stronger bitterness intensity when compared to the less hydrophobic dipeptides. However, unlike Leu, L-Phe dipeptides

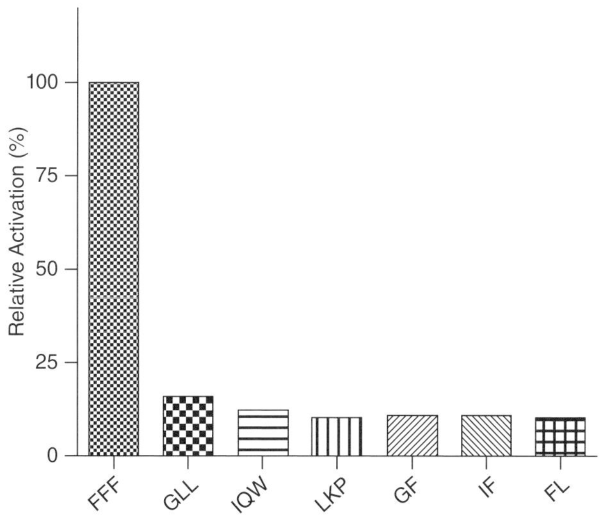

Figure 6.2 Relative activation rates of T2R1 in response to different peptides calculated based on their EC_{50} values. All values were normalized to the EC_{50} value of FFF. Adapted from Upadhyaya *et al.* (2010) with permission. Copyright 2015 Elsevier B.V.

with Gly had higher bitterness intensity but the threshold value is dependent on amino acid position. Phe at the C-terminal Gly-Phe has bitterness threshold value of 1.2 mM, which is similar to that of Phe-Phe but lower than the 6.0 mM value for Phe-Gly (Ishibashi *et al.*, 1987b). Therefore, similar to the Leu peptides, the presence of Phe at the C-terminal seems to be a critical determinant of bitterness intensity of the dipeptides. Similar results were also obtained for tripeptides where it was shown that the peptide containing Phe only had the lowest bitterness threshold value of 0.2 mM and peptides with Phe at the C-terminal had lowest threshold values than those with Phe at the N-terminal. Peptide bitterness intensity increased as the number of Phe residues in the tripeptides was increased. In contrast to Phe, other results suggest that the number of Tyr in a peptide was more important than the position in potentiating bitterness intensity. For example, Tyr-Gly and Gly-Tyr had similar bitterness threshold value of 3.0 mM, which decreased to 2.3 mM when both residues were tyrosine (Tyr-Tyr). For tripeptides, Tyr in the middle position (Gly-Tyr-Gly) had the least bitterness intensity with a threshold value of 19 mM when compared to Tyr-Gly-Gly (2.3 mM), Gly-Gly-Tyr (1.5 mM), Tyr-Tyr-Gly (0.6 mM), Tyr-Gly-Tyr (0.8 mM), Gly-Tyr-Tyr (0.4 mM) and Tyr-Tyr-Tyr (0.2 mM); there was a slight tendency towards higher bitterness intensity when Tyr was at the C-terminal (Ishibashi *et al.*, 1987b).

The structure-bitterness taste properties of Asn-Ala-Leu-Pro-Glu, a strongly bitter peptide isolated from soybean 11S glycinin has also been studied. This peptide has a hydrophocity index of 980 cal/mol (Kim *et al.*, 2008), which is less than the minimum 1400 cal/mol suggested for bitter peptides (Ney, 1979). Therefore, amino acid sequence of the peptide seems to be a more important determinant of bitterness intensity than the hydrophobicity. This peptide lacks a basic amino acid residue at the N-terminal while the C-terminal does not contain hydrophobic residues, which are the structural features that have been proposed for bitter peptides (Shinoda *et al.*, 1987). When the C-terminal Glu was replaced with aspartic acid (Asn-Ala-Leu-Pro-Asp), there was no change in hydrophobicity index but the peptide lost bitterness taste and became sour or astringent (Kim *et al.*, 2008). Replacement of Glu with arginine (Asn-Ala-Leu-Pro-Arg) led to increased hydrophobicity (1130 cal/mol) but bitterness intensity was reduced as evident by the higher minimum response threshold (MRT) value of 0.420 mM when compared to the 0.074 mM for Asn-Ala-Leu-Pro-Glu. Similarly, when the C-terminal Glu was replaced with serine (Asn-Ala-Leu-Pro-Ser), hydrophobicity decreased to 920 cal/mol while MRT increased to 0.250 mM, which means higher bitterness intensity when compared to Asn-Ala-Leu-Pro-Arg. In contrast, replacement of the C-terminal Glu with leucine (Asn-Ala-Leu-Pro-Leu) and tryptophan (Asn-Ala-Leu-Pro-Trp) led to increased hydrophobicity while MRT values were 0.149 and 0.105 mM, respectively. Thus since all the peptides contain the core hydrophobic Ala-Leu-Pro sequence, difference in bitterness intensity must be due to spatial arrangement of the amino acid residues. It has been postulated that proximity of the N-terminal C=O groups to hydrophobic regions may contribute to bitterness intensity (Kim *et al.*, 2008). Using computer simulations, the strongly bitter peptides (Asn-Ala-Leu-Pro-Glu, Asn-Ala-Leu-Pro-Trp, and Asn-Ala-Leu-Pro-Leu) had similar spatial structure, especially distance between the N-terminal C=O groups and hydrophobic regions (Kim *et al.*, 2008). However, the less bitter peptides (Asn-Ala-Leu-Pro-Arg and Asn-Ala-Leu-Pro-Ser) had similar spatial structure that showed longer distances between the N-terminal C=O groups and hydrophobic regions when compared to the intensely bitter Asn-Ala-Leu-Pro-Glu. Table 6.2 contains several amino acid sequences as examples of the structure-function properties discussed for bitter peptides.

6.3.4 Influence of amino acid isomeric configuration

Peptide bitterness intensity has also been shown to be strongly influenced by amino acid configuration in terms of the L- and D-stereoisomers. For example, the hexapeptide Arg-Arg-Pro-Pro-Phe-Phe with all amino acids in the L-configuration had a bitterness threshold value of 0.007 mM, which is similar to the 0.006 mM obtained for same peptide sequence but with C-2 Phe in the D-configuration (Shinoda *et al.*, 1986a). However, when both the Phe residues or the C-terminal Phe were switched to the D-configuration the threshold value was increased to 0.03 or 0.04 mM, respectively. Thus, the L-configuration of the C-terminal Phe residue seems to be critical for potentiating bitterness of this hexapeptide. Secondary structure measurement showed that the two Arg-Arg-Pro-Pro-Phe-Phe peptides with C-terminal Phe residue in the L-configuration had a positive peak in the 215-220 nm, which is different from Arg-Arg-Pro-Pro-Phe-Phe peptides with C-terminal Phe residue in the D-configuration (Shinoda *et al.*, 1986a). Therefore, the similarity in the spatial structure of peptides with similar bitterness intensity is consistent with a previous work (Shinoda *et al.*, 1985) and suggests that secondary structure conformation may be an important determinant. Even though D-Leu has a sweet taste, a dipeptide containing two D-Leu amino acids has an intense bitter taste with a threshold value of 2.5 mM, which is the same as dipeptides that contain two L-Leu residues or a combination of one L-Leu with one D-Leu, irrespective of the position.

6.4 PEPTIDE DEBITTERING METHODS

6.4.1 Peptide complexation

Activated carbon is highly hydrophobic and will form strong interactions with equally hydrophobic peptides. For example, a 5% (w/v) corn gluten hydrolysate solution was stirred with 10% activated carbon for 2 h, which was followed by filtration through a Whatman #2 paper or centrifugation at 3,000 x *g* for 15 min. Filtration and centrifugation protocols serve to remove the activated carbon (along

Table 6.2 Structure-taste properties of typical bitter peptides.

Amino acid sequence	Threshold value (mM)	References
Arg-Arg	8.000	Otagiri *et al.*, 1985
Arg-Arg-Arg	4.000	Otagiri *et al.*, 1985
Pro-Pro	4.500	Otagiri *et al.*, 1985
Pro-Pro-Pro	2.000	Otagiri *et al.*, 1985
Arg-Pro	0.800	Otagiri *et al.*, 1985
Pro-Arg	3.000	Otagiri *et al.*, 1985
Arg-Gly	0.130	Otagiri *et al.*, 1985
Gly-Arg	100	Otagiri *et al.*, 1985
Arg-Pro-Phe-Phe	0.040	Otagiri *et al.*, 1985
Arg-Arg-Pro-Phe-Phe	0.020	Otagiri *et al.*, 1985
Arg-Arg-Pro-Pro-Phe-Phe	0.007	Otagiri *et al.*, 1985
Arg-Arg-Pro-Pro-Pro-Phe-Phe-Phe	0.002	Otagiri *et al.*, 1985
Pro-Phe-Pro-Ile-Ile-Val	0.130	Shinoda *et al.*, 1985
Pro-Val-Leu-Gly-Pro-Val	0.500	Shinoda *et al.*, 1985
Arg-Gly-Pro-Phe-Pro-Ile-Ile-Val	0.004	Shinoda *et al.*, 1985
Pro-Val-Arg-Gly-Pro-Phe-Pro-Ile-Ile-Val	0.004	Shinoda *et al.*, 1985
Tyr-Pro-Phe-Pro-Gly-Pro-Ile	0.160	Shinoda *et al.*, 1986c
Pro-Phe-Pro-Gly-Pro-Ile-Pro	0.250	Shinoda *et al.*, 1986c
Pro-Phe-Pro-Gly-Pro-Ile	0.440	Shinoda *et al.*, 1986c
Tyr-Pro-Phe	0.400	Shinoda *et al.*, 1986c
Pro-Phe-Pro	0.400	Shinoda *et al.*, 1986c
Pro-Val-Leu-Gly-Pro-Val	0.500	Shinoda *et al.*, 1986c
Val-Val	25.00	Shinoda *et al.*, 1986b
Val-Gly	0.000	Shinoda *et al.*, 1986b
Gly-Val	4.500	Shinoda *et al.*, 1986b
Val-Val-Val	4.500	Shinoda *et al.*, 1986b
Val-Val-Val-Pro-Pro	3.750	Shinoda *et al.*, 1986b
Val-Val-Val-Pro-Pro-Phe-Leu	0.140	Shinoda *et al.*, 1986b
Val-Val-Val-Pro-Pro-Phe-Leu-Gln-Pro	0.380	Shinoda *et al.*, 1986b
Leu-Leu-Leu	1.200	Ishibashi *et al.*, 1987a
Leu-Leu-Gly	5.000	Ishibashi *et al.*, 1987a
Gly-Leu-Gly	10.00	Ishibashi *et al.*, 1987a
Leu-Gly-Gly	75.00	Ishibashi *et al.*, 1987a
Gly-Leu-Leu	1.500	Ishibashi *et al.*, 1987a

(*continued*)

Table 6.2 (*Continued*)

Amino acid sequence	Threshold value (mM)	References
Leu-Leu-Leu-Leu	0.600	Ishibashi *et al.*, 1987a
Asp-Ile-Lys-Gln-Met	6.000	Toelstede & Hofman, 2008
Glu-Ile-Val-Pro-Asn	0.430	Toelstede & Hofman, 2008
Met-Ile	0.420	Toelstede & Hofman, 2008
Tyr-Pro-Phe-Pro-Gly-Pro-Ile-His-Asn-Ser	0.050	Toelstede & Hofman, 2008
Leu-Val-Tyr-Pro-Phe-Pro-Gly-Pro-Ile-His-Asn	0.080	Toelstede & Hofman, 2008
Gly-Pro-Val-Arg-Gly-Pro-Phe-Pro	1.180	Toelstede & Hofman, 2008
Tyr-Pro-Phe-Pro-Gly-ProIle-Pro-Asn	0.230	Toelstede & Hofman, 2008

with bound bitter peptides) from solution. The obtained filtrate or supernatant was shown to possess reduced bitterness intensity when compared to the untreated control solution (Suh *et al.*, 2000). However, the authors showed that effectiveness of the debittering process was highly dependent on the type of hydrophobic matrix used for complexation and hydrolytic specificity of the protease. Bitterness intensity of the hydrolysates was shown to have positive correlation with peptide surface hydrophobicity, which confirms contribution of hydrophobic amino acids. A skim milk protein hydrolysate was also shown to be successfully debittered using activated carbon; sensory evaluation showed that a beverage containing 10% of the treated hydrolysate was as acceptable as regular orange juice but apple juice was preferred (Helbig *et al.*, 1980). In a follow-up work, it was shown that the debittered milk protein hydrolysate had reduced contents of Phe and Tyr, which may have contributed to the reduced bitterness after activated carbon treatment (Ma *et al.*, 1983).

Complexation with cyclodextrin (CD) has also been used to reduce bitterness of protein hydrolysates. CD is ideally suited for complex formation because its hydrophobic cavity can form strong interactions with hydrophobic bitter peptides, which become sequestered and unable to interact with bitter taste receptors during oral consumption. A 5% aqueous solution of an alcalase digest of soybean protein was mixed with 3 or 5% (w/v) CD and then subjected to sensory evaluation by a 6-member human taste panel (Linde *et al.*, 2010). Their results showed that the bitterness intensity of the soybean protein hydrolysate containing 3% (w/v) CD was about 50% less than the untreated sample. When the CD level was increased to 5% (w/v), bitterness intensity of the soybean protein hydrolysate was reduced by 90%. Complexation strength of CD may be increased at acidic pH values where the peptides carry no ionic charges and would interact better with the hydrophobic CD core. Cold temperature may also enhance CD complexation with peptides since there is decreased molecular vibration which reduces potential disruption of the CD-peptide complexes (Linde *et al.*, 2010). The advantages of acidic and cold environments make CD a good choice to formulate non-bitter or weakly bitter soft drinks and juices that are formulated with protein hydrolysates. Since the CD sequesters mostly the hydrophobic peptides, the remaining peptides will be rich in hydrophilic (and likely acidic) peptides. Coupled with the acidic pH of these beverages, the use of CD may allow incorporation of higher levels of peptides, especially to achieve therapeutic levels such as for antihypertensive and antioxidant effects.

6.4.2 Hydrophobic column adsorption

Since several bitter peptides have been shown to consist of hydrophobic amino acid residues, passage of a bitter peptide mixture through a hydrophobic column could enable separation of the hydrophobic peptides to produce a more hydrophilic peptide mixture with reduced bitterness intensity. A bitter casein hydrolysate was passed through a C8 or C18 column followed by elution with water to collect unbound hydrophilic peptide fractions as the debittered product (Lin *et al.*, 1997a,b). The column was then regenerated by washing with absolute ethanol to remove bound bitter peptides. A phenolic resin column was also used and the bitter casein hydrolysate passed through; the effluent collected before appearance of the highest peak at 280 nm was collected as the debittered fraction (Lin *et al.*, 1997a,b). Results showed that the C18 column and the phenolic resin column were the most effective in removing a broad range of hydrophobic peptides, whereas the C8 column removed mostly the strongly hydrophobic peptides. Amino acid composition also showed reductions in contents of Phe, Pro and Tyr in debittered fractions collected from the C18 and phenolic resin columns. But the phenolic resin column had a poorer yield of non-bitter peptides because it removed both hydrophobic and hydrophilic peptides. In contrast the C18 column seem to work best by producing a higher yield of non-bitter peptides since it binds mostly hydrophobic peptides. Overall, the C8 column products still retained bitterness attributes whereas the C18 and phenolic resin products had reduced bitterness intensities. Lin *et al.* (1997a,b) reported optimized conditions for the use of C18 adsorption columns to dibitter protein hydrolysates. Their results showed that a column diameter-to-height ratio of 1.33-5.0 coupled with a linear flow rate of 200-400 cm/h and feed concentration of up to 30% (w/v) had a debittered protein hydrolysate yield of up to 72% and was effective up to 70 debittering cycles.

A macroporous adsorption resin was used to separate whey protein hydrolysate (WPH) into various peptide fractions that were then evaluated for bitterness intensity. An aqueous WPH solution (2 mg/ml, pH 6.5) was pumped through a glass column (500 ml capacity) packed with styrene-based macroporous resin followed by washing with water (Cheison *et al.*, 2007). The adsorbed peptides were then eluted stepwise with 20, 40, and 75% (v/v) alcohol to give three fractions (F1, F2, and F3, respectively), which were then analyzed for amino acid composition and bitterness intensity. The contents of hydrophobic amino acids for F1, F2, and F3 were 27, 37, and 46%, respectively, which was directly correlated to the bitterness intensity. The F1 had no detectable bitter taste while F2 had some bitter but F3 had a very bitter taste. Therefore, F1 and F2 fractions contain peptides that may be used to formulate products with reduced or zero bitterness intensity. Moreover, a previous work has shown that acidic pH or acidic peptides could suppress human bitter taste receptor and mask the bitterness taste of foods (Sakurai *et al.*, 2009).

6.4.3 Enzyme treatment

Most bitter peptides contain amino acid residues at the N- or C-terminal positions and this locations have been shown by many researchers to potentiate peptide bitterness (Matoba & Hata 1972; Otagiri *et al.*, 1985; Ishibashi *et al.*, 1987a,b). In addition, peptide chain length seems to be directly related to bitterness intensity (Ishibashi *et al.*, 1987a,b). Therefore, application of exopeptidases could provide an effective means of reducing the amount of hydrophobic amino acids at the peptide chain terminals while at the same time result in reduced chain length. Arai *et al.* (1970a) and Fujimaki *et al.* (1970) performed batch treatments of soybean protein hydrolysates with carboxypeptidase A (removes amino acids from the peptide C-terminal) to reduce bitterness intensity. In one of the experiments, the carboxypeptidase A-treated soybean hydrolysate solution was dialyzed against water for 48 h at 5 °C, which allowed separation of the cleaved free amino acids from the debittered peptides (Arai *et al.*, 1970a). Wheat carboxypeptidase treatment was also shown to be effective in reducing bitterness intensity of casein with an inverse relationship between amount of released hydrophobic amino acids and bitterness intensity of the product (Umetsu *et al.*, 1983). A serine carboxypeptidase extracted from the Japanese common squid hepatopancreas was shown to be effective in eliminating the bitterness taste of a soybean digest prepared by pepsin digestion (Komai *et al.*, 2007). In contrast, the serine carboxypeptidase was not efficient in reducing bitterness intensity of tryptic digest of casein or peptic digest of corn. Therefore, the efficiency of some carboxypeptidases may be dependent on the amino acids involved in peptide bond formation. For example, the serine carboxypeptidase had the highest rate of hydrolysis when the peptide bond is the Phe-Leu type but low rates for Phe-Pro, Gly-Val, Gly-Phe, Gly-Met and Gly-Leu (Komai *et al.*, 2007). Gly-Pro, Gly-Lys and Pro-Pro were resistant to hydrolysis by the serine carboxypeptidase.

A casein hydrolysate was treated with an aminopeptidase, which removes amino acids from the peptide N-terminal (Minagawa *et al.*, 1989). Hydrolysis with aminopeptidase was carried out at pH 8.5 and 60 °C followed by sample evaluation for amino acid release up to 20 h. The results showed that Phe, Ala, Val, Tyr and Leu were the most released amino acids, which could account for the decreased or eliminated bitterness intensity of these protein hydrolysates. This is because peptide bitterness intensity is highly potentiated by the presence of hydrophobic amino acids. Bitter peptide fractions were collected by passing a pepsin digest of casein through Sephadex G-15; the fractions were then treated for 3 h with aminopeptidase II isolated from *Penicillium caseicolum* (Matsuoka *et al.*, 1991). Amino acid analysis showed that bitterness intensity was inversely related to level of cleaved amino acids, which were mainly Thr, Ser, Leu, Met, Phe, and Lys. A similar work also showed that an aminopeptidase N obtained from *Lactococcus lactis* could degrade a bitter tryptic β-casein digest to give a product with reduced bitterness and peptides with reduced hydrophobicity (Tan *et al.*, 1993). Interestingly, most of the non-bitter peptides were resistant to hydrolysis by the aminopeptidase N and overall, the bitter score was shown to be inversely related to incubation time, which indicates gradual degradation of the bitter peptides. A similar work used *Aeromonas caviae* aminopeptidase but found higher debittering activity with soybean protein hydrolysates than casein hydrolysate (Izawa *et al.*, 1997). Over 76% of the released free amino acids consisted of hydrophobic residues such as Phe, Tyr, Val, Ile, and Leu, which led to decreased peptide hydrophobicity and hence decreased bitterness intensity. The lower debittering efficacy against casein hydrolysate was believed to be due to the presence of several proline residues in the casein peptides. This is because the *A. caviae* aminopeptidase cannot remove amino acids that are adjacent to proline, which leads to fractional accumulation of bitter proline-rich peptides (Izawa *et al.*, 1997). A recent work showed that for a thermolysin whey protein hydrolysate, exopeptidase treatment was a more effective debittering method than aminopeptidase treatment (Cheung *et al.*, 2015). Therefore, choice of enzyme treatment will depend on the protein hydrolysate substrate and it may be necessary to test various debittering exopeptidases in order to determine the most effective.

Bitter peptides have been shown to consist of several proline (Pro) residues; Pro-Pro peptide bond is believed to be resistant to hydrolysis by most endoproteases, which could lead to accumulation of such peptides and increased bitterness intensity of the food product (Edens *et al.*, 2005). Therefore, the use peptidases such as prolyl endoprotease that have specificity for the Pro-Pro bond could enhance degradation of bitter peptides and reduce bitterness intensity of food products. A bitter casein hydrolysate obtained by treatment with thermolysin or subtilisin was subjected to proteolysis by a proline-specific protease that was isolated from *Aspergillus niger*. Human sensory evaluation of the hydrolysates showed a high bitterness intensity score of 4 for the thermolysin or subtilisin hydrolysate. In contrast, incubation of the thermolysin and subtilisin hydrolysates with *A. niger* prolyl endoprotease led to a low bitterness score of 1, which confirms debittering ability (Edens *et al.*, 2005). HPLC analysis of the prolyl endoprotease digest showed a substantial decrease in the hydrophobic peptides fraction, which was attributed to hydrolysis of the Pro-Pro bonds in the thermolysin and subtilisin hydrolysates. The decreased bitterness intensity of the prolyl endoprotease digests was attributed to decreased hydrophobicity that resulted from peptide degradation through hydrolysis of Pro-Pro peptide bonds.

6.5 CONCLUSIONS

Recent advances in the functional foods and nutraceuticals industry have led to increased production of protein hydrolysates for potential use as therapeutic agents. Associated with these protein hydrolysates is the presence of bitter peptides, which can have a negative influence on the taste and consumer acceptance of formulated foods. Therefore, there is need for additional research that will discover new proteases to produce non-bitter bioactive protein hydrolysates. Research is also required to produce novel bitter taste blockers to be used in suppressing bitterness properties of bioactive protein hydrolysates such that therapeutic uses can be achieved. These approaches are critical in producing next generation of protein hydrolysastes because current methods that involve complexation or exopeptidases are likely to inactivate bioactive peptides, which will prevent therapeutic use. Cheese is another food product that has the intrinsic problem of developing bitter peptides during manufacture or aging. Future research activities should build upon current knowledge of bacteria cultures that do not produce bitter peptides. Advances in bacteria biotechnology can lead to identification of species that can suppress bitter peptides formation during cheese aging, which will allow production of cheese products (especially Cheddar) with reduced bitter taste and enhanced eating quality.

ACKNOWLEDGEMENT

The research program of Dr. Rotimi Aluko is funded by grants from the Natural Sciences and Engineering Council of Canada (NSERC).

REFERENCES

Arai, S., Noguchi, M., Kurosawa, S., Kato, H., & Fujimaki, M. (1970a). Applying proteolytic enzymes on soybean. 6. Deodorization effect of Aspergillopeptidase A and debittering effect of Aspergillus acid carboxypeptidase. *J. Food. Sci.* **35**, 392–395.

Arai, S., Yamashita, M., Kato, H., & Fujimaki, M. (1970b). Applying proteolytic enzymes on soybean. Part V. Nondialyzable bitter peptide in peptic hydrolyzate of soybean protein and its bitterness in relation to the chemical structure. *Agric. Biol. Chem.* **34**, 729–738.

Aubes-Dufau, I., Seris, J-L., & Combes, D. (1995). Production of peptic hemoglobin hydrolysates: bitterness demonstration and characterization. *J. Agric. Food Chem.* **43**, 1982–1988.

Borsting, M., Qvist, K. B., Rasmussen, M., et al. (2012). Impact of selected coagulants and starters on primary proteolysis and amino acid release related to bitterness and structure of reduced-fat Cheddar cheese. *Dairy Sci. Technol.* **92**, 593–612.

Broadbent, J. R., Barnes, M., Brennand, C., et al. (2002). Contribution of *Lactococcus lactis* cell envelope proteinase specificity to peptide accumulation and bitterness in reduced-fat Cheddar cheese. *Appl. Environ. Microbiol.* **68**, 1778–1785.

Broadbent, J. R., Strickland, M., Weimer, B. C., Johnson, M. E., & Steele, J. L. (1998). Peptide accumulation and bitterness in cheddar cheese made using single-strain *Lactococcus lactis* starters with distinct proteinase specificities. *J. Dairy Sci.* **81**, 327–337.

Chaves-Lopez, C., Tofalo, R., Serio, A., et al. (2012). Yeasts from Colombian Kumis as source of peptides with Angiotensin I converting enzyme (ACE) inhibitory activity in milk. *Int. J. Food Microbiol.* **159**, 39–46.

Cheison, S. C., Wang, Z., & Xu, S-Y. (2007). Use of macroporous adsorption resin for simultaneous desalting and debittering of whey protein hydrolysates. *Int. J. Food Sci. Technol.* **42**, 1228–1239.

Cheung, L. K. Y, Aluko, R. E., Cliff, M. A., & Li-Chan, E. C. Y. (2015). Effects of exopeptidase treatment on antihypertensive activity and taste attributes of enzymatic whey protein hydrolysates. *J. Funct. Foods.* **13**, 262–275.

Cho, M. J., Unklesbay, N., Hsieh, F-H., & Clarke, A. D. (2004). Hydrophobicity o bitter peptides from soy protein hydrolysates. *J. Agric. Food Chem.* **52**, 5895–5901.

Edens, L., Dekker, P., van der Hoeven, R., et al. (2005). Extracellular prolyl endoprotease from *Aspergillus niger* and its use in debittering of protein hydrolysates. *J. Agric. Food Chem.* **53**, 7950–7957.

Engel, E., Septier, C., Leconte, N., Salles, C., & Le Quere, J-L. (2001). Determination of taste-active compounds of a bitter Camembert cheese by omission tests. *J. Dairy Res.* **68**, 675–688.

Exterkate, F. A., Alting, A. C., & Slangen, C. J. (1995). Conversion of α_{s1}-casein-(24-199)-fragment and β-casein under cheese conditions by chymosin and starter peptidases. *Syst. Appl. Microbiol.* **18**, 7–12.

Fallico, V., McSweeney, P. L. H., Horne, J., et al. (2005). Evaluation of bitterness in Ragusano cheese. *J. Dairy Sci.* **88**, 1288–1300.

Fujimaki, M., Yamashita, M., Okazawa, Y., & Arai, S. (1970). Applying proteolytic enzymes on soybean. 3. Diffusable bitter peptides and free amino acids in peptic hydrolysate of soybean protein. *J. Food. Sci.* **35**, 215–218.

Fujimaki, M., Yamashita, M., Okazawa, Y., & Arai, S. (1968). Diffusable bitter peptides in peptic hydrolysate of soybean protein. *Agric. Biol. Chem.* **32**, 794–795.

Gallagher, J., Kanekanian, A. D., & Evans, E. P. (1994). Hydrolysis of casein: a comparative study of two proteases and their peptide maps. *Int. J. Food Sci. Technol.* **29**, 279–285.

Gomez, M. J., Garde, S., Gaya, P., Medina, M., & Nunez, M. (1997). Relationship between level of hydrophobic peptides and bitterness in cheeses made from pasteurized and raw milk. *J. Dairy Res.* **64**, 289–297.

Heinio, R-L., Nordlund, E., Poutanen, K., & Buchert, J. (2012). Use of enzymes to elucidate the factors contributing to bitterness in rye flavor. *Food Res. Int.* **45**, 31–38.

Helbig, N. B., Ho, L., Christy, G. E., & Nakai, S. (1980). Debittering of skim milk hydrolysates by adsorption for incorporation into acidic beverages. *J. Food. Sci.* **45**, 331–335.

Hicks, C. L., Onuorah, C. O'Leary, J., & Langlois, B. E. (1986). Effect of milk quality and low temperature storage on cheese yield—a summation. *J. Dairy Sci.* **69**, 649–657.

Humiski, L., & Aluko, R. E. (2007) Physicochemical and bitterness properties of enzymatic pea protein hydrolysates. *J. Food. Sci.* **72**, S605–S611.

Ishibashi, N., Arita, Y., Kanehisa, H., Kouge, K., Okai, H., & Fukui, S. (1987a). Bitterness of leucine-containing peptides. *Agric. Biol. Chem.* **51**, 2389–2394.

Ishibashi, N., Sadamori, K., Yamamoto, O., et al. (1987b). Bitterness of phenylalanine- and tyrosine-containing peptides. *Agric. Biol. Chem.* **51**, 3309–3313.

Izawa, N., Tokuyasu, K., & Hayashi, K. (1997). Debittering of protein hydrolysates using *Aeromonas caviae* aminopeptidase. *J. Agric. Food Chem.* **45**, 543–545.

Karametsi, K., Kokkinidou, S., Ronningen, I., & Peterson, D. G. (2014). Identification of bitter peptides in aged Cheddar cheese. *J. Agric. Food Chem.* **62**, 8034–8041.

Kim, H-O., & Li-Chan, E. C. Y. (2006). Quantitative structure-activity relationship study of bitter peptides. *J. Agric. Food Chem.* **54**, 10102–101011.

Kim, M-R., Yukio, K., Kim, K. M., & Lee, C-H. (2008). Tastes and structures of bitter peptide, asparagine-alanine-leucine-proline-glutamate, and its synthetic analogues. *J. Agric. Food Chem.* **56**, 5852–5858.

Kodera, T., Asano, M., & Nio, N. (2006). Characteristic property of low bitterness in protein hydrolysates by a novel soybean protease D3. *J. Food. Sci.* **71**, S609–S614.

Kohl, S., Behrens, M., Dunkel, A., Hofman, T., & Meyerhof, W. (2013). Amino acids and peptides activate at least five members of the human bitter taste receptor family. *J. Agric. Food Chem.* **61**, 53–60.

Komai, T., Kawabata, C., Tojo, H., Gocho, S., & Ichishima, E. (2007). Purification of serine carboxypeptidase from the hepatopancreas of Japanese common squid *Todarodes pacificus* and its application for elimination of bitterness from bitter peptides. *Fish. Sci.* **73**, 404–411.

Kukman, I. L., Zelenik-Blatnik, M., & Abram, V. (1995). Isolation of low molecular-mass hydrophobic bitter peptides in soybean protein hydrolysates by reversed-phase high-performance liquid chromatography. *J. Chromatogr.* **704**, 113–120.

Larsson, M., Zakora, M., Dejmek, P., & Ardo, Y. (2006). Primary proteolysis studied in a cast cheese made from microfiltered milk. *Int. Dairy J.* **16**, 623–632.

Lee, K. D., Lo, C. G., & Warthesen, J. J. (1996). Removal of bitterness from the bitter peptides extracted from cheddar cheese with peptidases from *Lactococcus lactis* ssp. *cremoris* SK11. *J. Dairy Sci.* **79**, 1521–1528.

Lee, K-P. D., & Warthesen, J. J. (1996). Preparative methods for isolating bitter peptides from cheddar cheese. *J. Agric. Food Chem.* **44**, 1058–1063.

Lin, S-B., Nelles, L. P., Cordle, C. T., & Thomas, R. L. (1997a). Debittering casein hydrolysate with octadecyl-siloxane (C18) columns. *J. Food. Sci.* **62**, 665–670.

Lin, S-B., Nelles, L. P., Cordle, C. T., & Thomas, R. L. (1997b). Physical factors related to C18 adsorption columns for debittering protein hydrolysates. *J. Food. Sci.* **62**, 946-948, 1010.

Linde, G. A., Junior, A. L., de Faria, E. V., et al. (2010). The use of 2D NMR to study β-cyclodextrin complexation and debittering of amino acids and peptides. *Food Res. Int.* **43**, 187–192.

Ma, C.-Y., Amantea, G. F., & Nakai, S. (1983). Production of nonbitter, desalted milk hydrolysate for fortification of soft drinks and fruit juices. *J. Food. Sci.* **48**, 897–899.

Maehashi, K., Matsuzaki, M., Yamamoto, Y., & Udaka, S. (1999). Isolation of peptides from an enzymatic hydrolysate of food proteins and characterization of their taste properties. *Biosci., Biotechnol., Biochem.* **63**, 555–559.

Matoba, T., & Hata, T. (1972). Relationship between bitterness of peptides and their chemical structures. *Agric. Biol. Chem.* **36**, 1423–1431.

Matoba, T., Hayashi, R., & Hata, T. (1970). Isolation of bitter peptides from tryptic hydrolysate of casein and their chemical structure. *Agric. Biol. Chem.* **34**, 1235–1243.

Matsuoka, H., Fuke, Y., Kaminogawa, S., & Yamauchi, K. (1991). Purification and debittering effect of aminopeptidase II from *Penicillium caseicolum*. *J. Agric. Food Chem.* **39**, 1392–1395.

McSweeney, P. L. H. (1997). The flavor of milk and dairy products: III. Cheese: taste. *Int. J. Dairy Technol.* **50**, 123–128.

Minagawa, E., Kaminogawa, S., Tsukasaki, F., & Yamauchi, K. (1989). Debittering mechanism in bitter peptides of enzymatic hydrolysates from milk casein by amino peptidase T. *J. Food. Sci.* **54**, 1225–1229.

Minamiura, N., Matsumura, Y., Fukumoto, J., & Yamamoto, T. (1972). Bitter peptides in cow milk casein digests with bacterial proteinase. Part I. Isolation and determination of amino acid sequence of a bitter peptide. *Agric. Biol. Chem.* **36**, 588–595.

Ney K. H. (1979). Bitterness of peptides: amino acid composition and chain length. In *Food Taste Chemistry, ACS Symposium Series* **115**, pp 149–173. Boudreau J. C. (ed.), Washington DC: American Chemical Society.

Ono, I., Ishibashi, N., Kato, K., et al. (1988). Role of the hydrophobic amino acid residue in the bitterness of peptides. *Agric. Biol. Chem.* **52**, 91–94.

Otagiri, K., Nosho, Y., Shinoda, I., Fukui, H., & Okai, H. (1985). Studies on a model of bitter peptides including arginine, proline and phenylalanine residues. I. Bitter taste of di- and tripeptides, and bitterness increase of the model peptides by extension of the peptide chain. *Agric. Biol. Chem.* **49**, 1019–1026.

Sakurai, T., Misaka, T., Nagal, T., et al. (2009). pH-dependent inhibition of the human bitter taste receptor hTAS2R16 by a variety of acidic substances. *J. Agric. Food Chem.* **57**, 2508–2514.

Shakeel-Ur-Rehman, Farkye, N. Y., Considine, T., Schaffner, A., & Drake, M. A. (2003). Effects of standardization of whole milk with dry milk protein concentrate on yield and ripening of reduced-fat cheddar cheese. *J. Dairy Sci.* **86**, 1608–1615.

Shimamura, T., Nishimura, T., Iwasaki, A., Odake, S., & Akuzuwa, R. (2009). Degradation of a bitter peptide derived from casein by lactic acid bacterial peptidase. *Food Sci. Technol. Res.* **15**, 191–194.

Shinoda, I., Fushimi, A., Kato, H., Okai, H., & Fukui, S. (1985). Bitter taste of synthetic C-terminal tetradecapeptide of bovine β-casein, H-Pro^{196}-Val-Leu-Gly-Pro-Val-Arg-Gly-Pro-Pro-Ile-Ile-Val^{209}-OH, and its related peptides. *Agric. Biol. Chem.* **49**, 2587–2596.

Shinoda, I., Nosho, Y., Kouge, K., et al. (1987). Variation in bitterness potency when introducing Gly-Gly residue into bitter peptides. *Agric. Biol. Chem.* **51**, 2103–2110.

Shinoda, I., Nosho, Y., Otagiri, K., Okai, H., & Fukui, S. (1986a). Bitterness of diastereoisomers of a hexapeptide Arg-Arg-Pro-Pro-Phe-Phe containing D-phenylalanine in place of L-phenylalanine. *Agric. Biol. Chem.* **50**, 1785–1790.

Shinoda, I., Okai, H., & Fukui, S. (1986b). Bitter taste of H-Val-Val-Val-Pro-Pro-Phe-Leu-OH corresponding to the partial sequence (positions 82-88) of bovine β-casein, and related peptides. *Agric. Biol. Chem.* **50**, 1255–1260.

Shinoda, I., Tada, M., Okai, H., & Fukui, S. (1986c). Bitter taste of H-Pro-Phe-Pro-Gly-Pro-Ile-Pro-OH corresponding to the partial sequence (positions 61-67) of bovine β-casein and related peptides. *Agric. Biol. Chem.* **50**, 1247–1254.

Singh, T. K., Young, N. D., Drake, M., & Cadwallader, K. R. (2005). Production and sensory characterization of a bitter peptide from β-casein. *J. Agric. Food Chem.* **53**, 1185–1189.

Soeryapranata, E., Powers, J. R., Fajarrini, F., et al. (2002). Relationship between MALDI-TOF analysis of β-CN f193-209 concentration and sensory evaluation of bitterness intensity of aged Cheddar cheese. *J. Agric. Food Chem.* **50**, 4900–4905.

Sorensen, N. K., Qvist, K. B., & Mulholland, F. (1996). Significance of low molecular weight nitrogen constituents in cheese milk and of proteinase-negative variants of a *Lactococcus lactis* subsp. *cremoris* strain on the production and maturation of Danbo cheese. *Int. Dairy J.* **6**, 113–128.

Spellman, D., O'Cuinn, G., & FitzGerald, R. J. (2005). Physicochemical and sensory characteristics of whey protein hydrolysates generated at different total solids levels. *J. Dairy Res.* **72**, 138–143.

Sridhar, V. R., Hughes, J. E., Welker, D. L., Broadbent, J. R., & Steele, J. L. (2005). Identification of endopeptidase genes from the genomic sequence of *Lactobacillus helveticus* CNRZ32 and the role of these genes in hydrolysis of model bitter peptides. *Appl. Environ. Microbiol.* **71**, 3025–3032.

Suh, J. S., Bae, S. H., & Noh, D. O. (2000). Debittering of corn gluten hydrolysate with active carbon. *J. Sci. Food Agric.* **80**, 614–618.

Sullivan, J. J., Mou, L., Rood, J. I., & Jago, G. R. (1973). The enzymatic degradation of bitter peptides by starter streptococci. *Aust. J. Dairy Technol.* **28**, 20–26.

Taborda, G., Gomez-Ruiz, J. A., Martinez-Castro, I., et al. (2008). Taste and flavor of artisan and industrial Manchego cheese as influenced by the water-soluble extract compounds. *Eur. Food Res. Technol.* **227**, 323–330.

Tan, P. S. T., van Kessel, T. A. J. M., van de Veerdonk, F. L. M., et al. (1993). Degradation and debittering of a tryptic digest from β-casein by aminopeptidase N from *Lactococcus lactis* subsp. *cremoris* WG2. *Appl. Environ. Microbiol.* **59**, 1430–1436.

Toelstede, S., & Hofman, T. (2008a). Quantitative studies and taste re-engineering experiments toward the decoding of the nonvolatile sensometabolome of Gouda cheese. *J. Agric. Food Chem.* **56**, 5299–5307.

Toelstede, S., & Hofman, T. (2008b). Sensomics mapping and identification of the key bitter metabolites in Gouda cheese. *J. Agric. Food Chem.* **56**, 2795–2804.

Umetsu, H., Matsuoka, H., & Ichishima, E. (1983). Debittering mechanism of bitter peptides from milk casein by wheat carboxypeptidase. *J. Agric. Food Chem.* **31**, 50–53.

Upadhyaya, J., Prasad Pydi, S., Singh, N., Aluko, R.E., & Chelikani, P. (2010). Bitter taste receptor T2R1 is activated by dipeptides and tripeptides. *Biochem. Biophys. Res. Commun.* **398**, 331–335.

Wu, J., & Aluko, R. E. (2007). Quantitative structure-activity relationship study of bitter di- and tri-peptides including relationship with angiotensin I-converting enzyme inhibitory activity. *J. Pept. Sci.* **13**, 63–69.

Yamashita, M., Arai, S., & Fujimaki, M. (1969). Applying proteolytic enzymes on soybean. Part IV. A ninhydrin-negative bitter peptide in peptic hydrolysate of soybean protein. *Agric. Biol. Chem.* **33**, 321–330.

Zhang, S. B., Wang, Z., & Xu, S. Y. (2007). Downstream processes for aqueous enzymatic extraction of rapeseed oil and protein hydrolysates. *J. Am. Oil Chem. Soc.* **84**, 693–700.

SECTION III

ANALYTICAL TECHNIQUES FOR SEPARATING AND CHARACTERIZING BITTER COMPOUNDS

7 Sensory Evaluation Techniques for Detecting and Quantifying Bitterness in Food and Beverages

Donna Ryland, Erin Goldberg, and Michel Aliani

7.1 SCREENING METHODS

The ability of subjects to detect bitterness at a level that will be high enough to detect in food and beverage products is critical. Humans vary in their sensitivities toward bitterness as well as other compounds. Screening tools have been developed for this purpose. In many studies, researchers will initially screen for bitterness perception prior to recruitment for further sensory analysis. After the screening process, panelists used in this capacity are certainly not representative of a normal population wherein bitterness perception can differ greatly. At the same time, it is a useful tool to determine subtle differences between bitter products that some consumers cannot perceive, and it is inherent to test these products in those who are sensitive to bitterness and who it would ultimately affect. Bitterness sensitivity can be determined with triangle tests, a method that presents two blank samples with one sample containing the bitter stimuli. The bitter stimuli can be set at levels appropriate for the product being studied. Seo, Lee and Baek (2008) used this method for selecting 10 panelists to evaluate the bitterness in enzyme hydrolyzed soy protein isolates. Information was provided by Kranz *et al.* (2010) regarding performance levels required by panelists in order to participate in the sensory testing. Caffeine was among other solutions containing stimulants for the four basic tastes, and astringency that were presented in low concentrations.

Panelists needed to correctly identify at least 7 of the 10 solutions including at least one of the two containing caffeine. Secondly panelists were presented with aqueous samples containing 30 and 60 mg of the polyphenol solution contained in the olive leaf extract under study. Panelists with correct responses regarding the presence of bitterness and its ranking were allowed to proceed to further sensory studies. These screening methods are detailed in ISO standard methods ISO 4120 (2004) for triangle testing and, ISO 8587 (1998) for ranking. The ability to taste PROP, a bitter compound, is one way to determine that potential subjects can perceive this tastant. Whether it is related to an individual's ability to perceive

Bitterness: Perception, Chemistry and Food Processing, First Edition. Edited by Michel Aliani & Michael N. A. Eskin.

bitterness in other compounds or food matrices and at what level is difficult to determine. Kobue-Lekalake, Taylor, & de Kock, (2012) used this as part of their criteria for panelist participation but it is not noted as a common screening tool.

In addition to sensitivity of bitterness, other things need to be considered for successful sensory evaluation by human subjects. Subjects participated in a very thorough screening process for the study of the sensory characteristics of honey, including bitterness perception (González, Lorenzo & Perez, 2010). The first part included information gathered on age, gender (to maintain a panel with equal male and female numbers), health, motivation, performance in acuity testing, interest, and availability for at least 80% or more of the scheduled sessions. For subjects who passed part one, further selection was made according to specified standards (ISO 8586:1, 1993) which included performance in discrimination (paired comparison, triangle and rank testing) and descriptive flavor profile analysis. Scaling exercises were conducted for quantitative response scales (ISO 4121, 2003) and estimation of percentage of shaded portions for geometrical designs. Penci *et al.* (2013) used eight screening criteria in their study of the relationship between sensory, and chemical and physical parameters of pistachio nuts as follows: no food allergies, nonsmokers, complete natural dentition, aged 18 to 64 years, consume nuts, availability for sessions, interest, verbal communication skills regarding sampling. Other researchers conducted acuity testing to determine sensitivities to basic tastes (Heiniö *et al.*, 2012; Kreutzmann, Christensen & Edelenbos, 2008). Interest and availability may be the only criteria sought in the screening process (Fontoin *et al.*, 2008; Saenz-Navajas *et al.*, 2010).

Another type of selection criteria is to recruit those based on their experience. The experience could be in overall sensory testing (Yousfi, Cayuela & Garcia, 2008), tasting a wide variety of food products for a long period of time that is, >2000 hours (Miller & Chambers, 2013), the specific product being tested (Sokolowsky & Fischer, 2012) and experience with the test method being employed (Esti *et al.*, 2009). This would also be the case when using experts for panelists (Chira & Teissedre, 2013). Inarejos-Garcia *et al.* (2009) used panelists that had been trained by the International Olive Oil Council another form of screening. The many variations in selection criteria reflect objectives of the study.

7.2 TEST METHODS

Several studies utilize a trained, expert and/or consumer panel to evaluate bitterness in food and beverages. It is important that consumer panels be used to determine any possible differences in food and beverages, which could affect their palatability and acceptability. This is extremely critical in functional food development, as functional ingredients tend to decrease palatability, and many nutritious compounds are inherently bitter, like polyphenols. The completion of a consumer panel should be the first step in analyzing bitter food products. Thereafter, if differences in acceptability are found, the use of trained panels can be used to further understand the specific attributes in such products.

Based on principles outlined in Frank, Ottinger, & Hoffman (2001) *taste dilution analysis* (TD) uses a series of samples that are diluted to determine the threshold of tastants. This is particularly useful for compounds like bitterness that have very low thresholds. The dilution factor was calculated as the taste difference between the diluted fraction that could just be detected, and two blanks. Thus a higher dilution factor means the compound is detected in lower levels. Seo, Lee, & Baek (2008) used this technique to determine the bitterness of the degree of hydrolysis over time for six selected enzymes which could be used in soy protein isolate hydrolysates. This method has an element of screening to it as enzymes with high TD factors would exhibit high bitterness intensity and low detection threshold. Of the six enzymes tested Flavourzyme showed a TD factor of 0 which would warrant further investigation of this enzyme for reduction of bitterness in soy protein isolates. Frank, Ottinger, & Hoffman (2001) studied the thermal reactions that produce bitter taste. They concluded that quinizolate had a very low detection threshold and a very high TD factor compared to other compounds studied including caffeine and quinine hydrochloride.

Ranking as its name suggests is a method whereby a group of samples is placed in order from least to most bitter for example. Sums of the rankings are compared for the samples and non-parametric statistics applied to check for significance. Koprivnjak *et al.* (2009) used this method to determine which level of phospholipid to include in olive oil in order to decrease bitterness perception.

Quantitative Descriptive Analysis (QDA) provides a sensory description of the product expressed in numerical format. Bitter taste is a component in many food products and ingredients. The ability to quantify the level allows for correlation with other measures such as chemical composition and consumer acceptability. Decreasing the bitterness in a food matrix can alter the perception of other tastes and flavor as well. The effect can only be determined by measuring all of the existing product attributes. The QDA method (Stone, Bleibaum & Thomas, 2012) is measurement of sensory attribute intensities by a small group of individuals (10 to 12) that have been screened for particular abilities and characteristics. Development of the product sensory descriptors and definitions, use of the intensity scale, and protocols regarding sample handling and testing procedures are facilitated by a group leader during training sessions. Length of training varies depending on the study objective and can be done intensively by running a number of sessions over a short period of time, or spread over a longer period of time with sessions held intermittently. Screening of panelists and selecting those with specific ability such as low bitterness threshold for caffeine aqueous solutions can decrease training time. Selecting those with prior experience with the particular test method and/or product category can also lead to shorter training time as documented by Le Berre *et al.* (2013) who selected panelists with years of experience that were screened and had experience with the time intensity method. Bitterness was the only attribute measured. On the other hand, Penci *et al.* (2013) trained panelists for a complete sensory description of pistachio nuts during 24 sessions of two hours each spanning a three-month period.

Once training is complete attribute intensities are evaluated individually at personal workstations. This feature is the basic difference between QDA and the traditional flavor profiling sensory method whereby the group of trained individuals reaches consensus regarding the attribute intensities of food products (Keane, 1992).

The relation between bitter taste and phenolic compounds was studied in eight carrot genotypes (Kreutzmann, Christensen, & Edelenbos, 2008). A 10-member panel was used that was screened on basic taste, odor detection and color vision and also their ability to communicate the descriptions of sensory attributes as noted in the ISO 8586-1:1993 standard. Six attributes including bitterness were agreed on and defined by the group. An unstructured 15-point line scale was used to mark intensities from low (0) to high (15). Bitterness intensity for the cultivars ranged from 2.1 to 10.4 on the 15 cm scale but no information was given regarding the association between bitter amounts and the scale value. Some phenolic compounds were correlated with bitterness but not the one in the highest amount, which was present in amounts greater than detection level. This could possibly be due to a masking effect by high sugar levels.

The effect of roasting and salting methods on the sensory, and physical and chemical parameters of pistachio nuts was studied by Penci *et al.* (2013). Eight screening criteria for the 12 panelists included no food allergies, nonsmokers, complete natural dentition, aged 18 to 64 years, consume nuts, availability for sessions, interest, verbal communication skills regarding sampling. Two aroma attributes, four basic tastes, and texture and appearance attributes were defined during the training period and evaluated using a 10 cm unstructured scale from 0 to 10. All of the mean values for bitterness were less than 1 on the 10 cm scale. A value of 1.2 on the scale corresponded to the bitter taste of a 0.05% caffeine solution. From principal component analysis it was determined that drying of the nuts was associated with higher bitterness compared to roasting.

The QDA method has been used extensively for quantitating bitterness in a variety of food products with some alterations. In some cases a partial description of the product is analyzed. Bitterness was one of two attributes that were deemed important attributes of Pilsner beer samples (Da Silva *et al.*, 2012). Fifteen volunteers took part in a 5-day training period where they learned the low level of bitterness (diluted beer with an undisclosed amount of deionized water) 1 on a 9-point scale and full-scale bitterness (undiluted beer spiked with an undisclosed amount of caffeine) 9 on the 9-point scale. The resulting bitterness scores for the 32 beers tested ranged from 2.1 to 8.4, which the authors concluded would provide a good sample set for correlational data from headspace solid phase microextraction – gas chromatography with mass spectrophotometric detection.

Sanchez-Molinero & Arnau (2010) used QDA to determine bitterness intensity of dry cured ham varying in the type of atmosphere in the packaging. Selection and training of the six panelists was done according to ASTM and ISO methodologies. They had at least 10 years of experience in the evaluation of dry cured hams. Bitterness was defined as the taste perceived from caffeine and L-tryptophan and measured on a 10 cm non-structured scale from 0 (absence) to 10 (maximum intensity). Samples stored 289 days with reduced oxygen atmosphere in the packaging resulted in significantly higher bitterness (2.0 on the 10 cm scale) compared to samples stored 289 days with air or 214 days with air and 75 days with reduced oxygen. Bitterness perception again was not related to the measurement scale with a specified amount of caffeine or tryptophan.

Scaling methods used to measure bitterness intensity include 0 = low and 15 = high on an unstructured 15-point scale (Kreutzmann, Christensen, & Edelenbos, 2008); 0 = absence of bitterness and 10 = maximum intensity on a non-structured scale (Sanchez-Molinero & Arnau, 2010); 0 to 10 cm from non-bitter to extremely bitter (Kranz *et al.*, 2010); 0 = lowest intensity; 10 = highest intensity on a continuous non structured (Heiniö *et al.*, 2012); 0 = not noticeable to 10 cm = very strong (Sokolowsky & Fischer, 2012); 9-point scale where 1 = none, 5 = definite, 9 = pronounced (Agrawal & Hassan, 2007); 7-point scale (Chira & Teissedre, 2013); and a 6-point scale where 0 = no perception; 1 = scarce; 2 = light; 3 = middle; 4 = strong; 5 = intense (Koprivnjak *et al.*, 2009). It appears that the 0 to 10 scale is most common. The analysis for the majority of these results is based on calculation of a mean value of the intensity scores for each panelist. It was noted however that Inarejos-Garcia *et al.* (2009) used the median score for the analysis of data collected from a 10 cm non-structured scale. Generally panelists are asked to make two to three evaluations of the same samples on different days. Measures are deemed to be continuous and thus analysis of variance is performed to determine treatment differences followed by multiple comparison testing such as Tukey HSD, Duncan multiple range, Student-Newman-Keuls, and Fisher LSD.

General Labeled Magnitude Scale is another method suitable for measuring bitterness. Briefly the scale is a ratio scale that can be used to quantify taste in all forms and quantities. The vertical scale from 0 to 100 was developed using geometric means of magnitude estimates of six verbal descriptors. 1.5 = barely detectable; 6 = weak; 17 = moderate; 35 = strong, 52 = very strong and 100 = strongest imaginable (Green, Shaffer, & Gilmore, 1993). Keast (2008) used this method for evaluating the effect of four compounds – sodium gluconate, zinc lactate, sucrose and milk fat on the bitterness of caffeine. Sodium gluconate was the only compound that did not reduce bitterness perception of caffeine.

Time intensity methodology emphasizes the importance of bitterness duration and sample bitterness intensities at specified intervals or as a continuous tracking during the time frame associated with perception from the time the sample is placed in the mouth until the bitterness perception is gone which could be well beyond the point of swallowing. Various measures can be taken to provide a continuum of bitterness perception (Lawless & Heymann, 2010).

Le Berre *et al.* (2013) measured bitterness of theobromine, a bitter compound in cacao, during consumption of chocolate ice cream and 4 minutes after swallowing. Panelists were coordinated for number and size of bites (6 bites every 45 sec) followed by swallowing (4.5 min from start) and tracking of bitterness for a total test time of 10 minutes. Three measurements were taken – total area under the curve for the bitterness perception for the whole product over the total time of evaluation; the point over this period where bitterness peaked; the time where bitterness was no longer perceived. A more complete understanding of the bitter perception is provided as time passes.

Dual Attribute Time Intensity, first documented by Duizer, Bloom, & Findlay (1997), measured sweetness and peppermint flavor in chewing gum simultaneously over time. Advantages of the dual attribute over the single attribute method were reduction in testing time by half and additional information regarding interactions of the two attributes. Recognizing that bitterness and astringency are lingering sensations Kobue-Lekalake, Taylor & de Kock., (2012) used dual attribute time intensity for sorghum infusion analysis. The panelists participated in a previous study and completed an additional 10 hours of training to be familiarized with the time intensity method. Proficiency using the measuring scale for each one of the attributes was completed before the task of evaluating two attributes was undertaken. A line scale with 10 markings was labeled 'none' at 0 and 'extreme' at 100.

Bitterness was measured on a vertical line and astringency on a horizontal line. The panelist moved the 'marker' diagonally to the right as intensity increased and moved it to the left as the intensity decreased. Instructions for panelists were to hold the sample in the mouth while swirling for 15 seconds and to expectorate the sample. Bitterness and astringency were evaluated as soon as the sample entered the mouth and continued for 90 seconds. Significant differences between the tannin and tannin free sorghum infusions were similar between the maximum bitterness intensity and the one time bitterness evaluation of the original study. Bitterness maximum intensity appeared before the astringency maximum. In addition, the more bitter the sample the more astringent it was. The duration of bitterness and astringency attributes varied by individual.

Temporal Dominance of Sensations (TDS) records the perception of up to 10 sensory attributes in the same product over time. Whatever is perceived as dominant at the time is selected from the predetermined list and is recorded with the corresponding intensity (Pineau *et al.*, 2009).

Meillon, Urbano, & Schlich (2009) used this method to determine the effect of decreased alcohol content in red wines. Complete flavor profiling was done first to determine the attributes followed by TDS. Some products that were not found to be significantly different with profiling were found different for the TDS method including a decrease in bitterness due to astringency and a decrease in bitter due to an increase in fruity sensations.

More than one test method may be employed to satisfy the objectives of the study. In the study by Heiniö *et al.* (2012) subjects first took part in a *Difference from Control* sensory test. Samples of rye flour/water suspensions treated with enzymes were compared to a reference or control sample where samples higher in bitterness than the control could be assigned values up to +5 (clearly more bitter than control) and those lower in bitterness than the control assigned a value of −5 (clearly less bitter than control) with points between. Samples with bitterness greater than the control were quantified using QDA along with five other flavor attributes. It was found that the formation of small peptides as a result of enzymatic reaction contributed to the bitter flavor of the rye suspensions, information that will be critical for the development of healthy foods containing rye.

Esti *et al.* (2009) used both QDA and time intensity analysis to determine the relationship between bitterness and pungency of extra virgin olive oil containing different levels of phenolic compounds. At the lower levels of bitterness and pungency results from time intensity yielded differences in samples not found from QDA results.

Bitterness of white wine was studied using three sensory methods QDA, time intensity and temporal analysis (Sokolosky & Fischer, 2012). From the quantitative descriptive analysis of 28 different wines, 13 were selected for duration studies as they showed a variety of flavors as well as a range in bitterness. It was determined that information from all of the three methods was useful in characterization of the bitterness attribute for the wines.

Bitterness detection, recognition and increase in bitterness in smoothies with different levels of polyphenol from olive leaf extract was completed by 11 panelists who had detected bitterness in one or both of the following solutions: caffeine (30 mg/100mL) and tannin (30 mg/100 mL) (Kranz *et al.*, 2010). The sample was selected that was clearly bitter to test the effect of three different bitter masking agents. Subjects were asked to rank bitterness of smoothies that contained a bitter masking agent at increasing concentrations so that ideally the samples with increased levels of masking agent would be ranked lower for bitterness than those without the agent. As ranking only provides information regarding ordering of the samples for bitterness further sensory analysis was done to determine sample bitterness levels from non-bitter (0) to extremely bitter (10) on the 10 cm line scale. Sodium cyclamate in smoothies reduced bitterness significantly compared to sodium chloride which actually was perceived as more bitter than the smoothie with no masking agent.

7.3 TECHNIQUES TO MAXIMIZE BITTERNESS PERCEPTION

General panelist requests for palate preparation prior to testing include the request to refrain from eating, drinking, chewing gum or smoking 1 hour prior to sensory evaluation (Keast, 2008; Sun *et al.*, 2011). In addition, specific protocols to increase sensitivity to bitterness perception due to bitter compounds having a lingering sensation include no coffee one hour prior to testing (Sokolowsky & Fischer, 2012; Kranz *et al.*, 2010). To block interfering volatile perception, nose clips were applied to enhance the ability to detect bitter taste of bread crumb and crust (Bin *et al.*, 2012) and chocolate and coffee (Keast, 2008). Expectoration may be encouraged (Miller & Chambers, 2013; Heiniö *et al.*, 2012) to facilitate cleansing of the palate between samples and to avoid carry over from the previous sample (Lawless & Heymann, 2010). Time between samples also is advised to refresh the palate of bitterness and prepare for future tasting. Heiniö *et al.* (2012) requested a one minute wait time for samples containing rye, two minutes between samples was implemented for dry white wine (Sokolosky & Fischer, 2012) and, four minutes was requested for sorghum infusions (Kobue-Lekalake, Taylor & de Kock., 2012).

Palate cleansers after sample ingestion are also documented to mitigate the effects of lingering bitter taste in addition to preventing adaptation to a particular tastant which could affect intensity ratings of other tastants being evaluated. A wide variety of palate cleansers both singly and in combination are noted for various food products. Raw carrot and deionized water were used for sorghum infusions (Kobue-Lekalake, Taylor & de Kock., 2012); still water and unsalted crackers for dry white wine (Sokolosky & Fischer, 2012) and bread crust and crumb (Bin *et al.*, 2012); spring water for lentil sprouts (Troszynska *et al.*, 2011); odorless water and green apple for pistachio nuts (Penci *et al.*, 2013); a selection of one of the following four items to be used consistently for black walnuts-deionized reverse osmosis water, baby carrots, mozzarella cheese, skinless cucumber slices (Miller & Chambers, 2013); apple, crisp bread and water for virgin olive oil (Koprivnjak *et al.*, 2009); apple and water for olive oil (Garcia-Mesa *et al.*, 2008); tap water and unsalted crackers for olive leaf fortified fruit smoothies (Kranz *et al.*, 2010); warm water for ice cream (Esti *et al.*, 2009); and deionized water for bitter solutions (Keast, 2008) and tannin extracts (Fontoin *et al.*, 2008). The effectiveness of palate cleansers for cream cheese with increasing levels of caffeine was studied by Johnson & Vickers (2004). No differences were found in bitterness detection or residual mouth buildup for water, rinsing with water six times, carrot, cracker, cream cheese, or nothing. Sparkling water reduced bitterness perception for all of the samples compared to the other cleansers. They postulated that perhaps the activity of rinsing helped to reset the mental function for preparation for the next sample rather than the rinse agent itself.

7.4 USE OF STANDARDS

The majority of food products being tested for bitterness include beer, tea (black, oolong, yerba), vegetable oils (olive), nuts (peanuts, walnuts), cocoa products (nibs, liquor, chocolate), dairy (yogurt, UHT milk, ultrapasteurized milk, cheese), meat (chicken, pork), wine (red, white), various vegetables (carrots), and functional beverages, among others. With the wide variety of bitter products being tested for, which contain a variety of bitter compounds, in lies the predicament of standardization of standards used in training sessions. Typical standards used in research for this purpose include solutions of caffeine, quinine sulfate, quinine chloride, quinine dihydrochloride, L-tryptophan, glycyl-L-leucine and catechin. Concentrations of such standards also vary depending on the food or beverage being tested. In addition, reference standards of very bitter foods, typically that are being tested, are also used in many trials. For instance, in a sensory panel conducted in Jinap, Jamilah, and Nazamid (2004), in which various cocoa liquors were being tested, a very bitter Ghanaian cocoa liquor was used as the bitter standard to demonstrate this attribute. Of course, there are a variety of compounds and ingredients that are bitter, and the standard chosen is highly dependent on the food products that are being tested. For instance, in a trial testing wines rich in catechins, catechins (bitter) will likely be chosen as a reference standard over caffeine. There is also a certain amount of variation in concentration of reference solutions, or the strength of bitterness that these reference standards have. This variation will impact the conclusions that can be drawn from such studies, especially when differing methodologies are used, and comparing studies can be quite difficult because of this.

In some cases, bitter products may pose health problems if consumed. Many of the earlier sensory panels used quinine solutions as bitterness standards; however, concern over the potential of serious health problems, including possibility of death caused the United States Food and Drug Administration to ban over-the-counter quinine in 1994. However, many sensory panels continued to use quinine as standards in the early 2000s. The possibility that certain standards or ingredients may be deemed unacceptable for use in the future impacts the consistency of sensory testing across time.

7.5 CONCLUSION

The presence of bitterness in food products significantly affects consumer acceptability, which can result in a decrease in consuming products that benefit human health, which are typically bitter in taste. In order to detect and quantify bitterness, a number of sensory methods are available to test for bitterness of foods and beverages: taste dilution analysis, ranking, quantitative descriptive analysis, time intensity among others. Depending on the product being tested and the objective, the method of testing will change, as will the use of certain bitter compounds used as standards in trained panels. Sensory evaluation of bitterness is and will continue to be increasingly important with the creation of novel functional food products.

Table 7.1 Review of sensory panels with bitter products.

Author	Measurement	Panelist details	Test product
Vegetable oils			
Esti *et al.*, 2009	descriptive intensity to determine attributes for Time Intensity measurement of bitterness and pungency 15 cm line scale from 0 = no perception to 15 = very strong 150 sec time period strongest sample to start and used as the reference and the calibration sample during the testing of the other samples; samples swallowed	10 trained with previous experience in time intensity method	olive oil - extra-virgin
Beltran *et al.*, 2007	bitterness of olive oil, Off J Eur Communities 1991 referred method of the European Commission regulation no EEC/2568/91	9 trained	olive oil - virgin
Inarejos-Garcia *et al.*, 2009	10 cm non-structured scale for intensity of positive attributes of fruity, bitter and pungent recorded the median score of the panelists	official Spanish panels recognized by the International Olive Oil Council 1 - 10 panelists av age = 32 years; 2 - 9 panelists av age = 56 years	olive oil - virgin
Yousfi *et al.*, 2008	quality 9-pt: 1 = poorest quality possible; 9 = the best; bitterness intensity - 5-pt scale: 0 = absence of attribute; 2 = simple perception; 2 = slight presence; 3 = middle presence; 4 = strong intensity; 5 = highest intensity	8 trained ($\geq$5 years)	olive oil - virgin
Garcia-Mesa *et al.*, 2008	'Organoleptic Assessment of Virgin Olive Oil' IOOC procedure scored (0 to 10) for 'positive attributes' such as fruity, bitter and pungent and 'negative attributes' such as fusty, musty, rancid, metallic, wine-vinegary and others; extra = median = 0 for defects and >0 for fruity; virgin = medium for defects is between 0 and 2.5 and fruity is >0; lampante median for defects is >2.5. PLUS triangular tests for determination of varietal differences	8-10 trained	olive oil - virgin

Koprivnjak *et al.*, 2009	QDA - aromas and taste including bitter - six point scale sensory 0 = no perception; 1 = scarce; 2 = light; 3 = middle; 4 = strong; 5 = intense quality grading 1 = lowest quality to 9 = highest quality with greater than or equal to 6.5 rated as extra quality.	8 trained in VOO	olive oil (VOO) - virgin
Caporale, Policastro & Monteleone, 2004	10 cm line scale (quinine dihydrochloride as bitter standard)	31 females and 29 males (24–35 years)	olive oil
Sinesio, Moneta & Esti, 2005	15 cm line scale	10 panelists	olive oil - virgin
Gutierrez-Rosales, Rios & Gomez-Rey, 2003	5-point scale	12 panelists	olive oil - virgin
Mateos *et al.*, 2004	6-point scale	N/A	olive oil - virgin
Busch *et al.*, 2006	N/A	N/A	olive oil - virgin
Andrewes *et al.*, 2003)	taste threshold test	3 females and 1 male	olive oil polyphenols - virgin
Siliani *et al.*, 2006)	5-point scale	5 panelists	olive oil – extra virgin
Nuts/seeds			
Miller & Chambers, 2013	modified flavor profile; 0 to 15-point scale in 0.5 increments - 0 = none 15 = highest possible intensity	6 females 1 male extensive sensory experience	black walnuts
Colarič *et al.*, 2006	10 cm line scale	2 panels (young and old); 20 panelists each	walnuts

(*continued*)

Table 7.1 *(Continued)*

Author	**Measurement**	**Panelist details**	**Test product**
Penci *et al.*, 2013	descriptive analysis; 10 cm line scale	10 females 2 males screened (8 criteria) and trained; 24 sessions x 2 hours over 3 months	pistachio nuts
Talcott *et al.*, 2005	15 cm line scale	6 females and 4 males	peanuts
Hamada, Hashim, & Sharif, 2002	N/A	5 expert panelists	date pits
Dairy products			
Chapman, Lawless, & Boor, 2001	10-point scale	12 panelists	milk -ultrapasteurized
Le Berre *et al.*, 2013	time intensity bitterness; continually trace bitterness from beginning and up to 5 minutes after swallowing; measured total area under the curve (overall bitterness perception); overall maximum intensity for bitterness over the time period; the time taken for bitterness to disappear	8 trained with years of experience in many food products; screened; 2 training sessions	ice cream product
Jaworska *et al.*, 2005	10-point scale and 9-point hedonic scale	10 trained panelists; 68 females and 12 males (20–24 years)	yoghurt - natural
Stephan & Steinhart, 2000	6-point scale	20 panelists	soybean lecithins in ultrahigh temperature (UHT) milk
Poveda & Cabezas, 2006	10-point scale	10 panelists	goat cheese

Soeryapranata *et al.*, 2002	15 cm line scale (glycyl-L-leucine solution as bitter standard)	6 panelists	cheese - aged cheddar
Agrawal & Hassan, 2007	bitter and sour attributes 1 = none; 5 = definite; 9 = pronounced	4 experienced judges	cheese - reduced fat cheddar
Broadbent *et al.*, 2002	10-point scale (quinine sulfate as bitter standard)	14 panelists	cheese - reduced fat cheddar
Madsen & Ardö, 2001	7-point scale	8 panelists	cheese - Danbo
Izco *et al.*, 2000	7-point scale	8 panelists	cheese - Ossau-Iraty
Singh *et al.*, 2005	15-point scale (caffeine and quinine sulfate as bitter standards)	7 panelists	casein peptide from cheese
Pineau *et al.*, 2009	temporal dominance of sensations (pick most dominant attribute to score when it appears) compared to time intensity	16 panelists per test method; screened and trained	dairy products
Johnson & Vickers, 2004	bitterness on 150 mm line scale no bitterness to extreme bitterness	untrained; 4 males 16 females 18 to 65 years	cream cheese with caffeine
Grain products			
Heiniö *et al.*, 2003	10 cm line scale	4 and 5 expert panelists	rye flour and bread - milling fractions
Heiniö *et al.*, 2012	difference from control (phase 1 and 3) -5 = clearly less bitter than the reference and +5 clearly more bitter than the reference phase 2 QDA (9 attributes with descriptors on 10 cm scale)	10 trained panelists (7 = phase 1; 9 = phase 2; 10 = phase 3) who passed the basic taste test, odor test and color vision tests as well as checked continually using control card checks and also with rye samples in 'several' pre-sessions	rye - effect of enzyme activity on suspensions and crackers

(*continued*)

Table 7.1 (*Continued*)

Author	**Measurement**	**Panelist details**	**Test product**
Kobue-Lekalake, Taylor, & de Kock., 2007	1 = not intense; 10 = very intense 9-pt scale; sweet, sour, bitter tastes and astringency as well as other sorghum related tastes and texture (rice)	12 screened 6 males, 6 females (19-39 years); 1 hr/day for 3 wks; bitter standard = 1.0 g L^{-1} caffeine food grade	sorghum bran infusions and sorghum rice of 3 tannin and 3 tannin free cultivars
Kobue-Lekalake , Taylor, & de Kock., 2012	dual attribute time intensity for bitterness and astringency; 0 = none 100 = extreme with 10 markings; bitterness and astringency; caffeine 1 g/L 1,5 g/L and tannic acid 0.5 g/L alum (potassium aluminium sulphate; bitterness on vertical and astringency on horizontal at the same time	six men; six women screened for PROP sensitivity, trained previously on earlier sorghum study plus 10 h for use of Compusense	sorghum bran infusions of six cultivars
Watts *et al.*, 2012	0 = none; 15 = high 15 cm line scale; sweet, bitter, wheaty, yeasty, doughy (pita only),wheaty aftertaste, sour aftertaste, bitter aftertaste, doughy aftertaste (pita only)	10 trained panelists; ANOVA effects: cultivar, panelist and replication and interactions	whole wheat pan and pita bread of eight cultivars
Animal products			
Carrapiso, Bonilla, & García, 2003	10 cm line scale	18 panelists	ham
Sanchez-Molinero & Arnau, 2010	QDA; 0-10 non-structured scale 0 = absence of descriptor; 10 = maximum intensity; bitterness among other flavor and texture attributes; bitterness definition = fundamental taste sensation elicited by caffeine and L-tryptophan	six selected and trained (ASTM 1981; ISO8586-1,1993; ISO8586-2,1994) minimum 10 years experience	ham - dry cured
Nissen *et al.* 2004	15 cm line scale (quinine chloride as bitter standard)	5 females and 3 males	pork - patties with added rosemary, green tea, coffee and grape skin extracts
Dauksas *et al.*, 2004	5-point ranking test (caffeine as bitter standard)	12 panelists	fish - protein hydrolysates

Soy products			
Seo , Lee, & Baek., 2008	taste dilution analysis; bitterness of soy hydrolysates varied by time of enzyme hydrolysis (6 proteases); TD factor measured = bitterness int. & defined as the bitterness of diluted samples - blank; the higher the TD the higher the bitterness	10 - screened for bitterness sensitivity with triangle test; trained with triangle test for bitterness; final = 3 triangle tests	soy - protein isolates
Torres-Penaranda & Reitmeier, 2001	10 cm line scale	7 panelists	soymilk
Alcohol			
Da Silva *et al.*, 2012	QDA (Stone, 1992) - bitter (low - diluted beer; high - undiluted beer spiked with caffeine - full-scale); intensities scored from 1 to 9	15 trained (25 to 45 years old); 5 day training	beer
Fritsche & Shellhammer, 2007	bitterness intensity	trained	beer - lager with and without alpha-acids
Kolpin & Shellhammer., 2009	bitterness detection thresholds of iso-alpha-acids and tetrahydro-iso-alpha acids; ASTM 1432	14 volunteers, 3 training sessions and 6-13 testing sessions	beer - unhopped lager
Techakriengkrai *et al.*, 2004	0–100 line scale	14 panelists	beer - lager
Corzo & Bracho, 2004	5-point category scale	8 panelists	beer - canned
Chira & Teissedre, 2013	7-point scale	20 experts fully trained	wine

(*continued*)

Table 7.1 (*Continued*)

Author	Measurement	Panelist details	Test product
Chapman, Matthews, & Guinard, 2004	16-point scale	Panel 1 for 2000 wines: 4 females and 9 males (21–33 years) Panel 2 for 2001 wines: 8 females and 7 males (21–41 years)	wine - cabernet sauvignon
Schlosser *et al.*, 2005	10 cm line scale (caffeine as bitter standard)	14 panelists	wine - chardonnay
Preys *et al.*, 2006	10 cm line scale	4 trained panels	wine - red
Saenz-Navajas *et al.*, 2010	bitterness (quinine sulphate 0 to 10 mg per L) and other attributes 6 point scale 0 = absence; 1 = very low; 5 = very high	13 trained; 6 males, 7 females; 23 to 60 years; 10 training sessions over 2 months	wine - red
Fontoin *et al.*, 2008	scale from 0 to 10; low and high intensities of quinine sulphate for indication of low and high bitterness intensities	6 females and 10 males trained reference standards 0.15 g/L quinine sulphate = bitterness, 2.0 g/L aluminum sulphate = astringency; 1.0 g/L tartaric acid=sourness; 4 training sessions in total (2 practice sessions at the end)	wine (red) tannins 2 g/L in solution with NaOH (to adjust pH from 2.5 to 4) tartaric acid (0, 2, 4, 6 g/L), ethanol (0, 7, 11, 15%)
Sun *et al.*, 2011	tasted grape seed procyanidins, polyphenol extracts and polyphenols from different wine making stages for astringency and bitterness; scoring from at least 0 to 10 no details on this only shown on table	10 judges participating in at least one session on wine per week	wine components - red

Boselli *et al.*, 2004	10-point intensity scale (catechin as bitter standard)	7 panelists	wine - red
Meillon Urbano, & Schlich., 2009	temporal dominance of sensations; profile first for attributes; 10 cm unstructured line scale then 10 most often used terms for TDS measured on 30 cm unstructured scale from weak to strong	9 women and 7 men 29 - 65 years old; wine consumers; selected based on ranking and discrimination test performance; 16 one-hour training sessions	wine - red
Vidal *et al.*, 2003	general labeled magnitude scale (gLMS) (quinine sulfate as bitter standard)	2 females and 14 males (35 average age); 2 females and 13 males (35 average age)	grape and apple proanthocyanidins in wine-like medium; polysaccharides and anthocyanins in wine-like medium
Brossaud, Cheynier, & Noble, 2001	quantitative descriptive analysis (QDA) (quinine sulfate as bitter standard)	3 females and 9 males (26–52 years)	grape and wine polyphenol extracts
Gutiérrez Afonso, 2002	QDA	4 females and 11 males (23–35 years)	wine - white
Sokolowsky & Fischer, 2012	QDA compared with time intensity and temporal analysis for bitter intensity and bitter persistency detailed training described with standards and methods for tasting; 0 = not noticeable to 10 cm = very strong	9 males; 9 females experienced in wine tasting	wine - white
Nurgel, Pickering, & Inglis, 2004	15 cm line scale (quinine sulfate as bitter standard)	9 panelists (21–55 years)	ice wines

(*continued*)

Table 7.1 (*Continued*)

Author	Measurement	Panelist details	Test product
Tea / coffee			
Santa Cruz, Garitta, & Hough, 2002	0–100 scale	8 panelists	yerba mate
Drobna, Wismer, & Goonewardene, 2004	10 cm line scale (caffeine as bitter standard)	3 females and 5 males (19–35 years)	black tea
Yau & Huang, 2000	13.5 cm line scale and 9-point hedonic scale	15 trained panelists; 50 untrained panelists	oolong tea beverages
Frank *et al.*, 2007	triangle test	7 females and 5 males	coffee
Esteban-Diez, González-Sáiz, & Pizarro, 2004	10-point scale	11 panelists	green coffee and espresso
Vegetables			
Liu *et al.*, 2005	12.6 cm line scale	4 females and 4 males	oyster mushrooms
Kreutzmann *et al.*, 2008	0 = low 15 = high on 15-pt line scale; terpene flavor, carrot flavor, bitterness, green flavor, burning aftertaste, sweetness	5 males; 5 females (26-54 years old) screened for sensory taste, odor and color) and ability to communication of sensory descriptions; trained	carrots - raw

Varming *et al.*, 2004	15 cm line scale and 7-point category scale	8 panelists; 112 untrained panelists (25–65 years)	carrots - raw
Czepa & Hofmann, 2003	3-point bitterness scale (0 = not detectable; 3 = strong bitter)	N/A	carrots – fresh, stored and puree
Czepa & Hofmann, 2004	3-point bitterness scale (caffeine as bitter standard)	12 panelists	carrots - raw, puree, juice
Seljåsen *et al.*, 2001	9 cm line scale (caffeine as bitter standard)	10 panelists	carrot - puree
Chiwona-Karltun *et al.*, 2004	5-point scale	12 panelists	cassava root
Jones & Sanders, 2002	QDA and 8-point hedonic scale	9 trained panelists; 51 untrained panelists	turnip greens
Dinehart *et al.*, 2006	100-point gLMS	71 females and 39 males (18–60 years)	various vegetables
Miscellaneous			
Ly & Drewnowski, 2001	9-point scale	54 females (18–30 years)	caffeine, sucrose, neohesperidin dihydrochalcone and chocolate

(continued)

Table 7.1 (*Continued*)

Author	Measurement	Panelist details	Test product
Keast, 2008	gLMS vertical axis barely detectable = 1.5; 6 = weak; 17 = moderate; 35 = strong; 52 = very strong; 100 = strongest imaginable only adjectives no numbers for panelists; placed semi-logarithmically depending on experimentally determined intervals, similar to magnitude estimation, i.e., ratio scaling	33 trained to id five tastes given singular and mixed solutions	Exp 1 - effect of sodium gluconate and zinc lactate on caffeine bitterness; Exp 2 - effect of sucrose on caffeine bitterness; Exp 3 - effect of milk fat on caffeine bitterness; Exp 4 - effect of chocolate and coffee aromas on caffeine bitterness; used solutions containing chemicals (mM)
Kranz *et al.*, 2010	bitter detection and recognition of polyphenols; ranking test and asked the group to rank each set of 5 samples with sucrose, sodium cyclamate and sodium chloride for bitterness; 10 cm scale from non-bitter to extremely bitter for smoothies with sodium cyclamate, sucrose and sodium chloride at specific levels in smoothie containing 20 mg/100 g OLE polyphenol (level of OLE based on preliminary acceptance test)	11 with previous sensory evaluation experience - 3 males; 8 females (24 to 60 years old) screened with taste solutions and polyphenols in water	olive leaf extract (OLE) fortified fruit smoothie
Jinap, Jamilah, & Nazamid, 2004	10-point scale (Ghanaian cocoa liquor as bitter standard)	3 females and 5 males	cocoa liquor
González de Lorenzo, & Perez., 2010	QDA (Stone & Sidel, 2004) - bitter scale not bitter (water); slightly bitter (spinach plain); bitter (chocolate); very bitter (curly endive)	11 selected and trained over 4 months	honey

Dalev *et al.*, 2006	10 cm line scale (caffeine as bitter standard)	4 females and 2 males	probiotic beverages
Troszynska, Lamparski, & Kosinska, 2004	QDA and hedonic ratings	9 panelists (24–50 years)	sprouts - lentil, mung bean, pea and soybean
Troszynska *et al.*, 2011	QDA (Lawless & Heymann, 1999 and Stone & Sidel, 1993) ISO standards; unstructured, graphical scale of 10 cm with verbal anchors at each end 0 = none to 10 = very intense; 0.2% caffeine = bitter training standard	9 members 6 females, 3 trained males; 28–54 years	sprouts - lentil seeds
Wolejszo, Szymkiewicz, & Troszynska, 2007	QDA (Lawless & Heymann, 1999 and Stone & Sidel, 1993) ISO standards; unstructured, graphical scale of 10 cm with verbal anchors at each end none to very intense; bitter basic taste (reference sample: caffeine in water 0.5%)	9 selected and trained ISO standards	sprouts- lentil and mung bean
Heng *et al.*, 2006	15 cm line scale (quinine sulfate dehydrate as bitter standard)	8 females and 6 males (19–36 years)	saponin extracts from peas
Hance *et al.*, 2007	1 = least bitter; 4 = most bitter; AFNOR methodology	trained (# not given) in chicory root bitterness; related to 11β, 13 dihydrolactucin content; rank test used	chicory used for ingredients including roasted products (eaten as coffee substitute) and flour
Tamamoto, Schmidt, & Lee, 2010	16-point scale (naringin standard for fruit bitter taste; black tea for bitter tea taste)	9 females and 4 males (18–50 years)	energy drinks with added caffeine, ginseng and taurine
Kappes, Schmidt, & Lee, 2006	16-point scale (caffeine as bitter standard)	9 females and 3 males (18–50 years)	cola and lemon/lime carbonated drinks

(continued)

Table 7.1 (*Continued*)

Author	Measurement	Panelist details	Test product
Tang, Kälviäinen, & Tuorila, 2001	QDA and 9-point hedonic ratings	8 females and 2 males (23–49 years); 24 females and 16 males (22–59 years)	sea buckthorn juice
Duizer Bloom, & Findlay, 1997	dual attribute time intensity; peppermint and sweet; mouse moved from zero point depending on intensity over 15-minute period	10 trained; experience in time intensity evaluation; 8 one hour training sessions	chewing gum
Frank Ottinger, & Hofmann, 2001	taste dilution analysis; 21 fractions diluted stepwise 1+1 with water; increasing concentrations to panelists in triangle tests; taste dilution factor is the difference between the blanks and the diluted fraction that could be just detected	trained panel	chemicals from HPLC fractions

REFERENCES

Agrawal, P. & Hassan, A.N. (2007). Ultrafiltered milk reduces bitterness in reduced-fat cheddar cheese made with an exopolysaccharide-producing culture. *J. Dairy Sci.* **90**, 3110–3117.

Andrewes, P., Busch, J.L., de Joode, T., Groenewegen, A. & Alexandre, H. (2003). Sensory properties of virgin olive oil polyphenols: Identification of deacetoxy-ligstroside aglycon as a key contributor to pungency. *J. Agric. Food Chem.* **51**(5), 1415–1420.

Beltran, G., Ruano, M. T., Jimenez, A., Uceda, M., & Aguilera, M. P. (2007). Evaluation of virgin olive oil bitterness by total phenol content analysis. *Eur. J. Lipid Sci. Technol.* **109**(3), 193–197.

Bin, Q., Jiang, D., Cho, I.H., & Peterson, D.G. (2012). Chemical markers for bitterness in wheat bread. *Flavour Fragrance J.* **27**, 454–458.

Boselli, E., Boulton, R.B., Thorngate, J.H., & Frega, N.G. (2004). Chemical and sensory characterization of DOC red wines from Marche (Italy) related to vintage and grape cultivars. *J. Agric. Food Chem.* **52**(12), 3843–3854.

Broadbent, J.R., Barnes, M., Brennand, C., Strickland, M., Houck, K., Johnson, M.E., & Steele, J.L. (2002). Contribution of Lactococcus lactis cell envelope proteinase specificity to peptide accumulation and bitterness in reduced-fat Cheddar cheese. *Appl. Environ. Microbiol.* **68**(4), 1778–1785.

Brossaud, F., Cheynier, V., & Noble, A.C. (2001). Bitterness and astringency of grape and wine polyphenols, *Aust. J. Grape Wine Res.* **7**(1), 33–39.

Busch, J.L., Hrncirik, K., Bulukin, E., Boucon, C., & Mascini, M. (2006). Biosensor measurements of polar phenolics for the assessment of the bitterness and pungency of virgin olive oil, *J. Agric. Food Chem.* **54**(12), 4371–4377.

Caporale, G., Policastro, S., & Monteleone, E. (2004). Bitterness enhancement induced by cut grass odorant (cis-3-hexen-1-ol) in a model olive oil. *Food Qual Prefer* **15**(3), 219–227.

Carrapiso, A.I., Bonilla, F., & García, C. (2003). Effect of crossbreeding and rearing system on sensory characteristics of Iberian ham. *Meat Sci.* **65**(1), 623–629.

Chapman, D.M., Matthews, M.A., & Guinard, J. (2004). Sensory attributes of Cabernet Sauvignon wines made from vines with different crop yields. *Am. J. Enol. Vitic.* **55**(4), 325–334.

Chapman, K., Lawless, H., & Boor, K. (2001). Quantitative descriptive analysis and principal component analysis for sensory characterization of ultrapasteurized milk. *J. Dairy Sci.* **84**(1), 12–20.

Chira, K. & Teissedre, P. (2013). Extraction of oak volatiles and ellagitannins compounds and sensory profile of wine aged with French winewoods subjected to different toasting methods: behaviour during storage. *Food Chem.* **140**, 168–177.

Chiwona-Karltun, L., Brimer, L., Kalenga Saka, J.D., Mhone, A.R., Mkumbira, J., Johansson, L., Bokanga, M., Mahungu, N.M., & Rosling, H. (2004). Bitter taste in cassava roots correlates with cyanogenic glucoside levels. *J. Sci. Food Agric.* **84**(6), 581–590.

Colarič, M., Stampar, F., Hudina, M., & Solar, A. (2006). Sensory evaluation of different walnut cultivars (Juglans regia L.). *Acta Agric. Slov.* **87**, 403–413.

Corzo, O. & Bracho, N. (2004). Prediction of the sensory quality of canned beer as determined by oxygen concentration, physical chemistry contents, and storage conditions. *J. Food Sci.* **69**(7), 285–289.

Czepa, A. & Hofmann, T. (2004). Quantitative studies and sensory analyses on the influence of cultivar, spatial tissue distribution, and industrial processing on the bitter off-taste of carrots (Daucus carota L.) and carrot products. *J. Agric. Food Chem.* **52**(14), 4508–4514.

Czepa, A. & Hofmann, T. (2003). Structural and sensory characterization of compounds contributing to the bitter off-taste of carrots (Daucus carota L.) and carrot puree. *J. Agric. Food Chem.* **51**(13), 3865–3873.

Dalev, D., Bielecka, M., Troszyñska, A., Ziajka, S., & Lamparski, G. (2006). Sensory quality of new probiotic beverages based on cheese whey and soy preparation. *Pol. J. Food Nutr. Sci.* **15**(1), 71.

Da Silva, G. A. da, Maretto, D.A., Bolini, H.M.A., Teofilio, R.F., Augusto, F., & Poppi, R.J. (2012). Correlation of quantitative sensorial descriptors and chromatographic signals of beer using multivariate calibration strategies. *Food Chem.* **134**, 1673–1681.

Dauksas, E., Slizyte, R., Rustad, T., & Storro, I. (2004). Bitterness in fish protein hydrolysates and methods for removal. *J. Aquat. Food Prod. Technol.* **13**(2), 101–114.

Dinehart, M., Hayes, J., Bartoshuk, L., Lanier, S., & Duffy, V. (2006). Bitter taste markers explain variability in vegetable sweetness, bitterness, and intake. *Physiol. Behav.* **87**(2), 304–313.

Drobna, Z., Wismer, W.V., & Goonewardene, L.A. (2004). Selection of an astringency reference standard for the sensory evaluation of black tea. *J. Sens. Stud.* **19**(2), 119–132.

Duizer, L.M., Bloom, K., & Findlay, C.J. (1997). Dual-attribute time-intensity sensory evaluation: A new method for temporal measurement of sensory perceptions. *Food Qual Prefer* **8**, 261–269.

Esteban-Diez, I., González-Sáiz, J., & Pizarro, C. (2004). Prediction of sensory properties of espresso from roasted coffee samples by near-infrared spectroscopy. *Anal. Chim. Acta* **525**(2), 171–182.

Esti, M., Contini, M., Moneta, E., & Sinesio, F. (2009). Phenolics compounds and temporal perception of bitterness and pungency in extra-virgin olive oils; changes occurring throughout storage. *Food Chem.* **113**, 1095–1100.

Fontoin, H., Saucier, C., Teissedre, P.L., & Glories, Y. (2008). Effect of pH, ethanol and acidity on astringency and bitterness of grape seed tannin oligomers in model wine solution. *Food Qual Prefer* **19**, 286–291.

Frank, O., Blumberg, S., Kunert, C., Zehentbauer, G., & Hofmann, T. (2007). Structure determination and sensory analysis of bitter-tasting 4-vinylcatechol oligomers and their identification in roasted coffee by means of LC-MS/MS. *J. Agric. Food Chem.* **55**(5), 1945–1954.

Frank, O., Ottinger, H., & Hofmann, T. (2001). Characterization of an intense bitter-tasting 1H,4H-quinolizinium-7-olate by application of the taste dilution analysis, a novel bioassay for the screening and identification of taste-active compounds in foods. *J. Agric. Food Chem.* **49**, 231–238.

Fritsch, A. & Shellhammer, T. H. (2007). Alpha-acids do not contribute bitterness to lager beer. *J. Am. Soc. Brew. Chem.* **65**(1), 26–28.

Garcia-Mesa, J.A., Pereira-Caro, G., Fernandez-Hernandez, A., Civantos, C.G., & Mateos, R. (2008). Influence of lipid matrix in the bitterness perception of virgin olive oil. *Food Qual Prefer* **19**, 421–430.

González, M.M., de Lorenzo, C., & Perez, R.A. (2010). Development of a structured sensory honey analysis: application to artisanal Madrid honeys. *Food Sci. Technol. Int.* **16**, 19–29.

Green, B.G., Shaffer, G.S., & Gilmore, M.M. (1993). Derivation and evaluation of a semantic scale of oral sensation magnitude with apparent ratio properties. *Chem. Senses*, **6**, 683–702.

Gutiérrez Afonso, V. (2002). Sensory descriptive analysis between white wines fermented with oak chips and in barrels. *J. Food Sci.* **67**(6), 2415–2419.

Gutierrez-Rosales, F., Rios, J., & Gomez-Rey, M.L. (2003). Main polyphenols in the bitter taste of virgin olive oil. Structural confirmation by on-line high-performance liquid chromatography electrospray ionization mass spectrometry. *J. Agric. Food Chem.* **51**(20), 6021–6025.

Hamada, J., Hashim, I., & Sharif, F. (2002). Preliminary analysis and potential uses of date pits in foods. *Food Chem.* **76**(2), 135–137.

Hance, P., Martin, Y., Vasseur, J., Hilbert, J.L., & Trotin, F. (2007). Quantification of chicory root bitterness by an ELISA for 11β, 13-dihydrolactucin. *Food Chem.* **105**(2), 742–748.

Heiniö, R., Liukkonen, K., Katina, K., Myllymäki, O., & Poutanen, K. (2003). Milling fractionation of rye produces different sensory profiles of both flour and bread. *LWT-FOOD SCI TECHNOL* **36**(6), 577–583.

Heiniö, R.L., Nordlund, E., Poutanen, K., & Buchert, J. (2012). Use of enzymes to elucidate the factors contributing to bitterness in rye flavour. *Food Res. Int.* **45**, 31–38.

Heng, L., Vincken, J., van Koningsveld, G., Legger, A., Gruppen, H., van Boekel, T., Roozen, J., & Voragen, F. (2006). Bitterness of saponins and their content in dry peas. *J. Sci. Food Agric.* **86**(8), 1225–1231.

Inarejos-Garcia, A.M., Androulaki, A., Salvador, M.D., Fregapane, G., & Tsimidou, M.Z. (2009). Discussion on the objective evaluation of virgin olive oil bitterness. *Food Res. Int.* **42**, 279–284.

ISO. (1993). *Sensory analysis - General guidance for the selection, training and monitoring of assessors - Part 1: Selected Assessors. ISO 8586:1.* Geneva: International Organization for Standardization.

ISO. (1998). *Sensory analysis - Methodology. Ranking. ISO 8587.* Geneva: International Organization for Standardization.

ISO. (2003). *Sensory analysis - Guidelines for the use of quantitative response scales ISO 4121.* Geneva: International Organization for Standardization.

ISO. (2004). *Sensory analysis - Methodology. Triangle test. ISO 4120.* Geneva: International Organization for Standardization.

Izco, J., Irigoyen, A., Torre, P., & Barcina, Y. (2000). Effect of added enzymes on the free amino acids and sensory characteristics in Ossau–Iraty cheese. *Food Control* **11**(3), 201–207.

Jaworska, D., Wszkiewicz-Robak, B., Kolanowski, W., & Swiderski, F. (2005). Relative importance of texture properties in the sensory quality and acceptance of natural yoghurts. *Int. J. Dairy Technol.* **58**(1), 39–46.

Jinap, S., Jamilah, B., & Nazamid, S. (2004). Sensory properties of cocoa liquor as affected by polyphenol concentration and duration of roasting. *Food Qual Prefer* **15**(5), 403–409.

Johnson, E.A. & Vickers, Z. (2004). The effectiveness of palate cleansing strategies for evaluating the bitterness of caffeine in cream cheese. *Food Qual Prefer* **15**, 311–316.

Jones, G. & Sanders, O. (2002). A sensory profile of turnip greens as affected by variety and maturity. *J. Food Sci.* **67**(8), 3126–3129.

Kappes, S., Schmidt, S., & Lee, S. (2006). Descriptive analysis of cola and lemon/lime carbonated beverages. *J. Food Sci.* **71**(8), S583–S589.

Keane, P. (1992). The flavor profile. In: Hootman, R.C., ed. *ASTM manual on descriptive analysis testing for sensory evaluation*, Philadelphia, PA: ASTM. pp. 5–15.

Keast, R.S.J. (2008). Modification of the bitterness of caffeine. *Food Qual Prefer* **19**, 465–472.

Kobue-Lekalake, R.I., Taylor, J.R.N., & de Kock, H.L. (2012). Application of the dual attribute time-intensity (DATI) sensory method to the temporal measurement of bitterness and astringency in sorghum. *Int. J. Food Sci. Technol.* **47**, 459–466.

Kobue-Lekalake, R.I., Taylor, J.R.N., & de Kock, H.L. (2007). Effects of phenolics in sorghum grain on its bitterness, astringency and other sensory properties. *Journal of the Science of Food and Agriculture* **87**. 1940–1948.

Kolpin, K. M. & Shellhammer, T. H. (2009). The human bitterness detection threshold of iso-α-acids and tetrahydro-iso-α-acids in lager beer. *J. Am. Soc. Brew. Chem.* **67**(4), 200–205.

Koprivnjak, O., Skevin, D., Petricevic, S., Bubola, K.B., & Mokrovcak, Z. (2009). Bitterness, odor properties and volatile compounds of virgin olive oil with phospholipids addition. *LWT-FOOD SCI TECHNOL* **42**, 50–55.

Kranz, P., Braun, N., Schulze, N., & Kunz, B. (2010). Sensory quality of functional beverages: bitterness perception and bitter masking of olive leaf extract fortified fruit smoothies. *J. Food Sci.* **75**, S308–S311.

Kreutzmann, S., Christensen, L.P., & Edelenbos, M. (2008). Investigation of bitterness in carrots (Daucus carota L.) based on quantitative chemical and sensory analyses. *LWT-FOOD SCI TECHNOL* **41**, 193–205.

Lawless, H.T. & Heymann, H. (2010). *Sensory Evaluation of Food: Principles and Practices*, 2nd ed. Springer Science Business Media, New York, NY. p. 66.

Le Berre, E., Boucon, C., Knoop, M., & Dijksterhuis, G. (2013). Reducing bitter taste through perceptual constancy created by an expectation. *Food Qual Prefer* **28**, 370–374.

Liu, J., Vijayakumar, C., Hall, C.A., Hadley, M., & Wolf-Hall, C.E. (2005). Sensory and chemical analyses of oyster mushrooms (Pleurotus sajor-caju) harvested from different substrates. *J. Food Sci.* **70**(9), S586–S592.

Ly, A. & Drewnowski, A. (2001). PROP (6-n-propylthiouracil) tasting and sensory responses to caffeine,sucrose, neohesperidin dihydrochalcone and chocolate. *Chem. Senses* **26**(1), 41–47.

Madsen, J.S. & Ardö, Y. (2001). Exploratory study of proteolysis, rheology and sensory properties of Danbo cheese with different fat contents. *Int. Dairy J.* **11**(4-7), 423–431.

Mateos, R., Cert, A., Pérez-Camino, M.C., & García, J.M. (2004). Evaluation of virgin olive oil bitterness by quantification of secoiridoid derivatives. *J. Am. Oil Chem. Soc* **81**(1), 71–75.

Meillon, S., Urbano, C., & Schlich, P. (2009). Contribution of the temporal dominance of sensations (TDS) method to the sensory description of subtle differences in partially dealcoholized red wines. *Food Qual Prefer* **20**, 490–499.

Miller, A. & Chambers, D.H. (2013). Descriptive analysis of flavor characteristics for black walnut cultivars. *J. Food Sci.* **78**, S887–S893.

Nissen, L.R., Byrne, D.V., Bertelsen, G., & Skibsted, L.H. (2004). The antioxidative activity of plant extracts in cooked pork patties as evaluated by descriptive sensory profiling and chemical analysis. *Meat Sci.* **68**(3), 485–495.

Nurgel, C., Pickering, G.J., & Inglis, D.L. (2004). Sensory and chemical characteristics of Canadian ice wines. *J. Sci. Food Agric.* **84**(13), 1675–1684.

Penci, M.C. Martinez, M.L., Fabani, M.P., Feresin, G.E., Tapia, A., Ighani, M., Ribotta, P.D., & Wunderlin, D.A. (2013). Matching changes in sensory evaluation with physical and chemical parameters. *Food Bioprocess Technol.* **6**, 3305–3316.

Pineau, N., Schlich, P., Cordelle, S., Mathonniere, C., Issanchou, S., Imbert, A., Rogeaux, M., Etievant, P., & Koster, E. (2009). Temporal dominance of sensations: Construction of the TDS curves and comparison with time-intensity. *Food Qual Prefer* **20**, 450–455.

Poveda, J. & Cabezas, L. (2006). Free fatty acid composition of regionally-produced Spanish goat cheese and relationship with sensory characteristics. *Food Chem.* **95**(2), 307–311.

Preys, S., Mazerolles, G., Courcoux, P., Samson, A., Fischer, U., Hanafi, M., Bertrand, D., & Cheynier, V. (2006). Relationship between polyphenolic composition and some sensory properties in red wines using multiway analyses. *Anal. Chim. Acta* **563**(1), 126–136.

Saenz-Navajas, M., Ferreira, V., Dizy, M., & Fernandez-Zurbano, P. (2010). Characterization of taste-active fractions in red wine combining HPLC fractionation, sensory analysis and ultra performance liquid chromatography coupled with mass spectrometry detection. *Anal. Chim. Acta* **673**, 151–159.

Sanchez-Molinero, F. & Arnau, J. (2010). Processing of dry-cured ham in a reduced-oxygen atmosphere: Effects on sensory traits. *Meat Sci.* **85**, 420–427.

Santa Cruz, M., Garitta, L., & Hough, G. (2002). Sensory descriptive analysis of yerba mate (Ilex paraguariensis Saint Hilaire), a South American beverage. *Food Sci. Technol. Int.* **8**(1), 25–31.

Schlosser, J., Reynolds, A., King, M., & Cliff, M. (2005). Canadian terroir: sensory characterization of Chardonnay in the Niagara Peninsula. *Food Res. Int.* **38**(1), 11–18.

Seljåsen, R., Bengtsson, G.B., Hoftun, H., & Vogt, G. (2001). Sensory and chemical changes in five varieties of carrot (Daucus carota L) in response to mechanical stress at harvest and post-harvest. *J. Sci. Food Agric.* **81**(4), 436–447.

Seo, W.H., Lee, H.G., & Baek, H.H. (2008). Evaluation of bitterness in enzymatic hydrolysates of soy protein isolate by taste dilution analysis. *J. Food Sci.* **73**, S41–S46.

Siliani, S., Mattei, A., Innocenti, L.B., & Zanoni, B. (2006). Bitter taste and phenolic compounds in extra virgin olive oil: an empirical relationship. *J. Food Qual.* **29**(4), 431–441.

Sinesio, F., Moneta, E., & Esti, M. (2005). The dynamic sensory evaluation of bitterness and pungency in virgin olive oil. *Food Qual Prefer* **16**(6), 557–564.

Singh, T.K., Young, N.D., Drake, M. & Cadwallader, K.R. (2005). Production and sensory characterization of a bitter peptide from β-casein. *J. Agric. Food Chem.* **53**(4), 1185–1189.

Soeryapranata, E., Powers, J.R., Fajarrini, F., Weller, K.M., Hill, H.H., & Siems, W.F. (2002). Relationship between MALDI-TOF analysis of β-CN f193-209 concentration and sensory evaluation of bitterness intensity of aged Cheddar cheese. *J. Agric. Food Chem.* **50**(17), 4900–4905.

Sokolowsky, M. & Fischer, U. (2012). Evaluation of bitterness in white wine applying descriptive analysis, time-intensity analysis, and temporal dominance of sensations analysis. *Anal. Chim. Acta* **732**, 46–52.

Stephan, A. & Steinhart, H. (2000). Bitter taste of unsaturated free fatty acids in emulsions: contribution to the off-flavour of soybean lecithins. *Eur. Food Res. Technol.* **212**(1), 17–25.

Stone, H. (1992). Quantitative descriptive analysis. R.C. Hootman (Ed.), *Manual on descriptive analysis testing for sensory evaluation*, (ASTM Manual Series: MNL 13), Baltimore, pp. 15–21.

Stone, H., & Sidel, J.L. (1993). *Sensory evaluation of food.* Academic Press, Redwood City, CA.

Stone, H. & Sidel, J.L. (2004). *Sensory evaluation practices* (3rd ed). San Diego, CA: Elsevier Academic Press.

Stone, H., Bleibaum, R.N., & Thomas, H.A. eds. (2012). *Sensory evaluation practices* (4th ed) Academic Press, London, UK. pp. 250–275.

Sun, B., Neves, A.C., Fernandes, T.A., Fernandes, A.L., Mateus, N., Freitas, V.D., Leandro, C., & Spranger, M.I. (2011). Evolution of phenolic composition of red wine during vinification and storage and its contribution to wine sensory properties and antioxidant activity. *J. Agric. Food Chem.* **59**, 6550–6557.

Talcott, S.T., Passeretti, S., Duncan, C.E., & Gorbet, D.W. (2005). Polyphenolic content and sensory properties of normal and high oleic acid peanuts. *Food Chem.* **90**(3), 379–388.

Tamamoto, L.C., Schmidt, S.J., & Lee, S. (2010). Sensory profile of a model energy drink with varying levels of functional ingredients - caffeine, ginseng, and taurine. *J. Food Sci.* **75**(6), S271–S278.

Tang, X., Kälviäinen, N., & Tuorila, H. (2001). Sensory and hedonic characteristics of juice of sea buckthorn (Hippophae rhamnoides L.) origins and hybrids. *LWT-FOOD SCI TECHNOL* **34**(2), 102–110.

Techakriengkrai, I., Paterson, A., Taidi, B., & Piggott, J.R. (2004). Relationships of sensory bitterness in lager beers to iso-α-acid contents. *J. Inst. Brew.* **110**(1), 51–56.

Torres-Penaranda, A. & Reitmeier, C. (2001). Sensory descriptive analysis of soymilk. *J. Food Sci.* **66**(2), 352–356.

Troszynska, A., Lamparski, E.I., Hernandez, T., Amarowicz, R., & Pegg, R.B. (2011). Relationship between the sensory quality of lentil (Lens culinaris) sprouts and their phenolic constituents. *Food Res. Int.* **44**, 3195–3201.

Troszynska, A., Lamparski, G., & Kosinska, A. (2004). The sensory quality of sprouts obtained from the selected species of legume seeds. *Czech J. Food Sci.* **22**, 174–176.

Varming, C., Jensen, K., Møller, S., Brockhoff, P.B., Christiansen, T., Edelenbos, M., Bjørn, G.K., & Poll, L. (2004). Eating quality of raw carrots—correlations between flavour compounds, sensory profiling analysis and consumer liking test. *Food Qual Prefer* **15**(6), 531–540.

Vidal, S., Francis, L., Guyot, S., Marnet, N., Kwiatkowski, M., Gawel, R., Cheynier, V., & Waters, E.J. (2003). The mouth-feel properties of grape and apple proanthocyanidins in a wine-like medium. *J. Sci. Food Agric.* **83**(6), 564–573.

Watts, B., Ryland, D., Malcolmson, L.J., Ambalamaatil, S., Adams, K.M., & Lukow, O.M. 2012. Flavor properties of pan and pita breads made from red and white hard spring wheats. *J. Food Qual.* **35**, 60–68.

Wolejszo, A., Szymkiewicz, A., & Troszynska, A. (2007). Immunoreactive properties and sensory quality of lentil [Lens culinaris] and mung bean [Vigma radiata L.] sprouts. *Pol. J. Food Nutr. Sci.* **57**(4), 415–420.

Yau, N. & Huang, Y. (2000). The effect of membrane-processed water on sensory properties of Oolong tea drinks. *Food Qual Prefer* **11**(4), 331–339.

Yousfi, K., Cayuela, J.A., & Garcia, J.M. (2008). Reduction of virgin olive oil bitterness by fruit cold storage. *J. Agric. Food Chem.* **56**, 10085–10091.

8 Analysis of Bitterness Compounds by Mass Spectrometry

Geraldine Dowling

8.1 INTRODUCTION

The sensation of bitterness caused by "bitter compounds" can be equated with dietary danger. In nature, microbiological fermentation products, hydrolysed proteins, plant-derived alkaloids and other toxins usually have a bitter taste. Plant based foods are very diverse and foods such as beans and peas, potatoes, yams, cabbage, pumpkins, cucumbers, lettuce, spinach and kale contain bitter compounds. Due to the wide distribution of plant based bitter toxins, efforts to develop less bitter cultivars of common plant foods may have been conducted primarily based on their safety concerns rather than their taste. It has been the ultimate goal of science to explain how so many structurally unrelated compounds can give rise to a bitter taste. There are a wide variety of bitter compounds that can impart bitter taste. Bitter tasting compounds (bitter agonists) can be present in a multitude of foods not just plant based foods and consumer products. Although in some cases these compounds can also contribute to the formation of desirable sensory attributes with a specific contribution to their characteristic taste (e.g., caffeine in tea or coffee, quinine in bitter-lemon drinks or bitter compounds such as humulones or iso-α-acids obtained from hop in beer), in most cases, the presence of these compounds lower food values as bitterness can be considered an undesirable sensory attribute in some foods.

Several bitter compounds in foods have been associated with health benefits. Certain virgin olive oils have very beneficial health effects. Studies have shown consumers have preferences for olive oils with low or moderate levels of bitterness (Mateos *et al.*, 2004). Chemopreventive phytonutrients are naturally present in plant foods and can beneficially improve human health. During food production and food processing, methods are developed to enhance and preserve these phytonutrients. In addition, functional foods are created by adding phytonutrients to foods at high concentration levels. It is well known that a number of phytonutrients such as phenolic compounds, flavonoids, terpenes and glucosinolates are bitter

Bitterness: Perception, Chemistry and Food Processing, First Edition. Edited by Michel Aliani & Michael N. A. Eskin.

or astringent (Drewnowski & Gomez-Carneros, 2000). This causes difficulties in enrichment because consumers mostly do not tolerate the bitter taste in food.

Therefore, strategies are needed to understand the unpleasant taste of bitter compounds. There is also a need to develop and identify bitter taste masking substances. Sensory methods are usually performed to identify bitter taste masking compounds, that is, by equating the taste of a mixture of bitter tasting compound and bitter antagonist with the taste of the bitter compound alone, and by screening in the presence of a bitter compound with/without a bitter antagonist. Sensory screening is both time and work consuming and generally performed on toxicologically harmless compounds. Determination of bitter compounds by means of sensory evaluation techniques using human panels and instrumental counterparts such as the electronic tongue are described in Chapters 10 and 11, respectively. The development of quantitative chemical analysis methods for potential bitter compounds in different foods, combined with sensory analysis is necessary to identify key compounds responsible for bitterness.

Rapid screening methods such as immunoassays or conventional high performance liquid chromatography (HPLC) or gas chromatography (GC) in combination with a variety of detectors are useful tools in modern laboratories but technologies are rapidly evolving. Nowadays, analytical instruments which gained substantial ground in food testing are the hyphenated techniques of GC-MS and LC-MS. Over a number of years, the strategies adopted to determine trace levels of compounds in food has changed dramatically moving away from the use of GC with various detectors to the sensitivity and specificity offered by mass spectrometry (MS). The combination of MS with either liquid chromatography (LC) or GC has been well recognized and recommended for quantitative and semi-quantitative screening. The technique of GC-MS is suitable for the analysis of volatile and semi-volatile bitter compounds and for compounds that can be derivatized in order to increase their thermal stability and/or volatility. The combination of the GC with the MS technique utilizing electron impact ionization (EI) gives rise to a high chromatographic resolution with detection in a targeted way, thus allowing bitter compound quantitation and the identification of non-targeted or "unknown" screening approaches in a single sample. The advantages of using EI are that it is highly reproducible when compared to other ionisation techniques and less affected by ion suppression. The consistent generation of fragmentation patterns in EI allows the development of spectral libraries and the searching of established libraries such as NIST (n.d.) to identify specific components.

LC-MS is particularly suitable for the analysis of non-volatile compounds or compounds that are thermally unstable and not suitable for GC-MS. In the analysis of bitter compounds with a wide range of polarities within a food matrix both techniques are complementary. LC-MS has clear advantages over more conventional techniques such as HPLC or GC-MS with reduced sample preparation and analysis time, higher sensitivity and specificity. In recent years the advances in LC-MS technology have meant that instruments are capable of reaching even lower concentrations in biological samples previously not detectable. There are some concerns when utilising LC-MS technology; however, such as the unavailability of standard spectral libraries (i.e., NIST), thus the analysis of unknown compounds is much more challenging compared with data obtained from GC-MS. In addition, the presence of matrix effects can hinder LC-MS analysis.

However, the recent advances in both LC and MS have given rise to highly sophisticated and powerful instrumentation for sensitive detection. In addition, the innovations made in chromatography allow for more rapid, highly efficient LC separations (Guillarme *et al.*, 2010; Nunez *et al.*, 2012), thus allowing opportunities to analyse ionic and polar compounds (Li *et al.*, 2008; McCalley, 2010; West *et al.*, 2010).

Different ionization techniques can be utilized in the practice of LC-MS but electrospray ionization (Yamashita & Fenn, 1984) remains one of the widespread ionization techniques employed for the determination of chemical constituents in food by LC-MS. The use of atmospheric pressure chemical ionization (APCI) (Bruins, 1991) for the analysis of ingredients in food (Santini *et al.*, 2009; Kaklamanos *et al.*, 2009) seems to be much less popular than ESI. This could be due to the improvements in source and probe design for ESI which has not yet been paralleled in APCI. In retrospect, the most important change in the past decade has been on the increase in the variety of mass analyzers for LC-MS and how these innovations have affected the approaches undertaken to monitor compounds of wide ranging polarity in a wide variety of scientific disciplines. The use of LC-MS in food analysis can have important advantages in the analysis of bitter compounds due to the high separation power of MS as identification and confirmation strategy. In addition, LC-MS has the ability to cover a wider range of compounds in a greater variety of matrices using a combination of targeted and non-targeted data collecting approaches.

In other disciplines, LC-MS has become a routine analytical tool and a mainstream technique for the detection of a wide range of polar and non-volatile compounds not compatible with GC analysis. Likewise, its application to the analysis of bitter compounds which are mostly non-volatile compounds dispersed in a large variety of food matrices could be very useful. A wide variety of substances require estimation which are non-volatile and a bitter tasting compound could be in a range of diverse foods or food additive/supplement ingredient, nutraceutical ingredients or pharmaceutical substances or formulation.

LC-MS may therefore be considered an extremely beneficial tool in determination of potential bitter compounds. This chapter will provide an overview of LC-MS considerations and its use as a powerful technique applied to non-volatile bitter compounds.

The most widely used ionization techniques in LC-MS analyses are electrospray ionization (ESI), (Dole *et al.*, 1968; Aleksandrove *et al.*, 1984; Horning *et al.*, 1973) atmospheric pressure chemical ionization (APCI) (Horning *et al.*, 1974; Robb *et al.*, 2000) and atmospheric pressure photoionization (APPI)

(Syage *et al.*, 2000; Niessen, 2006). These techniques provide user friendly coupling of the LC to a mass spectrometer. Due to the complexity of the ionization process and many factors that affect mass spectrometric sensitivity and chromatographic performance, obtaining optimal LC-MS conditions is not an easy task. As a highly sensitive method, LC-MS is considered an important analytical technique suitable for the analysis of bitter compounds which can also help to provide structural information of the analyte in a range of complex biological matrices. The choice of mobile phase solvents for bitter compound ionization can be further complicated because solvents do not often provide optimal retention time and resolution in the chromatography approach undertaken. Often a compromise must be made with selection of solvent in order to achieve sufficient ionization and chromatographic efficiencies when choosing this technique for test compounds. These techniques provide stable performance, good repeatability and high sensitivity. The applicability of ESI, APCI and APPI is different (Ma *et al.*, 2012). ESI can be used for small polar organic molecules and is the most widely used API technique, and therefore, this chapter will focus on its use in LC-MS analysis of bitter compounds in a range of biological matrices. Although ESI is the most widely used ionization technique and has significant advantages in analysis of bitter compounds in food, a disadvantage of ESI is that the ionization can be poor for non-polar organic compounds. In general, APCI and APPI can be utilized for non-polar organic compounds.

In addition to the ionization mode used as a part of an LC-MS methodology, the LC separation method plays a major role on the sensitivity and the selectivity of compounds. The most commonly used approach is reverse phase LC; however, other techniques such as ion pair, ion exchange, affinity and size exclusion chromatography have also been adopted for use. The chemistry, length and diameter of analytical columns used in LC-MS methods have an important impact on the separation efficiency. The 100 to 200 mm long with internal diameter of 3-4.6 mm analytical columns are widely used. Shorter columns with similar internal diameter may be used with reduced analysis time compared to longer counterparts. High separation power and sensitivity can be achieved using capillary columns with internal diameter of 0.05–0.3 mm, but the analysis can be time consuming.

Ultra high performance liquid chromatography (UHPLC) columns and a monolithic column can provide good chromatographic resolution with a shorter analysis time. Ma *et al.* (2012) evaluated the use of different columns (Gemini C_{18} or the Synergy RP) and mobile phases (e.g., water-acetonitrile and water-methanol) for the determination of constituents in dietary supplements containing bitter melon and the best results were obtained using Synergy RP column with water:acetonitrile and 10 mM ammonium formate and 1% formic acid. The 1% formic acid is a higher than would be expected concentration for LC-MS however was necessary in the analysis of compounds in bitter melon. Kenny *et al.* (2013) developed a quantitative UPLC-MS method for the determination of phenolic compounds in fenugreek seeds and bitter melon. The column utilized was a Waters Acquity UPLC

HSS T3 (2.1mm x 100 mm, 1.8 μm) with a gradient of mobile phase consisting of acetonitrile, formic acid and water.

The coupling of various technologies in LC-MS gives a wide variety of options for the analysis of bitter compounds. However, it should be noted that the appropriate implementation of LC-MS in a laboratory requires educated personnel and significant resources.

8.2 OVERVIEW OF LC-MS

8.2.1 Electrospray ionisation

Electrospray was presented as a method of ionization and an analytical technique four decades ago after Dole and co-workers (1968) studied the ionization mechanism by ion mobility spectrometry. In addition, another research group (Yamashita & Fenn, 1984) successfully combined ESI and MS. A further group identified that ESI would be suitable for large biomolecules and their study was rewarded a Nobel prize in chemistry in 2002 (Meng *et al.*, 1988; Mann *et al.*, 1989). In ESI a compound will be dissolved in LC effluent and channeled through a small capillary which is set to a high voltage usually (3-5 kV). Due to the high electrostatic field at the tip of the capillary negative counter ions (when positive ions are encountered) move away from the liquid surface towards the wall of the capillary where they are neutralized and the positive ions drift downfield towards the liquid front. The results are the formation of a liquid cone, so-called a Taylor cone (Taylor, 1964). The positive ions drift towards the surface of the liquid. When the electrostatic repulsion at the surface devastates the surface tension of the liquid at the cone tip, the jet breaks apart and small electrically charged droplets are formed. The droplets subsequently migrate towards the interface plate in the API source and while transitioning through the source the surface area of the droplet starts to reduce due to evaporation of the solvent in the droplet. Furthermore, the charge density at the surface increases. A set radius within the droplet is reached called the Rayleigh limit (Rayleigh, 1982). Consequently, the charge density at the surface becomes elevated and the repulsion forces on the surface exceed the surface tension of the droplet (Gomez & Tang, 1994). Ultimately a charged smaller droplet is formed and the procedure is repeated until the droplet size is small enough to obtain gas phase ions. The generation of gas phase ions has been proposed by two models namely the charge residue model (Dole *et al.*, 1968; Schmelzeisen *et al.*, 1989) and the ion evaporation model (Iribarne & Thomson, 1976; Thomson & Iribarne, 1979). The theory behind the electrospray ionization process is outlined in more detail in other reviews (Kebarle & Ho, 1997; Bruins, 1998; Cole, 2000; Kebarle, 2000). It is necessary in both principles that the test compound is already in the liquid phase. During the process of electrospray ionization, not only charged compounds in the gas phase are generated, but also charged eluent species at a high concentration are obtained from solvents and additives and can act as reagent ions in gas-phase ion-molecule reactions. Charged eluent molecules can be deprotonated or protonated solvent or additive molecules which can ionize neutral compounds present in the gas-phase by proton transfer reactions. It should be noted that the overall process of ESI is extremely complex with a variety of parameters requiring consideration such as volatility, surface tension, viscosity, conductivity, ionic strength, dielectric constant, electrolyte concentration and pH. Gas phase ion molecule reactions influence the ionization process also and the signal intensity. Other factors like chemical and physical properties of the compound including pKa, hydrophobicity, surface activity, ion solvent, ion energy, proton affinity and parameters such as solvent flow rate, temperature and ESI voltage also are equally important. The choice of LC mobile phase is hindered in LC-ESI/MS/MS as only polar solvents and volatile additives can be used in practice and the ESI response and LC separation efficiency must be finely tuned.

8.2.2 Solvents

Ma *et al.* (2012) optomized the LC-ESI-MS/MS conditions of five cucurbitane-type triterpene and triterpene glycoside in bitter melon. Individual standards were injected with different buffers in ESI and APCI in both positive and negative modes. ESI in positive mode with 10 mM ammonium formate and 1% formic acid buffer gave suitable sensitivity.

The conductivity of solvent must be sufficient in order to achieve high sensitivity and good stability. Solvents of choice for ESI vary from polar to medium polar with the most common combination being water and acetonitrile. Organic solvents such as methanol, acetonitrile and dichloromethane are better solvents for ESI compared to water alone (Kostiainen & Bruins, 1996). A comprehensive discussion on solvent use in ESI is outlined in various literature (Kostiainen & Bruins, 1996; Kebarle & Tang, 1993; Zhou & Cook, 2000; Cole *et al.*, 1993; Hiraoka & Kudaka, 1992).

The majority of LC-ESI/MS based strategies in literature across a variety of disciplines have been carried out by reverse phase LC with a non-polar C_{18} or C_8 bonded silica stationary phases (Hemström & Irgum, 2006). The mobile phase is generally water and organic modifier and is a balance between obtaining satisfactory chromatographic performance and ESI sensitivity. (Straub & Voyksner, 1993; Dams *et al.*, 2002; Needham *et al.*, 2000). Organic modifiers most widely utilized are methanol and acetonitrile in LC-ESI/MS. Methanol has been shown to offer better ESI efficiency than acetonitrile (Temesi & Law, 1999) and better peak shape (McCalley, 1996) but it is compound specific.

8.2.3 Additives

In the analysis of ginsenosides (Zhao *et al.*, 2013), ammonia, water, formic acid and ammonium chloride were tested in order to enhance the ionisation and to improve the collision induced dissociation (CID) efficiency in mass spectrometry. The fragmentation of the 15 ginsenosides tested depended on the nature of additives utilized and the most abundant fragment ions were obtained when 0.02% formic acid was added to the mobile phase. 0.1 mm ammonium chloride was an additive that enhanced the sensitivities and dynamic range for the tested compounds as well as the precision; however, fewer fragment ions were observed. They concluded that 0.01 mM ammonium chloride was the best mobile phase additive for quantitative analysis and 0.02% formic acid and 0.02% acetic acid were the most suitable ones for qualitative analysis. It is worth noting that the composition of the mobile phase might not be suitable for other compounds and studies should be performed for specialist bitterness compounds as an ideal composition of the mobile phase is compound dependent (Zhao *et al.*, 2013).

The resolution and reproducibility in LC mobile phases are generally improved by addition of additives and buffers. The concentration of the additives, their chemical properties as well as their pH values have significant effects on analyte response in ESI. The majority of additives and buffers used in LC is not compatible with ESI MS/MS. Non-volatile buffers such as phosphate and borate cause increased background signal, signal suppression and rapid contamination of the ion source resulting in reduced sensitivity and stability. Strong acids such as trifluoroacetic acid (TFA) are commonly used as ion pairing agents in the LC analysis of molecules such as peptides and proteins but may cause significant signal suppressions in ESI (Eshragi & Chowdhury, 1993; Appel *et al.*, 1995; Kuhlmann *et al.*, 1995). The most widely used reagents in the LC-MS analysis of polar compounds tends to be formic acid, ammonium hydroxide, ammonium acetate and ammonium formate (Gao *et al.*, 2005). In practical terms the additive concentration should not exceed 10 mM otherwise it might suppress the ionization and reduce the sensitivity for the tested compounds. Additives present in LC mobile phases at concentrations of 100 mM can be too high for ESI.

8.2.4 pH

Improved sensitivity in ESI can be achieved by adjusting the pH of the liquid phase in which the compound is ionized. This can be achieved by pH adjustment to an acidic mobile phase (two pH units below pKa of the analyte) for basic analytes such as amines and to basic conditions for acidic analytes such as carboxylic acids and phenols (two pH units above pKa of the analyte) (Zhou *et al.*, 1990). Good chromatographic performance in reverse-phase LC is achieved by altering the pH so that acidic or basic analytes are altered in the mobile phase. Alternatively, satisfactory chromatographic performance can be achieved if the mobile phase has sufficient interaction between the hydrophobic moiety of the compound and the reverse phase material. Another approach is to alter the pH so that the compound is not ionized in the mobile phase.

8.2.5 Adduct formation

Ma *et al.* (2012) studied cucurbitane-type and triterpene glycosides in bitter melon in positive ESI mode with 10 mM ammonium formate and 1% formic acid which provided adequate sensitivities for the corresponding $[M+NH_4]+$ adduct ions. Sodium or lithium salts have been used in literature to improve the ionization and repeatability of compounds such as trichothecenes (Dall'Asta *et al.*, 2004), carbohydrates (Dall'Asta *et al.*, 2004; Guignard *et al.*, 2005; Assam & Glish, 1997; Harvey, 2000) and lipids. Chloride, formate and acetate anions were utilized in negative ion mode to promote the formation of adducts ([M+Cl]-, [M+HCOO]-, $[M+CH_3COO]$-) for analytes that do not easily undergo deprotonation (Zhu & Cole, 2000). In reality, only very low concentrations of salts (below 0.1 mM) are added to facilitate ionization in ESI via adduct ion formation as significant concentrations may lead to strong background interference and rapid contamination of the source. The influence of the above parameters on APCI and APPI have been summarized previously (Kostianen & Kauppila, 2009).

Neutral compounds that are polar cannot be ionised by protonation or deprotonation in liquid phase but can be ionized by adduct formation, examples can be with ammonium, lithium, sodium in positive ion mode ([M+Na]+, [M+Li]+ and with chloride, acetate or formate ions in negative mode ($[M+Cl]^-$, ($[M+HCOO]^-$, ($[M+CH_3\ COO]^{-\cdot}$ The use of buffers such as ammonium acetate, ammonium formate and ammonium hydroxide can result in an ammonium adduct formation instead of the protonated molecule. This phenomenon is common for compounds having a proton affinity close to ammonia. Adducts of sodium [M+Na]+ can be generated in addition to [M+H}+ ions since sodium is always in the mobile phase at concentrations ranging from 0.01-0.1 mM due to sample vial derived impurities and LC solvents. The relative abundance of the [M+Na]+ may vary, thus reducing the precision of the analysis. The formation of sodium adducts can be reduced by adding formic acid to the eluent post-column.

8.2.6 Ion-Pairing and ion exchange

In reverse-phase, LC-ESI/MS ion-pairing can be used to improve the retention and resolution of polar ionic compounds. Generally, in the analysis of basic compounds, volatile ion pairing solvents such as pentafluoropropanoic acid (PFPA), trifluorobutanoic acid (HFBA) and TFA have commonly been used in the analysis of these polar compounds (Appfel *et al.*, 1995; Gustavsson *et al.*, 2001; Petritis *et al.*, 2002; McCalley, 2004; Häkkinen *et al.*, 2007). In the analysis of compounds the use of ion-pairing LC-MS/MS has been limited to date possibly due to the challenges cited above. Relatively stable ion-pairs with basic compounds can be formed thus reducing secondary interactions with free silanols on the stationary phase which could result in poor chromatography. It is important to note that acidic ion-pairing agents can suppress ionization as outlined in the use of fluorinated carboxylic acid as ion-pairing agents at concentrations of a few mM where shown to decrease the ESI signal by 30-89% compared to the intensity of the signal with formic acid-ammonium formate buffer (Gustavsson *et al.*, 2001). For instance the use of TFA is controversial as it can cause a suppression effect in LC-ESI-MS (Eshragi & Chowdhury, 1993; Kuhlmann *et al.*, 1995; Mirza & Chat, 1994). Formic acid is generally preferred. As formic acid provides satisfactory chromatographic performance without suppression, it has been more utilized in LC-ESI/MS.

8.3 DATA ACQUISITION IN LC-MS

LC-MS provides a multitude of data. analysis that can be performed by acquiring the data in a targeted compound screening approach and/or a non-targeted compound (retrospective) screening approach.

8.3.1 Targeted compound screening

LC-MS instruments which have a triple quadrupole analyser, operated in selected reaction monitoring (SRM) mode, achieve the selectivity and sensitivity necessary in this monitoring. The approach is widely used for targeted multi-component determination of constituents in food. The fast monitoring of numerous transitions is easily achievable with good sensitivity and precision thus allowing rapid method development covering a wider array of compounds with wide-ranging polarities. The ability to monitor a wider range of compounds is made possible due to retention-time window based SRM acquisition of data. The instrument users must only enter the masses and the retention times for each SRM and peak width and/or data points across the peak. Subsequently, the software sets acquisition windows for each SRM transition for each compound yet to elute.

8.3.2 Non-targeted compound (retrospective) screening

In the non-targeted compound (retrospective) screening approach no pre-programming of masses is required thus the selectivity is provided by the high mass resolving power. The benefits of non-targeted analysis are that it provides greater scope than a targeted approach. For instance, monitoring can be extended to certain metabolites or other transformation products where no reference standards are available. It is important to note that all LC-MS instruments can carry out full spectral acquisition but not all can attain sufficient sensitivity in this mode for non-target analysis. Approaches for evaluating data obtained from non-targeted acquisition using HRMS have been developed using two different processes: exact mass filtering and searching databases relating to molecular formulae. This approach concentrates more on detectability rather than reaching unequivocal confirmatory criteria. To be useful, these data handling processes must be automated and quick; however, there is a need for powerful computing power and data management/storage. In addition, the data processing is more time consuming than data traditionally required with LC-MS analyses using QQQ analysers. Molecular formulae databases containing information on exact mass and isotopic patterns are generally developed in-house but are available commercially or via the internet. If information on retention time is available then the search window can be narrowed. A 'hitlist' is generated as a result of the search with or without a chromatographic peak. Utilising orbitrap technology (Alder *et al.*, 2011) or when TOF analysers are in use, more careful optimization of the accurate-mass window tolerances (generally 2–50 ppm) is necessary to ensure adequate selectivity as resolving power varies considerably between instrument types (Mezcua *et al.*, 2009). If too narrow a search mass window is chosen around the exact mass this can cause loss of signal when the measured mass lies outside of the defined tolerance, the result is false negatives are obtained (Hird, 2008).

The use of advanced and powerful techniques like HRMS for acquiring full scan data is advantageous. This technique facilitates exact mass analysis of both MS and MS/MS ions, therefore allowing the detection of many compounds in a single sample thus providing useful information to identify the structures of test compounds. Statistical methods can then be carried out to select a list of molecular compounds whose levels are significantly altered in a test sample versus a control sample. The compounds are subjected to precursor ion (PI) scans in order to obtain MS/MS data. The MS/MS data in conjunction with the PIs and the retention times are utilized to obtain structural information. Another possibility is to obtain the MS/MS data "on the fly" by either data independent acquisition (DIA) or data dependent acquisition (DDA). DDA involves a survey scan followed by MS/MS acquisition. The MS during survey scan automatically selects PI above an abundance threshold and this triggers the instrument to start fragmentation of the PIs followed by subsequent MS/MS fragmentation of the product ions. In the case of DIA all ions within the m/z window are subjected to fragmentation instead of choosing a particular PI. Xu *et al.* (2013) developed a LC-QTOF-MS method for the analysis of phenolic compounds in liquorice. This group used a combination of data independent and data dependent acquisition. Fifteen flavonoids and one triterpenoid were investigated. The combined use of DIA and DDA in the study provided accurate and specific MS/MS spectra for co-eluting peaks. Dorta *et al.* (2014) identified 30 phenolic compounds including gallates, gallatannins, flavonoids, xanthones, benzophenones, gallic acid and derivatives from the peels and seeds of three mango varieties. The group processed the MS and MS/MS spectra acquired through Mass Hunter Work station software (version B.04.00, Agilent Technologies, Waldbronn, Germany) which gave a list of feasible elemental molecular formulas utilising the Generate Molecular Formula editor due to the accurate mass and isotopic pattern. The molecular formula obtained with the highest percentage score would give an indication that there is a closer familiarity between the formula generated by the software and the real molecular formula of the compound. There are two other factors that affect the overall correlation score which are the mass accuracy of the observed fragment ions and the overall percentage of fragment ion intensity that might be explained by substructures (Agilent Mass Hunter Molecular Structure Correlator Software). Other tools that were utilized in the study to identify the unknown phenolic compounds in order to interpret the observed MS/MS spectra were those available in the literature or in a online database such as ChemSpider, Mass Bank, METLIN, LIPID MAPS, MetaboAnalyst, Spectral Database for Organic Compounds. A comparison of the mass spectral data and the chromatographic behavior generated utilising authentic standards was performed when available.

8.3.3 Ion annotation

This is a strategy to identify a group of ions likely to originate from the same compound. In LC-MS-based compound analysis there is potential for a compound to be shown as numerous peaks in LC-MS data with m/z values at similar retention times due to adducts, isotopes and neutral loss fragments. It is important that the scan rate on the mass spectrometer is appropriately set so enough data points are acquired to define the points across the chromatographic peak. The ions obtained from the same compound share similar shaped elution profiles which can be shown by extracted ion chromatograms (EIC). Therefore, ion annotation can be achieved by grouping similar elution profiles together thus allowing compound identification.

Zhang *et al.* (2014) developed a new LC-MS data processing platform for metabolite compound feature extraction and annotation called MET-COFEA for use in metabolomics. This software can detect and compartmentalize relevant chromatographic peak features for each test compound based on the retention time and peak shape criteria and then annotate the relationship between each peak's identified m/z values with the appropriate test compound molecular mass. MET-COFEA integrates with a number of innovative algorithms such as mass trace based EIC extraction, compound associated peak annotation and compound alignment. In the study when MET-COFEA was compared with numerous open-source software such as MAVEN the software achieved superior performance in analysis of chromatographic peaks. The advantage of the software as a tool in LC-MS data analysis is the capability to significantly reduce the number of possible compound candidates in library searching and also improve compound quantification accuracy. This data processing approach could be used for the analysis of bitter compounds by LC-MS in the future.

8.3.4 Mass-based identification

In ion annotation, the peaks are grouped together and the monoisotopic exact masses can be calculated based on mass difference of the isotopes/adducts from their monoisotopic neutral forms. The masses calculated are subsequently compared against commercial databases. Bitter compounds having molecular mass within a pre-specified tolerance of the query are retrieved from the database. Generally, mass-based identification does not allow for identification of unique compounds. Automated LC-MS system performance software is available in other fields (Xu & Freitas, 2009). Analysis of MS/MS spectra of the ions is carried out and results from the mass based identification can further be investigated through the following steps.

8.3.5 Spectral interpretation

Spectral interpretation identifies the possible structure or sub-structure of an unknown molecular ion by comparing its MS/MS data with hypothetical spectra predicted through, for example, insilico fragmentation approaches. There are two well-known ways to predict fragmentation ions of a given molecule. One approach is to use a rule-based predictor which utilizes fragmentation patterns collected from the literature. Predictors available are, for example, Mass Frontier (High Chem, LtD). These software programs are commercially available. The advantage of a rule-based approach is its potentially high specificity. However, it is important to note that the particular fragment ion cannot be predicted if a particular fragmentation rule is not included in the database. Other insilico fragmentation tools such as MetFrag TM are also available (Wolf *et al.*, 2010).

Such software programs generate a list of possible fragments through combinational disconnection of chemical bonds. The internal energy of each cleaved bond is calculated. This approach does not require any type of knowledge base and therefore reduces the often time consuming data collection step and avoids possible bias from a limited set of fragmentation rules. After the hypothetical MS/MS spectra is generated, and the insilico fragmentation is compared against the experimental database to calculate a similarity score the results are ranked accordingly. Some caution should be undertaken, however, as the prediction of low-resolution electron ionization is found to exhibit bias towards certain categories of structures depending on the program setting. More work needs to be carried out to improve this and it is expected that high resolution mass spectrometry spectrum data might be in a position to achieve this in the future.

8.3.6 Spectral matching

Spectra matching mimics manual verification of the compounds identification using the MS/MS spectrum. Instead of acquiring the MS/MS spectrum of a standard each time, previously acquired MS/MS spectra from standards are amalgamated into a spectral library and compared with test sample spectra. An appropriate scoring mechanism measures the similarity between two MS/MS spectra. If spectra are extremely similar then this represents the same compound. There have been several spectra matching approaches for spectra obtained from GC-MS and LC-MS. The majority of the algorithms calculate the similarity between the query spectrum and the library spectra by treating the two spectra as vectors. A substantial difficulty with MS/MS spectra matching is that the acquired spectra depend significantly on the machines used and the acquisition settings utilised. One approach to overcome this is to expand the spectral libraries under different experimental settings (different experimental settings or collision energy conditions or design improved scoring functions by considering expanded aspects of spectra being similar). A further limitation of spectral matching is the minimal coverage in spectral libraries of metabolites. This can be difficult in fields like metabolomics or new designer drugs in forensic toxicology where no information on metabolism is available. In bitterness analysis if a new component is found and no information is available a similar problem exists. Yu *et al.*, (2013) expressed some difficulties with database building and matching as some compounds can share the same molecular composition hence m/z ratio. Without additional information, one cannot identify a one-to-one correspondence between the identified features and compounds. As extra information becomes available the potential of each feature could be reduced and the corresponding database entered can be split based on parameters such as retention time.

8.3.7 Compound identification

Compound identification in particular structural elucidation should be performed vigilantly. Various computational approaches can be used to prioritize assumed identifications. If the availability of authentic standards is limited then it will be necessary to identify via prioritizing assumed identifications so efforts can focus on the likely candidates. However, it is important to note that in order to verify the identity of the unknown compound it is vital to obtain an authentic standard and inject on the same instrument with a test biological sample in order to compare MS/MS spectra and retention times. It should be noted that for some compounds, it is not enough to generate only a MS/MS spectra in order to identify the compound uniquely. In this situation, MS^3 or MS^4 is required in order to obtain further fragmentation information from the desired test compounds. The ability to perform MS^3 and subsequent fragmentation of a compound can only be achieved on the ion trap mass spectrometer, for example, linear ion trap (LIT) can give fragmentation of precursor ions and fragment ions. The importance of MS^n is that it allows the elucidation of very similar molecules and more confidence in compound identification. The QTRAP is an ideal analytical platform for class-targeted bitter compound analysis as the structure-specific precursor ion (PI) (Wen *et al.*, 2008; Sandra *et al.*, 2004), neutral loss (NL) (Scholz *et al.*, 2005), or multiple reaction monitoring (MRM) (Wagner *et al.*, 2007; Steimer & Sjöberg, 2011) can be utilized as a survey scan to trigger sensitive enhanced product ion (EPI) spectra so that qualitative results can be obtained within the run time of the chromatographic analysis for the reliable characterization of low-level compounds. Within the range of different scanning modes MRM is often the most favoured due to its superior sensitivity, selectivity and wide linear range (Wagner *et al.*, 2007; Steimer & Sjöberg, 2011; Yao *et al.*, 2008).

The application of LC-MS to several bitter compounds in various foods is described in next section.

8.4 LC-MS APPLICATION OF BITTERNESS COMPOUNDS

One of the aims in bitter compound analysis is the quantitation of test compounds in order to evaluate changes in response to experimental conditions. Bitter compound quantitation will be discussed in relation to common LC-MS analysers such as triple quadrupole selected reaction monitoring (SRM), ion trap and full-scan HRLC-MS-based analysis.

8.4.1 Bitter compound quantitation by triple quadrupole and selected ion monitoring

In the triple quadrupole instrument the precursor ion is selected in the first quadrupole (Q1MS) and is dissociated into fragment ions in the collision cell (Q2MS) and only a specific fragment ion (daughter ion) is selected in the third quadrupole (Q3M3). This ion selection method is called selected reaction monitoring (SRM) and is affected by molecular weight and specific to the structure of the selected compound. SRM can identify the real concentration through absolute quantitation by correlating signal intensities in calibration curve with spiked stable isotope labeled analogues. This method has been utilized for over three decades (Baty & Robinson, 1977).

Ding *et al.* (2006) developed a reverse phase LC-ESI method for determination of bilobalide, ginkgolides A, B, C, quercetin, kaempferol, isorhamnetin, rutin hydrate, quercetin-3-β-D-glucoside and quercetin hydrate in Ginkgo biloba. The sensitivity of different mass spectrometry modes (full scan, selected ion monitoring (SIM) and selected reaction monitoring (SRM) were compared and quantitation were achieved with/without internal standard. Quantitation in the method was performed using negative mode ESI-MS in selected ion monitoring (SIM) mode. In SIM mode, a mass window of 0.5 Da was utilized in order to specify the ion monitored during SRM scanning mode and the same window was utilized for both the precursor ion and the product ion selection. The study showed that in most cases the signals obtained in full-scan mode were comparable except in the case of the signals obtained for bilobalide, ginkgolide, quercetin-3-β-D-glucoside. Theoretically, SRM analysis should result in significantly better sensitivity by reducing the background noise in the mass spectrometer but this was not the case in this study. The authors expressed the unexpected outcome may have been due to the use of an ion trap analyser for acquiring SIM and SRM data in the study as the ion trap performs MS/MS analysis in a single space over time rather than between different mass analysers (as in triple quadrupole instruments) and therefore a potential sensitivity loss.

In triple quadrupole analysis, effective sample preparation and chromatographic separation are important, but no single liquid chromatography method or sample preparation method is capable of separating or purifying all classes of bitter compounds. More research needs to be carried out to improve LC separation capabilities in order to detect bitter compounds in a variety of food matrices.

Regardless of the development of LC-SRM/MS/MS in targeted bitter compound analysis, there are a number of disadvantages that limit its application for bitter compound analysis. Firstly, a pre-defined SRM transitions can lack the flexibility of using a different product ions for quantitation which can be affected by cross talk among substances with the same structures and masses (same retention times and fragment ions) or subject to interference with endogenous isobaric interferences from matrix. In order to overcome this, other major product ions which are unique to the tested bitter compound should be chosen in the SRM. There are some software tools in MS data acquisition available which have the function to "scramble" transitions with identical product ions to avoid monitoring these transitions one after the other. Tools are also available which increase the inter channel delay between each SRM transition in order to give enough time for the collision cell to empty prior to loading ions for the next SRM transition. In addition, providing chromatographic resolution of bitter compounds prior to MS detection provides a further solution. In SRM, most ions are filtered out with loss of ions and therefore qualitative information which is necessary for identification or structural elucidation of bitter compounds.

This limits the use of triple quadrupole (QQQ) for targeted analysis based on full scan MS/MS. The triple quadrupole SRM quantitative method is limited also then to compounds already tuned for by the instrument and omits information relating to other compounds present in the sample as are invisible due to the specificity of the target analysis. Another point to note is that certain analytes have non-specific transitions that are common for matrix interference (e.g., neutral loss of H_2O or CO_2). This compromises the targeted approach and causes inaccurate quantitation. Targeted quantitation of theobromine and caffeine was achieved using a triple quadrupole MS/MS in electrospray positive ionization operated in multiple reaction mode. ^{13}C isotopically labeled caffeine standard was included as internal standard in order to improve accuracy and precision (Ptolemy *et al.*, 2010). Hofte *et al.* (1998) described a negative mode LC-ESI/MS/MS application for analysis of six major bitter acids in extracted hop. The fragmentation under negative ESI conditions occurred. It was not sufficient in the study that only one product ion was monitored and would not be sufficient to confirm unequivocally the structural difference of the homologs and analogs (Hofte *et al.*, 1998). Zhang *et al.* (2004) characterised bitter acids in crude hop using high- performance liquid chromatography tandem mass spectrometry. Structural information was obtained by collision-induced dissociation (CID). Analysis of the fragmentation patterns of the major α and β- bitter acids was undertaken and minor bitter acids were detected using selected reaction monitoring (SRM) utilising a qualitatively relevant selected precursor-product ion transition for each bitter acid in a single LC run. Utilising LC-MS/MS in this way, six minor bitter acids, including "adprelupulone" were identified for the first time (Zhang *et al.*, 2004).

More recently in the literature developments in LC provide highly resolved peaks with a narrow peak width. Higher resolution combined with narrower peaks provide a challenge for mass spectrometers. The compatibility of the acquisition rate of mass spectrometer (dwell time for SRM transitions) with chromatographic elution in a short time period is complicated as accurate quantitation by LC-MS requires a minimum number of data points across a peak (>20). An inadequate number of data points results in poor peak resolution and affects the sensitivity of the SRM. A study by Ortega *et al.* (2010) involved developing an ultra-performance liquid chromatography tandem mass spectrometry (UPLC-MS/MS) method to quantify procyanidins, monomers to oligomers, alkaloids, theobromine and caffeine in cocoa samples. This study reported a comprehensive comparison with an HPLC method in terms of speed, sensitivity, selectivity, peak efficiency, linearity, reproducibility, detection limits and quantification limits. The results showed that UPLC-MS/MS methodology allowed detection of procyanidins at low concentration levels in short time, that is, less than 12.5 mins (Ortega *et al.*, 2010). A rapid quantitative and qualitative method was developed for 17 phenolic acids in different beverages utilising UHPLC coupled with tandem mass spectrometry. The compounds were detected in MRM mode and quantified using internal standards of deuterium labeled 4-hydroxybenzoic (2,3,5,6-D_4) and salicylic (3,4,5,6-D_4) acids [88] (Gruz *et al.*, 2008).

A mass spectrometer instrument with higher acquisition speed (e.g., LIT) and high-mass resolution can be an additional aid for simultaneous quantitation of targeted bitter compounds and identification of non-target bitter compounds. A 3D-ion-trap MS with unit–mass resolution can provide reasonable quantitative results by extracting selected ions from full-scan data. It is important to note that the sensitivity of such a quantitative approach is not comparable with true SRM and selectivity in addition is reduced in scope (Zhang *et al.*, 2009).

A recent evaluation between LC-LIT-MS and LC-QQQ-MS identified that QQQ based LC-SRM-MS/MS methods are better options for analysis of small molecules in relation to limit of detection, lower limit of quantification and precision (Dai & Herman, 2010).

8.4.2 Quantitation of bitter compounds by LC-IT-MS and LC-HRMS

The triple quadrupole LIT hybrid instrument is designed so the manufacturer uses the Q3 analyser simultaneously as quadruple and LIT. The LIT has a fast-duty cycle allowing full scan on product ions. The mass spectrometer gives the same proficiencies of neutral loss (NL), scanning, PI scanning and SRM acquisition compared with triple quadrupole for unknown metabolite screening and known metabolite quantitation. The QTRAP can, in addition, trigger a survey scan to trigger information-dependent acquisition (IDA) of enhanced product ion (EPI) spectra. MRM to EPI spectra [MRM] is used by for SRM and provides better selectivity and more sensitivity than NL-EPI and PI-EPI (Wen *et al.*, 2008; Zheng *et al.*, 2007). The MRM to EPI can be set up to monitor up to 100 SRM transitions and also maintain the qualitative performance of SRM methods without significant loss of sensitivity. The advantage of utilising MS/MS spectra generated from MRM-EPI is in the identification of false-positive peaks displayed in the SRM ion chromatograms. There the QTRAP can be a suitable alternative to triple quadrupole SRM methodology as allows simultaneous quantitation of bitter compounds and verification of their identities by MS/MS. Yan *et al.* (2014) performed a generic MRM based strategy for flavonoids profiling in plants using a hybrid triple quadrupole linear ion trap (QTRAP) mass spectrometer. The strategy adopted the four following steps: (1) preliminary profiling of major aglycones by MRM triggering an EPI scan, (MRM-EPI) (2) glycones were profiled by precursor ion-triggered EPI scan (PI-EPI) of major aglycones, (3) aglycones profiling by combining MRM-EPI and neutral loss triggered EPI (NL-EPI) scan (NL-EPI) of major glycone and (4) in-depth flavonoids profiling by MRM-EPI with elaborated MRM transitions. As the group incorporated MRM, PI and NL scanning as a possible strategy, the approach not only makes full use of the sensitivity of MRM but also surmounts the obstacles of limited compound coverage and low throughput. The mass spectrometer used in the study was a QTRAP 4000. Recently, a newer model was introduced (QTRAP 6500) which gives a 30- to 40-fold better sensitivity with a much faster scanning speed thus allowing PI-EPI/NL-EPI scans or 500-800 MRM transitions in one injection. Other authors used QLIT technology to monitor anthocyanins in red cabbage (Arapitsas *et al.*, 2008).

HRLC-MS encompasses FT-MS (eg FT-ICR and Orbitrap), TOF, QTOF provide MS detection and provide solutions to the limitations in SRM analysis. HRLC-MS especially FT-MS in full-scan mode can identify virtually a limitless number of bitter compounds in a biological sample due to the high resolving power, high mass accuracy and the broad dynamic range. Faraq *et al.* (2012) studied 13 hop cultivars in Humulus lupulus analysed by nulear magnetic resonance (NMR), LC-MS and fourier transform ion cyclotron resonance (FTICR)-MS in parallel. Under experimental conditions the group identified 46 metabolites including 18 bitter acids, 12 flavonoids, 3 terpenes, 3 fatty acids and 2 sugars. Hop bitter acids studied were alicyclic phenolic acids categorized as α acids (humulone, cohumulone and adhumulone) and β acids (lupulone, colupulone and adlupulone). One difficulty of FTMS was the lack of quantification and the necessity of acquiring spectra in both positive and negative mode in order to obtain comprehensive fingerprints of samples. In addition, the technique could not distinguish between constitutional isomers such as humulone and adhumulone. Regueiro *et al.* (2014) analysed tannins, flavonoids and phenolic acids in walnuts using LC coupled to LIT-orbitrap mass spectrometer and 120 compounds were identified on the basis of their retention times, accurate mass measurements and subsequent mass fragmentation data or by comparing with reference substances within the literature. The group reported the presence of eight polyphenols that have never been reported in walnuts such as malabathrin A, stenophyllanin C, eucalbanin A, cornusiin B, heterophyllin E, pterocarinin B, reginin A and alienanin B. Diaz *et al.* (2013) developed a UHPLC –LTQ-Orbitrap method for determination of non-anthocyanin flavonoids quantification in Euterpe oleracea juice.

Current HRLCMS allow rapid scan rates allowing the acquisition of an appropriate number of data points across a chromatographic peak and utilizes EICs for accurate quantitation by centering a narrow mass window on the theoretical m/z value of the analyte. This strategy for quantitation avoids pre-selection of SRM transitions for target compounds and offers identification of non-targeted compounds at the same time. The hybrid configuration in HRMS (e.g., LIT-FT-ICR, LTQ-Orbitrap of QTOF) gives information-dependent MS/MS acquisition on full-scan product-ion spectra to assistance in the confirmation of compound characteristics.

FT-MS orbitrap-based mass spectrometers and FT-ICR offer high-mass resolution and mass accuracy (e.g., above 1,000000 FWHM at m/z 400 and sub-ppm for FT-ICR, 100,000 FWHM at m/z 400 and 1-2 ppm for Orbitrap (Cortes-Franciso *et al.*, 2011). The advantage of HRLC-MS is the high resolving power facilitates bitter compound identification as accurate mass measurements help to determine the elemental formula and high mass resolution generate an exact isotopic pattern. The advantages of FT-MS are useful to eliminate some 'putative identification' with similar mono-isotopic mass but different isotopic distributions. Such instruments also allow the quantitation of bitter compounds using EICs by centering the narrow mass window on the theoretical m/z value of target analyte and excluding overlapping isobaric signals, whereas the mass accuracy is maintained throughout the acquisition (Junot *et al.*, 2010; Kamleh *et al.*, 2008; Koulman *et al.*, 2009; MacIntyre *et al.*, 2011; Abello *et al.*, 2008).

In HRLC-MS untargeted quantitative approaches to the analysis of bitter compounds a wide amount of efforts have been made in order to improve sample preparation techniques, improve LC separation prior to LC-MS detection and improve sensitivity which expands the array of bitterness compounds that can be analysed and further improves quantitation.

LC separation developments similar to SRM based experiments have been made in LC-HRMS quantitative bitter compound analysis (using a number of different column chemistries in a number of LC platforms in order to achieve monitoring of a broad range of compounds with wide ranging polarities).

An analytical platform such as FT-ICR presents extremely high mass accuracy and high resolving power. However, it is not widely used in monitoring of bitter compounds because its costs are high, it is difficult to maintain and it is complicated to couple with LC compared to Orbitrap and TOF mass spectrometers (e.g., 15000 FWHM at m/z 400 and 5 -10 ppm for TOF) (Junot *et al.*, 2010).

In recent years, Orbitrap-focused HRLC-MS has become more prevalent in order to perform a more integrated qualitative and quantitative analysis in the full-scan mode. In the early days the Orbitrap suffered from reduced acquisition speeds in MS/MS scans, limited dynamic range and limited sensitivity. In recent times modern Orbitrap instruments such as the Benchtop Exactive Orbitrap have shown more feasible advantages in relation to cost, sensitivity and mass accuracy and linear dynamic range (Koulman *et al.*, 2009; Bateman *et al.*, 2009; Clasquin *et al.*, 2010). Lopez-Gutierrez *et al.*. (2014) reported an UHPLC single-stage orbitrap high-resolution mass spectrometer method for the determination of isoflavones, but in addition, phytochemicals such as favones, flavonols, flavanones and phenolic acids were detected and quantified. Work was undertaken by Quifer-Rada *et al.* (2015) to characterize beer polyphenols by ESI hybrid linear ion trap quadrupole orbitrap mass spectrometry with confirmation by MS_2 (Quifer-Rada *et al.*, 2015).

In the literature there are limited papers that report bitter compounds using TOF- or QTOF-based quantitative MS. In recent times with the advancement of TOF technology this could expand exponentially and be further alternative for the analysis of bitter compounds. The limited dynamic range in typical TOF instruments relying on TDC detectors has been improved by analog-to-digital converter technology. Tremendous improvements have been made in mass resolution and mass accuracy of TOF. At present, 40,000 in resolution FWHM m/z 922 and accuracy of < 1 ppm in TOF are available (Stroh *et al.*, 2007). Haseleu *et al.* (2009) screened for bitter compounds in wort boiling using LC-TOF from hop-derived precursors. The group identified that the bitterness of ethanolic hop extract containing the β-acids was enhanced upon boiling. In addition, the study identified previously unreported bitter-tasting colupulone degradation products (Haseleu *et al.*, 2009).

State of the art QTOF instruments (e.g., Agilent 6540 Ultra High Definition Accurate Mass QTOF) allows both accurate mass measurements for compound confirmation and molecular formula generation, but in addition, provide accurate isotope ratios. These options allow users to reduce down the list of possible molecular formulas and increase confidence in the results allowing these instruments to compete with the Orbitrap technology. Bondia-Pons *et al.* (2014) presented a non-targeted LC-QTOF-MS metabolite profiling showing the diversity of flavonoid and phenolic derivatives in Goji berries. The group identified that Mongolian Goji berries were particularly rich in flavonoids and phenolic acids. Identified flavonoids were quercetin, isorhamnetin derivatives rutin and narcissin respectively (Bondia-Pons *et al.*, 2014). Fabani *et al.* (2013) studied the chemical profile, mineral content and antioxidant activities of three cultivars of Pistachio (Pistacia vera cv Kerman). The total phenolic content flavonoids and anthocyanins were measured. LC-ESI-QTOF-MS was utilised in the study. Gallic acid and (+)- catechin were present in higher amounts, however, in addition, the study reported the presence of myricetin, isoquercitrin and a dimer of prostaglandin for the first time in pistachio. Confirmation of the identity of these compounds and ions monitored were shown in the paper by Fabani *et al.* (2013).

A review by Krauss *et al.* (2010) of each type of major mass analyser discusses overall performance in quantitative analysis of small molecules with detailed evaluations on dynamic range, sensitivity, resolving power and mass accuracy. The specifications in the paper apply to the majority of instruments, however, furthermore, newer mass spectrometers can achieve better performance of dynamic range and sensitivity. It has been reported that the Applied Biosystems Sciex Triple TOF 5600 has an equivalent dynamic range and limit of quantitation to triple quadrupole instruments. In addition, the sensitivity of each mass analyzer depends on the ionization capabilities of the compound in the ion source. In addition, mass spectrometers can provide higher resolution and mass accuracy depending on the m/z range and scan speed of the specific instrument.

Additional comparison studies (Lu *et al.*, 2008; Kreutzmann *et al.*, 2007; Drewnowski & Gomez-Carneros, 2000) are noted elsewhere. In general, the QQQ SRM-based approach to bitter compound detection and quantitation gives superior results in parameters such as linear dynamic range and sensitivity. However, there are HRLC-MS instruments available with high mass accuracy that gives comparable results. In addition, HRLC-MS gives favorable results in untargeted quantitative studies.

8.5 CHALLENGES AND FUTURE PERSPECTIVES

Application of the same analytical method to potential bitter compounds in different food matrices is a formidable task. Indeed, even within a particular food group such as vegetables, the peel or the main part of the vegetable can be an analytical challenge. The reason for this is that substances can be distributed in different parts of plant leaves which may require different extraction methods compared to the plant root. In the identification of bitter compounds in carrots, Kreutzman *et al.* (2007)showed that falcarindiol and di-caffeic acid derivative were highly related to bitterness in comparison to falcarinol. The falcarindiol and di-caffeic acid derivative were mainly present in the peel, whereas the falcarinol was more evenly distributed in the root.

Potentially, different food matrices will require significant modification of the sample preparation procedures for similar compounds which is time consuming. This is primarily due to the presence of varying matrix components and the wide variety of high and low molecular weight compounds with a wide range of polarities. This poses a challenge in method development and quantification. Another feature for consideration is that sporadic bitterness can occur when a foodstuff such as a vegetable is exposed to stress during growing, harvesting, transportation, storage and processing. At these various stages the ability to develop and apply rapid monitoring methods to quantify bitter compounds is highly advantageous.

There are a number of issues that need to be considered in order to utilize LC-MS in bitter compound analysis. The quality of the MS spectrum is of vital importance for confirmation but can be affected by the analytical platform used and the collision energy. It is vital to spend necessary time developing experimental conditions for generation of good-quality spectra.

A list of compounds and the analytical methods employed for their determination is provided in Table 8.1.

8.6 OPTIMISATION OF MASS SPECTRA PARAMETERS

This is a vital stage for recording useful and desired spectra. Some modern LC-MS instruments have the capability of collision energy ramping, as well as auto polarity switching which can be investigated to obtain maximum information in a single LC-MS injection. A challenge during MS runs is utilising atmospheric pressure ionization sources as adduct peaks can be encountered of molecular ions with sodium (+22 Da), potassium (+38 Da) and/or ammonium (+17 Da) in positive ion mode or in negative ion mode can obtain chloride (+36 Da), acetate (+60 Da) and formate (+46 Da). Buffer solutions like triethylamine (+101 Da) or water (+18 Da) can also be identified in spectrum. Adducts can be a challenge if present but additionally can be exploited to identify molecular ion peak. For instance a study by Singh *et al.* (2012), the total ion count (TIC) of two peaks were present in spectrum of a compound that had a acetate moiety in the structure). Their mass difference was equivalent to 60 Da, indicating that either formation of acetate adducts or fragment was yielded upon neutral loss of acetic acid. Subsequently the same sample was analyzed in formate buffer. A higher mass ion appeared with a difference of 46 Da due to formic acid adduct. This identified that the appearance of higher mass ion was the molecular ion peak. It can be challenging but very important to evaluate spectra for adducts and optomise instrument parameters in spectra acquisition in bitter compound detection.

8.7 RECORDING OF MS^n PROFILE

In MS^n experiments the fragments generated are captured one by one, held in a trap and fragmented further yielding information on the presence or absence

Table 8.1 The list of compounds and the analytical methods employed for their determination in various foods.

Compounds	Method	Source	Reference
Bilobalide, ginkgolides A, B, C, quercetin, kaempferol, isorhamnetin, rutin hydrate, quercetin-3-β-D-glucoside and quercetin hydrate	LC-MS; ESI–	Ginkgo biloba	Ding *et al.* (2006)
Theobromine and caffeine	LC- MS/MS ESI+	Saliva, plasma and urine	Ptolemy *et al.* (2010)
Cohumulone, humulone, adhumulone, colupulone, lupulone and adlupulone	LC-MS/MS ESI–	Hop	Hofte *et al.* (1998)
major α and β- bitter acids six minor bitter acids, adprelupulone*	LC-MS/MS ESI	Hop	Zhang *et al.* (2004)
(-)-Epicatechin, (+)-catechin, theobromine and caffeine	UPLC-MS/MS ESI +/–	Cocoa	Ortega *et al.* (2010)
Gallic acid, 3,5-dihydroxybenzoic acid, protocatechuic acid, chlorogenic acid, gentistic acid, 4-hydroxybenzoic acid, caffeic acid, vanillic acid, syringic acid, 3-hydroxybenzoic acid, 4-coumaric acid, sinapic acid, ferulic acid, 3-cournaric acid, 2-coumaric acid, salicylic acid and trans-cinnamic acid	UPLC-MS/MS ESI –	White wine, grapefruit juice and green tea infusion.	Gruz *et al.* (2008)
cyanidin,glucoside, triglucoside, sophoroside, caffeoyl,feruloyl, coumaroyl,sinapoyl, glucopyranosyl, benzenoyl, oxaloyl succinoyl.	QTRAP Linear Ion Trap MS ESI +	Red cabbage	Arapitsas *et al.* (2008)

Table 8.1 (*Continued*)

Compounds	**Method**	**Source**	**Reference**
Adlupulone, adhumulone, humulone, lupulone, cohumulone, α-humulene, xantholhumol, β-glucose, myracene*	LC-MS/MS, NMR, FTICR-MS ESI +/−	13 hop cultivars	Faraq *et al.* (2012)
Tannins, flavonoids and phenolic acids* and the following polyphenols that have never been reported in walnuts: malabathrin A, stenophyllanin C, eucalbanin A, cornusin B, heterophyllin E, pterocarinin B, reginin A and alienanin B.	Linear ion trap-orbitrap MS ESI −	Walnuts	Regueiro *et al.* (2014)
(+)-Catechin, rutin, Isovitexin, quercetin, chrysoeriol, luteolin, eriodictyol, homoorientin, orietin, kaempferol-3-rutinoside, quercetin-3-glucoside and (+)-dihydrokaempferol	UHPLC –LTQ-Orbitrap ESI −	Euterpe oleracea juice	Diaz *et al.* (2013)
Naringenin, isokuranetin, biochanin, hesperidin, gentisin, syringic acid, sinapic acid, eriodictyol, daidzin, apigenin-7-o-glucoside, ferulic acid, sakuranetin, vitexin, glycitin, luteolin-7-o-glucoside, luteolin-4-glucoside*	UHPLC orbitrap MS ESI +/−	Soy based nutraceutical products (tablets and capsules)	Lopez-Gutierrez *et al.* (2014)
4-hydroxybenzoic,caffeic acid,catechin, epicatechin,chlorogenic, ferulic acid, kaempferol-*O*glucoside, *p*-coumaric acid, protocatechuic acid, quercetin-3-*O*-glucoside, sinapic and vanillic acids	LC-orbitrap MS ESI −	Beer	Quifer-Rada *et al.* (2015)

(*continued*)

Table 8.1 (*Continued*)

Compounds	Method	Source	Reference
Colupulone, lupulone, xanthohumol, cohulupone, humulone, adlupulone, adhumulone, nortricyclocolupone*	LC-TOF-MS ESI –	Wort	Haseleu *et al.* (2009)
Dicaffeoylquinic acid, kaempferol glucoside, isohamnetin, myristic acid, chlorogenic acid, citric acid, quercetin*	LC-QTOF-MS ESI –	Goji berries	Bondia-Pons *et al.* (2014)
Gallic acid, (+)- catechin, myricetin, isoquercitrin, eriodictyol, cyaniding-o-galactoside*	LC-ESQTOF-MS ESI +/–	Pistachio	Fabani *et al.* (2013)
Quinic acid, 3-caffeoylquinic acid, caffeic acid, 4-caffeoylquinic acid, ferulic acid, p-coumaric acid, 6-methoxymellin, falcarinol*	LC-MS ESI – APCI +/-	Carrots	Kreutzman *et al.* (2007)

- More compounds mentioned in publication but not mentioned in the table; see cited reference for further information.
- ESI= Electrospray ionisation
- APCI = Atmospheric Pressure Chemical Ionisation

of connectivity of each observed fragment. It should be understood that online tandem mass studies beyond MS2 may not always be possible especially if concentration is low or fragments of interest have low relative abundance. In practical terms this would require recording more data within the same timeframe as the bitter compound is not in the source for sustained period of time thus less number of data points will be obtained leading to loss in sensitivity. This can be possibly resolved through multiple injections or data dependent analysis where specific transition(s) can be targeted using SRM or MRM mode. It is sensible to initially propose fragmentation pathways of an unknown bitter compound based on HR-MS data collected under collision energy ramping (where available).

In addition, the automated generation of spectra can be achieved which can improve method development and sample throughput for bitter compounds. The heterogeneity of spectral data provides a significant problem for the active usage of spectral libraries. In certain studies promising results showed a degree of reproducibility of MS/MS spectra using different instruments from different laboratories (Oberacher *et al.*, 2009). There is a need to carefully design experiments so appropriate spectral algorithms can be achieved which will aid compound identification times.

8.8 CHALLENGES IN THE COLLECTION OF HRMS DATA

Calculation of the molecular formula for precursors, fragments and losses can be difficult in bitter compound analysis. Some HRMS instruments are available with errors as low as 1 ppm. This is maintained throughout the analysis by use of calibrants and external control of temperature. In some instruments a calibrant can be added. The benefit is that calibrants with different masses can be added at the place where the analyte of interest is resolved. Some HRMS instruments utilise a lock mass where a single calibrant is continuously introduced throughout the time the mass spectrum is being acquired. This approach has been stated to give much higher accuracy without strict temperature control. The challenge with using this approach is that sometimes the calibrant can occasionally suppress ionization, or interfere with bitter compound ions of close exact mass. In addition, the abundance of the calibrant may not always match with that of the analyte and if the mass is significantly different from the analyte the same accuracy may not be extrapolated throughout the range. An interesting discussion on practices using HRMS equipment is previously discussed (Bristow, 2006; Webb *et al.*, 2004). Another consideration is upon acquiring HRLCMS data, molecular formula are generated for the parent, fragments and neutral losses. During this calculation it is important to be aware of the charge and that the correct value is fed into elemental composition calculator (see related query in references). Singh *et al.* (2012) stated that feeding 0 against 1 as the value of charge leads to a difference of 0.00055 Da (055 mmu) in theoretical mass value equivalent to the mass of one electron. This could possibly lead to the rejection of possible elemental composition due to falling outside of the predefined tolerance of error, defined as the difference between experimental accurate and theoretical exact mass value. Evaluating isotopic abundances of peaks of interest and mass fragments can also be challenging. Certain instrumental parameters affect the isotopic abundance ratio. Structures containing chlorine and sulphur, combined isotopic abundance with contribution due to sulfur could be minimal compared to chlorine and instrument parameters could have overwhelming influence and results might be misleading unless other HRMS data is considered. In order to address this compare the theoretical and observed values against a reference compound.

Other considerations such as incomplete collection of qualitative information such as using a single tool and not a combination can be a challenge in already acquired data. An example of this would be using a TOF instrument but collecting only molecular ion data but not fragmentation data (Dongre *et al.*, 2009).

The identification of unknown compounds in HRLCMS can take a significant amount of resources. The most reliable way to identify an unknown compound unmistakably with confidence is to make a comparison between retention time, its fragmentation spectrum and its mass with those of authentic standards. It has been outlined in the literature that obtaining two independent measurements analysed under identical experimental conditions are necessary to identify the compound (Sumner *et al.*, 2007).

It is vitally important with LC-MS data of bitter compounds to evaluate results against the published literature with respect to completeness and correctness of mass fragmentation patterns to ensure that incorrect structure is not assigned. It is not uncommon in other fields to have indicated the elucidation of a test compound merely based on the mass difference from parent ion without looking at the characteristic isotopic abundance which indicated the presence of a chlorine atom. An example of this is in the case of clopidogrel as its identification was elucidated merely on mass difference from ion but the characteristic isotopic abundance was overlooked which indicated a chlorine atom. Later another group identified this mistake and suggested the correct structure (Danikiewicz & Swist, 2007). In correct interpretation of raw data can also occur. It is vitally important that in-depth literature search and at times there can be necessity of advanced 2D-NMR experiments is carried out to minimize the risk of wrongly establishing structure.

A number of computational tools are available which help to prioritize the number of possible compounds possibly present in a sample thus improving sample handling time. The collation of knowledge of bitter compounds can be significant challenges for developing computational tools due to the fact that information can be available from numerous different sources and the spectra for example can be collected under different experimental conditions.

In silico fragmentation patterns aids the deduction of the compound during spectra interpretation. If the MS/MS spectra of different bitter compounds are available in different databases with spectral libraries then MS/MS spectra obtained from experimental samples are compared with library spectra to identify the bitter compound. This strategy does not give the same unambiguous confirmation as achieved with authentic standards, however, it employs important guidance for the determination of bitter compounds because it helps to reduce the number of possible substances that require investigation.

Generally, there is a need to develop better in-silico fragmentation models also in order to identify complex ion molecular interactions encountered in compound fragmentation. With the improvement of the specificity of these models they will assist with the identification of unknown bitter compounds when no spectral library coverage is available.

In general, the identification of a compound using a mass only gives limited information about the elemental composition of bitter compound and possible structure of the metabolite which can rarely give definite identification of the bitter compound. Adducts, isotopes and fragments of the same bitter compound should be initially identified. In addition, evaluation of the isotopic pattern of the MS spectrum a more confident interpretation of the elemental composition but still with limited knowledge about its structures particularly when there are isomers.

It is important to note that bitter compound metabolites would not exist alone but within certain context such as metabolic pathways and the integration of such contextual information would be extremely beneficial to reduce ambiguity.

8.9 CONCLUSIONS

LC-MS has secured a major role in bitter compound analysis due to its high selectivity, sensitivity, precision and accuracy. The choice of ionization mode has a significant effect on the results that are obtained. Some bitter compounds are ionized more efficiently in one ionization mode or polarity while some compounds show higher efficiency in a different mode. Electrospray ionization is the preferred ionization mode and generally spectra are obtained in positive and negative mode. The selection of the ionization mode will have a significant effect. To date there are studies using a combination of UPLC with different mass spectrometry analysers. LC-MS is an extremely versatile technology and can be used with several different modes or MS analyses in bitter compound research. The standardization of data obtained is a difficulty as a result and so bitter compound researchers should ensure that experiments are not designed where they are carried out on different instruments and stay with the same instrumental configuration. Bitter compound study experimenters should pay careful attention prior to performing experiments to include the number of samples to ensure statistical significance with the methods of sample handling and the selection of technology. It is good practice to perform a pilot study to identify possible sources of error particularly in the case of plant tissue samples or food matrices under study. Such difficulties as formation of emulsions during extraction can reduce precision of results. It is also important in LC-MS method development to be wary of solvents, vials, filters or tubing as these may introduce contaminants such as surfactants causing major interference in MS detection. Sample stability is another important consideration. In the development of non-targeted LC-MS methods for bitter compounds the length of the analytical run is also an important consideration. Factors such as the stability of the column, the potential contamination of the ion source and the stability of the samples in the autosampler are vital. It is essential that the retention time repeatability is within acceptable ranges. In the case of scheduled MRM experiments of bitter compound, it is essential that MS/MS transitions are not monitored in the wrong timeframe or information on test bitter compounds can be lost. It is also important to consider putting protocols in place that can identify possible instrument failures in bitter compound analysis using LC-MS. A widely used practice is quality control (QC) samples. These samples are analysed a number of times throughout the batch and assessed against certain criteria to check validity of the measurements (e.g., repeatability of bitter compound peak areas and retention times). It is required that a few injections of the QC sample at the beginning of the analytical run is performed to ensure correct "conditioning" or that the equilibration phase is satisfactory for bitter compound analysis prior to running the LC-MS system.

LC-MS-based bitter compound analysis will progress rapidly in the next years and LC-MS will become the key analytical methodology to aid this. There is great potential in the food chemistry field due to the ever increasing capabilities and technological advancements in chromatography and mass spectrometry. It is important to note that LC-MS platforms generate a huge amount of data and as the informatics field advances this field will be equally highly important. Maximizing the efficiency of advanced tools in informatics made available by LC-MS vendors for data mining will be vital for researchers of bitter compounds. There will also be a need for accepted reporting schemes and harmonization of method validation approaches in the measurement of bitter compounds by LC-MS. An important consideration in bitter compound analysis is the incorporation during validation, of studies to identify matrix effects especially when working in electrospray ionisation mode.

Associated health benefits of bitter compounds in food is highly important in today's society. During both food production and food processing it is beneficial to ascertain the benefits to the consumer. Methods to detect bitter compounds in high and low concentrations are essential. LC-MS is perfectly positioned within the current technological revolution as a powerful tool for the future investigation of established and novel bitter compounds.

REFERENCES

Abello, N., Geurink, PP., Toorn, M., Oosterhout, A.J.M., Lugtenburg, J., Marel, G.A., Kerstjens, H.A.M., Postma, D.S., Overkleeft, H.S., & Bischoff, R. (2008). Poly(ethylene glycol)-based stable isotope labeling reagents for the quantitative analysis of low molecular weight metabolites by LC-MS. *Anal. Chem.* **80**, 9171–9180.

Alder, L., Steinborn, A., & Bergelt, S. (2011). Suitability of an orbitrap mass spectrometer for the screening of pesticide residues in extracts of fruits and vegetables. *J. AOAC Int.* **94**, 1661–1673.

Aleksandrov, M.L L., Gall, L.N., Krasnov, V.N., Nikolaev, V.I., Pavlenko, V.A., & Shkurov, V.A. (1984). *Dokl. Akad. Nauk USSR* **277**, 379–383.

Apffel, A., Fisher, S., Goldberg, G., Goodley, P.G., & Kuhlmann, F.E. (1995). Enhanced sensitivity for peptide mapping with epectrospray liquid chromatography mass spectrometry in the presence of signal suppression due to trifluoroacetic acid containing mobile phases. *J. Chromatogr. A.* **712**, 177–190.

Arapitsas, P., Sjöberg, P.J.R., & Turner, C. (2008). Characterisation of anthocyanins in red cabbage using high resolution liquid chromatography coupled with photodiode array detection and electrospray ionisation-linear ion trap mass spectrometry. *Food Chem.* **109**(1), 219–226.

Asam, M.R. & Glish, G.L. (1997). Tandem mass spectrometry of alkali cationized polysaccharides in a quadrupole ion trap. *J. Am. Soc. Mass Spectrom.* **8**, 987–995.

Bateman, K.P., Kellmann, M., Muenster, H., Papp, R., & Taylor, L. (2009). Quantitative-qualitative data acquisition using a benchtop Orbitrap mass spectrometer. *J. AM. SOC. MASS. SPECTR.*, **20**, 1441–1450.

Baty, J.D. & Robinson, P.R. (1977). *Biol. Mass Spectrom.*, **4**(36), 36–41.

Bondia-Pons, I., Savolainen, O., Törrönen, R., & Bristow, A.W.T. (2006). Accurate mass measurement for the determination of ele-mental formula—a tutorial. *Mass Spectrom. Rev.*, **25**, 99–111.

Bruins A.P (1991). Mass-spectrometry with ion sources operating at atmospheric pressure. *Mass Spectrom. Rev.*, **10**, 53–77.

Bruins, A.P. (1998). Mechanistic aspects of electrospray ionization. *J. Chromatogr. A.*, **794**, 345–357.

Clasquin, W., Melamud, M.F., Amador Nuguez, D., Caudy, A.A., & Rabinowitz, J.D. (2010). Metabolomic Analysis via reverse phase ion pairing liquid chromatography coupled to a stand alone orbitrap mass spectrometer. *Anal. Chem.*, **82**, 3212–3221.

Cole, R.B. (2000). Some tenets pertaining to electrospray ionization mass spectrometry. *J. Mass. Spectrom.*, **35**, 763–772

Cole, R.B., Harrata, A., & Kamel, K. (1993). Solvent effect on analyte charge state, signal intensity and stability in negative ion electrospray mass spectrometry; implication for mechanism of negative ion formation. *J. Am. Soc. Mass Spectrom.*, **4**, 546–556.

Cortes-Franciso, N., Flores, C., Moyano, E., & Caixach, J. (2011). Accurate mass measurements and ultra high-resolution: evaluation of different mass spectrometers for daily routine analysis of small molecules in negative electrospray ionization mode. *Anal. Bioanal. Chem.*, **400**(10), 3595–3606.

Dai, S.Y. & Herman T.J. (2010). Evaluation of two liquid chromatography tandem mass spectrometry platforms in quantification of monesin in animal feeds and milk. *Rapid Commun Mass Spectrom.*, **24**, 1431–1438.

Dall'Asta, C., Sforza, S., Galaverna, G., Dossena, A., & Marchelli, R., (2004). Simultaneous determination of type A and type B trichothecenes in cereals by liquid chromatography electrospray ionization mass spectrometry using NaCl as cationisation agent. *J. Chromatogr. A* **1054**, 389–395.

Dams, T., Benijts, W., Günther, W., Lambert, A., & de Leenheer, A.(2002). Influence of the eluent composition on the ionization efficiency for morphine of pneumatically assisted electrospray, atmospheric pressure chemical ionsiation and sonic spray. *Rapid Commun. Mass Spectrom.*, **16**, 1072–1077.

Danikiewicz, W. & Swist M. (2007). The fragmentation reactions of anti-platelet drugs: clopidogrel and ticlopidine. *JMS, Letters, J. Mass Spectrom.*, **42**, 405–406.

Diaz, A.L., Rozet, E., Larondelle, Y., Hubert., P. Rogez, H., & Quetin-Leclercq, J. (2013). Development and validation of an UHPLC-LTQ-Orbitrap MS method for non-anthocyanin flavonoids quantification in Euterpe oleracea juice. *Anal. Bioanal. Chem.*, **28**, 9235–9249.

Ding, S., Dudley, E., Plummer, S., Tang, J., Newton, R.P., & Brenton, A.G. (2006). Quantitative determination of major active components In Ginkgo biloba dietary supplements by liquid chromatography/mass spectrometry. *Rapid Commun in Mass Spectrometry*, **20**, 2753–2760.

Dole, M., Hines, R.L., Mack, L.L., Mobley, R., Ferguson, L.D., & Alice, M.B. (1968). Molecular beams of macroions. *J. Chem. Phys.*, **49**, 2240–2249.

Dongre, V.G., Ghugare, P.D., Karmuse, P., Singh, D., Jadhav A. & Kumar A. (2009). Identification and characterisation of process related impurities in chloroquine and hydroxychloroquine by LC-ITMS, LC-TOFMS and NMR. *J. Pharm. Biomed. Anal.*, **49**, 873–879.

Dorta, E., González, M., Lobo, G., Sanchez-Moreno, C., & de Ancos, B. (2014). Screening of phenolic compounds in by-product extracts from mangoes (Mangifera indica L.) by HPLC-ESI-QTOF-MS and multivariate analysis for use as a food ingredient. *Food Res. Int.*, **57**, 51–60.

Drewnowski, A. & Gomez-Carneros, C. (2000). Bitter taste, phytonutrients, and the consumer: a review. *Am. J. Clin. Nutr.*, **72**, 1424–1435.

Eshraghi, J. & Chowdhury, K. (1993). Factors affecting electrospray ionization of effluents containing trifluoroacetic acid for high performance liquid chromatography mass spectrometry. *Anal. Chem.*, **65**, 3528–3533.

Fabani, M.P., Luna, L., Baroni, M.V., Monferran, M.V., Ighani, M., Tapia, A., Wunderlin, D.A., & Feresin, G.E. (2013). Pistachio (Pistacia vera var Kerman) from Argintinean cultivars. A natural product with potential to improve human health. *J. Funct. Foods*, **5**(3), 1347–1356.

Faraq M.A., Porzel, A., Schmidt J., & Wessjohann L. (2012). Metabolite profiling and fingerprinting of commercial cultivars of Humulus lupulus L (hop); a comparision of MS and NMR methods in metabolomics. *Metabolomics.*, **8**, 492–507.

Gao, S., Zhang, Z.P., & Karnes, H.T. (2005). Sensitivity enhancement in liquid chromatography atmospheric pressure ionization mass using derivatisation and mobile phase additives. *J. Chromatogr. B*, **825**, 98–110.

Gomez, A. & Tang, K. (1994). Charge and fission of droplets in electrospray. *Phys. Fluid.*, **6**, 404–417.

Gruz. J., Novák O., & Strnad M. (2008). Rapid analysis of phenolic acids in beverages by UPLC-MS/MS. *Food Chem.*, **111**(3), 789–794.

Guignard, C., Jouve, J., Bogeat-Triboulot, M.B., Dreyer, E., Hausman, J.F., & Hoffmann, L. (2005). Analysis of carbohydrates in plants by high performance anion chromatography coupled with electrospray mass spectrometry. *J. Chromatogr. A.*, **1085**, 137–142.

Guillarme, D., Ruta, J., Rudaz, S., & Veuthey J.L. (2010). New trends in fast and high-resolution liquid chromatography: a critical comparison of existing approaches. *Anal. Bioanal. Chem.*, **397**, 1069–1082.

Gustavsson, S.Å. Samskog, J., Markides, K.E., & Långström, B., (2001). Studies of signal suppression in liquid chromatography electrospray ionization mass spectrometry using volatile ion pairing reagents. *J. Chromatogr. A.*, **937**, 41–47.

Häkkinen, M.R., Keinänen, T.A., Vepsäläinen, J., Khomutov, A.R., Ahonen, L., Jänne, J., & Auriola, S. (2007). Analysis of underivatised polyamines by reverse phase liquid chromatography with electrospray tandem mass spectrometry. *J. Pharm. Biomed. Anal.*, **45**, 625–634.

Harvey, D.J. (2000). Collision induced fragmentation of underivatised N-linked carbohydrates ionized by electrospray. *J. Mass Spectrom.*, **10**, 1178–1190.

Haseleu, G., Intelmann, D., & Hofmann T. (2009). Structure determination and sensory evaluation of novel bitter compounds formed from hop (Humulus lupulus L) upon wort boiling. *Food Chem.*, **116**, 71–81.

Hemström, P. & Irgum, K. (2006). Hydrophilic interaction chromatography. *J. Sep. Sci.*, **29**, 1784–1821.

Hiraoka, K. & Kudaka, I. (1992). Negative-mode electrospray-mass spectrometry using non aqueous solvents. *Rapid Commun. Mass Spectrom.*, **6**, 265–268.

Hird, S.J. (2008). Analysis of pesticides by chromatographic techniques coupled with mass spectrometry. In: Tadeo, J.L., (Ed.), *Analysis of Pesticides in Food and Environmental Samples*. CRC Press, Boca Raton, pp. 59–94.

Hofte, A.J.M. & van der Hoeven, R.A.M. (1998). Characterization of hop acids by liquid chromatography with negative electrospray ionization mass spectrometry. *J. Am. Soc. Brew. Chem.*, **56**, 118–122.

Horning, E.C., Carroll, D.I., Dzidic, I., & Stillwell R.N. (1973). New picogram detection systems based on mass spectrometer with a external ionization source at atmospheric pressure. *Anal. Chem.*, **45**, 936–943.

Horning, E.C., Carroll, D.I., Dzidic, I., Haegele, M.G., Horning, M.G., & Stillwell, R.N. (1974). Atmospheric pressure ioinisation (API) mass spectrometry solvent-mediated ionization of samples introduced in solution and in a liquid chromatography effluent stream. *J. Chromatogr. Sci.*, **12**, 725–729.

Iribarne, J.V. & Thomson, B.A. (1976). On the evaporation of small ions from charged droplets. *J. Chem. Phys.*, **64**, 2284–2294.

Junot, C., Madalinski, G., Tabet, J.C. & Ezan, E. (2010). Fourier Transform Mass Spectrometry for Metabolome Analysis. *Analyst*, **135**(9), 2203–2219.

Kaklamanos, G., Theodoridis, G., & Dabalis, T. (2009). Determination of anabolic steroids in muscle tissue by liquid chromatography-tandem mass spectrometry. *J. Chromatogr. A.*, **1216**, 8072–8079.

Kamleh, A., Barett, M.P., Wildridge, D., Burchmore, R.J.S., Scheltema, R.A. & Watson, D.G. (2008). Metabolomic profiling using orbitrap fourier transform mass spectrometry with hydrophilic interaction chromatography: a method with wide applicability to analysis of biomolecules. *Rapid Commun Mass Spectrom.*, **22**, 1912–1918.

Kebarle, P. (2000). A brief overview of the present status of the mechanisms involved in electrospray mass spectrometry. *J. Mass Spectrom.*, **35**, 804–817.

Kebarle, P. & Ho, Y. (1997). On the mechanism of electrospray mass spectrometry. In: R.B. Cole (Ed.), *Electrospray Ionization Mass Spectrometry*, Wiley, New York, p. 3.

Kebarle, P. & Tang, L. (1993). From ions in solution to ions in gas phase- the mechanism of electrospray mass spectrometry. *Anal. Chem.*, **65**, 972A–986A.

Kenny, O., Smyth, T.J., Hewage, C.M., & Brunton, N.P., (2013). Antioxidant properties and quantitative UPLC-MS analysis of phenolic compounds from extracts of fenugreek (Trigonella foenum-graecum) seeds and bitter melon (Momordica charantia) fruit. *Food Chem.*, **141**, 4295–4302.

Kostiainen, R. & Bruins, A.P. (1996). Effect of solvent on dynamic range and sensitivity in pneumatically assisted electrospray (ionspray) mass spectrometry. *Rapid Commun. Mass Spectrom.*, **10**, 1393–1393.

Kostianen, R. & Kauppila, T.J. (2009). Effect of eluent on the ionization process in liquid chromatography mass spectrometry. *J. Chrom A.*, **1216**, 685–699.

Koulman, A., Woffendin, G., Narayana, V.K., Welchman, H., Crone, C., & Volmer, D.A. (2009). High-resolution extracted ion chromatography is a new tool for metabolomics and lipidiomics using second generation orbitrap mass spectrometer. *Rapid Commun. Mass Spectrom.*, **23**, 1411–1418.

Krauss, M., Singer, H., Hollender, J. (2010). LC high resolution MS in environmental analysis: from targeted screening to the identification of unknowns. *Anal. Bioanal. Chem.*, **397**, 943–951.

Kreutzmann, S., Christensen, L.P., & Edelenbos, M. (2007). Investigation of bitterness in carrots (Daucus carota L.) based on quantitative chemical and sensory analyses. *LWT*, **41**, 193–205

Kuhlmann, F.E., Apffel, A., Fisher, S.M., Goldberg, G., & Goodley, P.C. (1995). Signal enhancement for gradient reverse phase high performance liquid chromatography electrospray ionization mass spectrometry analysis with trifluoroacetic acid and other strong acid modifiers by post columns addition of propionic acid and isopropanol. *J. Am. Mass Spectrom.*, **6**, 1221–1225.

Li, J., Shao, S., Jaworsky, M.S., & Kurtulik, P. (2008). Simultaneous determination of cations, zwitterions and neutral compounds using mixed-mode reversed-phase and cation-exchange high-performance liquid chromatography. *J. Chromatogr. A.*, **1185**, 185–193.

López-Gutiérrez, N., Romero-González, R., Garrido Frenich, A., Luis, J., & Vidal, M. (2014). Identification and quantification of the main isoflavones and other phytochemicals in soy based nutraceutical products by liquid chromatography-orbitrap high resolution mass spectrometry. *J. Chrom. A.*, **1348**, 125–136.

Lu, W., Bennett, B.D., Rabinowitz, J.D. (2008). Analytical strategies for LC-MS based targeted metabolomics. *J. Chromatogr. B.*, **871**, 236–242.

Ma, J., Krynitsky, A.J., Grundel E., & Radar, J.I., (2012). Quantitative determination of cucubitane-type triterpenes and triterpene glycosides in dietary supplements containing bitter melon (Momordica charantia) by HPLC-MS/MS. *J. AOAC. Int. Vol.*, **95**(6), 1597–1608.

MacIntyre, L., Zheng, L., Scullion, L., Keating, P. & Watson, D.G. (2011). Metabolomic profiling of biomarkers of liver X receptor-induced toxicity in mouse liver tissue. *Metabolomics*, **7**, 54–70.

Mann, M., Meng, C.K., & Fenn, J.B. (1989). Interpreting mass spectra of multiply charged ions. *Anal. Chem.*, **61**, 1702–1708.

Martinez, J., Poutanen, K., & Hanhineva, K. (2014). Metabolic profiling of Goji berry extracts for discrimination of geographical origin by non-targeted liquid chromatography coupled to quadrupole time-of-flight mass spectrometry. *Food Res. Int.*, **63**, 132–138.

Mateos, R., Cert, A., Perez-Camino, M. C., & Garcia, J. M. (2004). Evaluation of virgin olive oil bitterness by quantification of secoiridoid derivatives. *J. Am. Oil Chem. Soc.*, **81**, 71–75.

McCalley, D.V. (1996). Effect of organic solvent modifier and nature of the solute on. *J. Chromatogr. A.*, **738**, 169–179.

McCalley, D.V. (2004). Comparision of a polymeric column and a silica based reverse phase for the analysis of basic peptides by high performance liquid chromatography. *J. Chromatogr. A.*, **1038**, 77–84.

McCalley, D.V. (2010). Study of the selectivity, retention mechanisms and performance of alternative silica-based stationary phases for separation of ionised solutes in hydrophilic interaction chromatography. *J. Chromatogr. A.*, **1217**, 3408–3417.

Mezcua, M., Malato, O., Garcia-Reyes, J.F., Molina-Diaz, A., & Fernandez-Alba, A.R. (2009). Accurate-mass databases for comprehensive screening of pesticide residues in food by fast liquid chromatography time-of-flight mass spectrometry. *Anal. Chem.*, **81**, 913–929.

Meng, C.K., Mann, M., & Fenn, J.B. (1988). *The 36th ASMS Conference on Mass Spectrometry and Allied Topics*, June 5–10, San Francisco, CA, pp. 771.

Mirza, U.A. & Chait, B.T. (1994). Effects of anions on the positive ion electrospray. *Anal. Chem.*, **66**, 2898–2904.

Needham, S.R., Brown, P.R., Duff, K., & Bell, D. (2000). Optomised stationary phases for the high performance liquid electrospray ionization mass spectrometric analysis of basic pharmaceuticals. *J. Chromatogr. A.*, **869**, 159–170.

Niessen, W.M.A. (2006). *Liquid Chromatography–Mass Spectrometry*, third ed., CRC Press, Taylor and Francis, Boca Raton.

NIST. (n.d.). Available from: http://webbook.nist.gov/chemistry [Accessed 19 September 2015].

Nunez, O., Gallart-Ayala, H., Martins, C.P., & Lucci, P. (2012). New trends in fast liquid chromatography for food and environmental analysis. *J. Chromatogr. A.*, **1228**, 298–323.

Oberacher, H., Pavlic, M., Libiseller, K., Schubert, B., Sulyok, M., Schuhmacher, R., Csaszar, E., & Kofeler, H.C. (2009). On the interinstrument and inter-laboratory transferability of a tandem mass spectral reference library: 1 Results of an Austrian multicenter study. *J. Mass Spectrom.*, **44**, 485–493.

Petritis, K., Brussaux, S., Guenu, S., Elfakir, C., & Dreux, M. (2002). Ion pair reverse phase liquid chromaography electrospray mass spectrometry for the analysis of underivatised small peptides. *J. Chromatogr. A.*, **957**, 173–185.

Quifer-Rada, P., Vallverdú-Queralt, A., Martínez-Huélamo, M., Chiva-Blanch, G., Jáuregui, O., Estruch, R., & Lamuela-Raventós R. (2015). A comprehensive characterisation of beer polyphenols by high resolution mass spectrometry (LC-ESI-LTQ-Orbitrap-MS). *Food Chem.*, **169**, 336–343.

Rayleigh, L. (1982). The equilibrium of liquid conductibity masses charged with electricity. *Philosophical. Mag.*, **14**, 184–186.

Regueiro, J., Sánchez-González, C., Vallverdú-Queralt, A., Simal-Gándara, J., Lamuela-Raventós, R., & Izquierdo-Pulido., M. (2014). Comprehensive identification of walnut polyphenols by liquid chromatography coupled to linear ion trap–Orbitrap mass spectrometry. *Food Chem.*, **152**, 340–348.

Robb, D.B., Covey, T.R., & Bruins, A.P. (2000). Atmospheric Pressure Photoionisation: An ionization method for liquid chromatography mass spectrometry. *Anal. Chem.*, **72**, 3653–3659.

Sandra, K., Devreese, B., Van Beeumen, J., Stals, I., & Claeyssens, M. (2004). The Q-Trap mass spectrometer, a novel tool in the study of protein glycosylation. *J. Am. Soc. Mass Spectr.*, **15**, 413–423.

Santini, A., Ferracane, R., Somma, M.C., Aragon, A., & Ritieni A. (2009). Multitoxin extraction and detection of trichothecenes in cereals: an improved LC-MS/MS approach. *J. Sci. Food Agric.*, **89**, 1145–1153.

Schmelzeisen-Redeker, G., Bütfering, L., & Röllgen, F.W., (1989). Desolvation of ions and molecules in thermospray mass spectrometry. *Int. J. Mass Spectrom. Ion Proc.*, **90**, 139–150.

Scholz, K., Dekant, W., Völkel, W., & Pähler, A. (2005). Rapid detection and identification of N-acetyl-L-cysteine thioethers using constant neutral loss and theoretical multiple reaction monitoring combined with enhanced product-ion scans on a linear ion trap mass spectrometer. *J. Am. Soc. Mass Spectr.*, **16**, 1976–1984.

Singh, S., Handa, T., Narayanam, S.A., Junwal, M., & Shah, R.P. (2012). A critical review on the use of modern sophisticated hyphenated tools in the characterization of impurities and degradation products. *J. Pharm. Biomedical Anal.*, **69**, 148–173.

Steimer, S. & Sjöberg, P.J. (2011). Anthocyanin characterization utilizing liquid chromatography combined with advanced mass spectrometric detection. *J. Agric. Food Chem.*, **59**, 2988–2996.

Straub, R.F. & Voyksner, R.D. (1993). Negative ion formation in electrospray mass spectrometry. *J. Am. Soc. Mass Spectr.*, **4**, 578–587.

Stroh, J.G., Petucci, C.J., Brecker, S.J., Huang, N., Lau, J.M. (2007). Automated sub ppm mass accuracy on the ESI TOF for use with drug discovery compound libaries. *J. Am. Soc. Mass Spectrom.*, **18**, 1612–1616.

Sumner, L., Amberg, A., Barrett, D., Beale, M., Beger, R., Daykin, C., Fan, T., Fiehn, O., Goodacre, R., Griffin, J., Hankemeier, T., Hardy, N., Harnly, J., Higashi, R., Kopka, J., Lane, A., Lindon, J., Marriott, P., Nicholls, A., Reily, M., Thaden, J., & Viant, M. (2007). Proposed minimum reporting standards for chemical analysis Chemical Analysis Working Group (CAWG) Metabolomics Standards Initiative (MSI). *Metabolomics*, **3**, 211–221.

Syage, J.A., Evans, M.D., & Hanold, K.A. (2000). Photoionisation mass spectrometry. *Am. Lab.*, **32**, 24–29.

Taylor, G.I. (1964). Distintegration of water drops by electric fields. *Proc. R. Soc. Lond. A.*, **280**, 383–397.

Temesi, D. & Law, B. (1999). The effect of LC eluent composition on MS responses using electrospray ionization. *LC–GC Int.*, **12**, 175.

Thomson, B.A. & Iribarne, J.V. (1979). Field induced ion evaporation from liquid surfaces at atmospheric pressure. *J. Chem. Phys.*, **71**, 4451–4463.

Wagner, S., Scholz, K., Sieber, M., Kellert, M., & Voelkel, W. (2007). Tools in metabonomics: an integrated validation approach for LC-MS metabolic profiling of mercapturic acids in human urine. *Anal. Chem.*, **79**, 2918–2926.

Webb, K., Bristow, T., Sargent, M. & Stein, B. (2004). *Methodology for Accurate Mass Measurement of Small Molecules*. Best Practice Guide, LGC, Limited, Teddington (UK) and EPSRC National Mass Spectrometry Centre, Swansea (UK).

Wen, B., Ma, L., Nelson, S.D., & Zhu, M.S. (2008). High-throughput screening and characterization of reactive metabolites using polarity switching of hybrid triple quadrupole linear ion trap mass spectrometry. *Anal. Chem.*, **80**, 1788–1799.

West, C., Elfakir, C., & Lafosse, M. (2010). Porous graphitic carbon: a versatile stationary phase for liquid chromatography. *J. Chromatogr. A.*, **1217**, 3201–3216.

Wolf, S. Schmidt, M. Müller-Hannemann, S., & Neumann, S. (2010). In silico fragmentation for computer assisted identification of metabolite mass spectra. *BMC Bioinform.*, **11**, 148.

Xu, H. & Freitas, M.A. (2009). Data and text mining Automated diagnosis of LC-MS/MS performance. *Bioinformatic Notes.*, **25**(10), 1341–1343.

Xu T., Yang, M., Li, Y., Chen, X., Wang, Q., Deng, W., Pang, X., Yu, K., Jiang, B., Guan, S., & Guo, D. (2013). An integrated exact mass spectrometric strategy for comprehensive and rapid characterisation of phenolic compounds in licorice. *Rapid Commun. Mass Spectrom.*, **27**, 2297–2309.

Yamashita, M. & Fenn, J.B. (1984). Electrospray ion source – another variation on the free-jet theme. *J. Phys. Chem.*, **88**, 4451–4459.

Yan, Z., Lin, G., Ye, Y., Wang, Y., & Yan, R. (2014). A generic multiple reaction monitoring based approach for plant flavonoids profiling using a triple quadrupole linear ion trap mass spectrometry. *J. Am. Soc. Mass Spectr.*, **25**(6), 955–965.

Yao, M., Ma, L., Humphreys, W.G., & Zhu, M.S. (2008). Rapid screening and characterization of drug metabolites using a multiple ion monitoring– dependent MS/MS acquisition method on a hybrid triple quadrupole- linear ion trap mass spectrometer. *J. Mass Spectrom.*, **43**, 1364–1375.

Yu, T., Pak, Y., Li, S., & Jones, D.P. (2013). Hybrid feature detection and information accumulation using high resolution LC-MS metabolomics data. *J. Proteome Res.*, **12**, 1419–1427.

Zhang, W., Chang, J., Lei, Z., Huhman, D., Sumner, L.W., & Zhao, P.X. (2014). MET-COFEA: A liquid chromatography/mass spectrometry data processing platform for metabolite compound feature extraction and annotation. *Anal. Chem.*, **86**, 6245–6253.

Zhang, X., Liang, X., Xiao, H., & Xu, Q. (2004). Direct characterization of bitter acids in a crude hop extract by liquid chromatography-atmospheric pressure chemical ionization mass spectrometry. *J. Am. Soc. Mass Spectr.*, **15**, 180–187.

Zhang, N. R., Yu, S., Tiller, P., Yeh, S., Mahan, E., & Emary, W.B. (2009). Quantitation of small molecules using high resolution accurate mass spectrometers-a different approach for analysis of biological samples. *Rapid Commun. Mass Spectrom.*, **23**, 1085–1094.

Zhao Y., Liu, G., Liu, Y., Yuan, L., Hawthorne, D., Shen, J.X., Guha, M., & Aubry, A. (2013). Improved ruggedness of an ion-pairing liquid chromatography/tandem mass spectrometry assay for the quantitative analysis of the triphosphate metabolite of a nucleoside reverse transcriptase inhibitor in peripheral blood mononuclear cells. *Rapid Commun. Mass Spectrom.*, **27**, 481–488.

Zheng, J., Ma, L., Xin, B., Olah , T., Humphreys, W.G. & Zhu, M. (2007). Screening and identification of GSH trapped reactive metabolites using hybrid triple quadrupole linear ion trap mass spectrometry. *Chem. Res. Toxicology*, **20**, 757–766.

Zhou, S. & Cook, K.D. (2000). Fluorescence of a solvatochromatic dye. *Anal. Chem.*, **72**, 963–969.

Zhou, Z., Ogden, S., & Leary, J.A. (1990). Linkage position determination in Oligiosaccharides: MS/MS study of lithium cationized carbohydrates. *J. Org. Chem.*, **55**, 5444–5446.

Zhu, J.H. & Cole, R.B. (2000). Formations and decompositions of chloride adduct ions [M+CL]$^-$ in negative ion electrospray ionisation mass spectrometry. *J. Am. Soc. Mass Spect.*, **11**, 932–941.

9 Evaluation of Bitterness by the Electronic Tongue: Correlation between Sensory Tests and Instrumental Methods

Michel Aliani, Ala'a Eideh, Fatemeh Ramezani Kapourchali, Rehab Alharbi, and Ronak Fahmi

9.1 INTRODUCTION

A range of analytical tools are available for studying the nature and compounds imparting bitterness in foods including sensory evaluation (see Chapter 7) and instrumental methods such as chromatography and mass spectrometry based techniques (see Chapter 8). While sensory studies are important for correlating the sensory characteristics of compounds identified by instrumental analyses they are costly and time-consuming and impractical for monitoring changes, such as bitterness, during the production of foods and beverages. This chapter will discuss the importance of the electronic tongue as an invaluable rapid and reliable tool for assessing the bitterness of foods and beverages.

9.2 THE ELECTRONIC TONGUE

Human physiology, as it relates to taste perception, has inspired scientists to create an effective tool "the so called electronic tongue" that can measure and compare tastes. Electronic tongues are simply devices designed to analyze taste in a similar way the human tongue analyzes taste. The principal role of the electronic tongue is to analyze different compounds dissolved in a solution using an array of low selective, nonspecific chemical sensors paired with a chemometric tool for data analysis (Di Natalca *et al.*, 2000). It can be also defined as a multisensor device devoted to analyze complex composition samples by recognizing their characteristic properties (Ciosek & Wroblewski, 2007). The electronic tongue is composed of various selective arrays of chemical sensors serving as tools to obtain the signals and transferring them to computer software with an advanced mathematical system for processing the data and generating results (Ghasemi-Varnamkhasti, Mohtasebi & Siadat, 2010). The following is a detailed description of its two essential components:

Bitterness: Perception, Chemistry and Food Processing, First Edition. Edited by Michel Aliani & Michael N. A. Eskin.

9.2.1 Sensor arrays

Sensor arrays are selected based on the chemical nature of the samples of choice to be examined. There are different types of sensor arrays such as the optical sensors and biochemical sensors, which include potentiometric, voltammetric and impedimetric sensors. Each of these sensors has different characteristics, advantages and disadvantages as follows.

a. *Potentiometric sensors:* In 1993, the first electronic tongue system (which was a typical model of a potentiometric sensor) was designed in Japan by Kiyoshi Toko and his co-workers. Toko, a professor of information science and electrical engineering at Kyushu University, created a taste sensor based on artificial lipid membranes and called it "taste sensor" (Rudnitskaya, Rocha, Legin, Pereira & Marques, 2010). The main principal of this method is using lipids as the main material to transform chemical information into electrical signals (Kobayashi *et al.*, 2010; Tahara & Toko, 2013). This type of the electronic tongue system is the most widely used, particularly the ion selective electrodes (ISEs) (Escuder-Gilabert & Peris, 2010). The old version of this system was able to detect the five main tastes: salty, sweet, acid, bitter and umami. The improved version is used in food quality control, beverage classifications and in environmental analysis (Codinachs *et al.*, 2008). The main advantages of potentiometric the electronic tongue are their well-knowing operation protocols, low cost and easy to set up fabrications, also the sensors in this system are selective to various species. However, the adsorption of solution components and the temperature dependence in this system affect membrane potential, which are considered the main disadvantages. Controlling the temperature and washing the electrodes can minimize these factors (Escuder-Gilabert & Peris, 2010; Tahara & Toko, 2013).
b. *Voltammetric sensors:* Another widely used model uses electrochemical sensors. These devices are used in food and beverage classifications and mixture quantifications of oxidizing substances (Del Valle, 2012). This type of electronic tongue consists of four working electrodes made of the platinum, rhodium, metals gold, silver metal, silver chloride electrode (Ag/AgCl) as a reference electrode and a stainless steel counter electrode (Ciosek & Wroblewski, 2007). In this device, the current between the encounter electrode and the metal working electrode is measured when a voltage pulse between the reference electrode and the working electrode is applied. A chain of pulses can be formed to extract as much of a possible information from the solution (Ciosek & Wroblewski, 2007). According to several studies, this technique in the electronic tongue is preferable in multicomponent measurements due to their low detection limits, high selectivity and several methods of measurements. Moreover, the ability to modify the electrode surface with several chemosensitive materials makes the sensors more sensitive and selective towards a various species (Del Valle, 2012). However, the temperature dependence and the drift caused by the large surface shift in sensors limits the

applicability of this system. To overcome the drift in the electrode reading, a cleaning process and mechanical polishing can be performed (Escuder-Gilabert & Peris, 2010).

c. *Impedimetric electronic tongues:* This system has been mainly used in detecting the basic tastes of food and beverages especially mineral water due to its high sensitivity (Escuder-Gilabert & Peris, 2010). The main parts of this system are one coated interdigitated electrode with different chemosensitive substances and another bare interdigitated electrode deposited by different methods of Langmuir–Blodgett technique. In contrast to other sensors system, this technique does not require a reference electrode (Escuder-Gilabert & Peris, 2010; Kumar *et al.*, 2012).
d. *Optical sensors:* This system is employed mainly for biomedical analysis and some food analysis. Optical electronic tongues offer several methods of operation such as absorbance, reflectance, and fluorescence (Di Natalea *et al.*, 2000; Escuder-Gilabert & Peris, 2010). This technique is applied when the analysis with electrochemical sensors is difficult or not feasible. However, sensor preparation, signal interference and durability are all factors limiting the use of this technique (Jain, Panchal, Pradhan, Patel & Pasha, 2010).

9.2.2 Data processing

Signal processing is one of the important aspects of the electronic tongue. To analyze the data from sensor arrays, several pattern recognition approaches are applied, primarily artificial neural networks (ANN) and principal component analysis (PCA). Also, multivariate calibration approaches use the primarily partial least square regression (PLS) and ANN (Di Natalea *et al.*, 2000; Escuder-Gilabert & Peris, 2010). These advanced mathematical signal processing techniques help in analyzing the response of the sensor array because the result is very complex and cannot be described by theoretical mathematic equations (Ciosek & Wroblewski, 2007); however, advanced mathematical methods cannot improve or alter the results (Jain *et al.*, 2010).

The electronic tongue system has been widely applied in food and flavor evaluation; sometimes it is paired with the electronic nose to provide wider complementary taste analysis information. The electronic tongue is a useful tool of great benefit when human panelists are not applicable. By understanding the structure of the different types of the electronic tongue systems, the applications, uses, advantages and disadvantages are to follow.

9.3 THE ELECTRONIC TONGUE AND FOOD PRODUCTION

Electronic tongues are analytical instruments patterned after the biological sensory system, mainly the sense of taste in human, as a new method to discriminate between complex samples. Recent research indicates that the electronic tongue was first applied to the food industry but has since found application in nutraceuticals, herbal medicines, pharmaceuticals, safety, environmental monitoring and medical diagnostics (Woertz *et al.*, 2011).

A taste sensing system can predict whether the raw food ingredients or the finished food products are fresh or rotten. The accuracy of the e-tongue has shown it to be a simple, rapid and inexpensive way to assess the shelf-life time and freshness of foods, a hot topic of concern for marketing foods over a long time period (Gil *et al.*, 2008). Moreover, the electronic tongue is considered a perfect method to monitor continually the changes occurring during the processing of food products such as fermentation (Claire Turner, 2003; Peris & Escuder-Gilabert, 2013). These devices can be applied as a quality control tool for such foods as milk, tea, beer and wine (Ciosek *et al.*, 2006; Palit *et al.*, 2010; Parra *et al.*, 2006; Rudnitskaya *et al.*, 2009). The applications of the electronic tongue for detection of compounds such as gluten in the foods for special dietary use (Miyanaga *et al.*, 2003) or unwanted ingredients such as copper, cadmium and iron in wine are also very beneficial (Simões da Costa *et al.*, 2014). In terms of the food industry, quantitative and qualitative analysis of food products can also be done using the electronic tongue. Sensors used in electronic tongues do not need any information about ingredients in samples as they use a number of non-specific, ion selective and low-selective chemical sensors with wide capability, high sensitivity and stability (Hruskar *et al.*, 2010; Martina *et al.*, 2007; Di Natalea *et al.*, 2000).

Nowadays, taste as a quality characteristic of food products plays a pivotal role in the different aspects of the food industry. A rapid response to the changing tastes of consumers is of primary importance to a modern and sustainable food industry. To meet such demands, minimal sample preparation and economical (cost effective) and prompt analysis of food are required.

The main methods for food analysis are either subjective or objective. Subjective approaches are sensory evaluation techniques using human panels to assess the sensory characteristics of foods. This approach is considered the gold standard by the food industry (Hruskar *et al.*, 2010). Artificial tongues mimic the human sensory response to food (Martina *et al.*, 2007). The different types of electronic tongues commercially available have shown acceptable results that correlate with human organoleptic scores (Escuder-Gilabert & Peris, 2010; Pein *et al.*, 2013) and are capable of predicting human taste panel scores. For example, electronic tongues can be used to reduce or replace the subjective methods of sensory analysis and distinguish the different gustatory sensations of sourness, saltiness, umami taste (Japanese term for deliciousness), sweetness and bitterness. This is particularly advantageous as human taste panels are expensive and time-consuming with the possible toxicity of samples being a problem for tasting (Cram *et al.*, 2009; Davies & Tuleu, 2008). Moreover, the variability in the physiological and physical conditions of evaluators as well as their preference can affect the result of test panels, despite the fact that they are highly trained for this purpose (Cram *et al.*, 2009). In contrast, artificial sensors are easy-to-use tools that can be repeated as needed as they are unaffected by adaptation and side effects associated with the evaluation of samples such as human fatigue or cramp (Escuder-Gilabert & Peris, 2010; Smyth & Cozzolino, 2013). Taste sensors are applicable for analyzing a wide range of taste and toxic samples (Kirsanov *et al.*, 2014).

Artificial sensors can mimic the taste in the food, as well as "olfaction". They are developed to recognize all kinds of dissolved compounds in the food including volatile compounds that cause odor after evaporation. Due to such properties, they are more sensitive than the natural senses to search for substances in food that are undetectable by humans (Legin *et al.*, 2002). In conclusion, these devices can and are being used to evaluate the different amounts of ingredients in foods (Gallardo, Alegret & Del Valle, 2005). Further research combining the electronic tongue systems with other available artificial sense technologies, such as the electronic nose, will better replicate the human sensing system with more reliable and precise results.

A wide range of methodologies are available to quantify and characterize food compounds physically and chemically using non-subjective techniques. Moreover, they can also assess the taste, aroma, texture and color and provide an overall sensory evaluation of the food. Although the results of these methods are accurate and reliable, they are time-consuming. A particular drawback to the existing objective measurements is that they are not cost effective since they require very expensive equipment (Escuder-Gilabert & Peris, 2010; Smyth & Cozzolino, 2013). Many of these quantitative methods are destructive and cannot be applied *in situ* or as monitoring devices, which is further disadvantage of the non-subjective techniques. Association between sensory evaluation and measured food characteristics cannot be achieved by analytical chemistry (Smyth & Cozzolino, 2013). A particular advantage of the electronic tongues is that they are environmentally friendly methods. They provide quick screening tools to assess food quality parameters. This device can be used as an in-line and at-line process analyzer generating rapid and low cost results using a small sample size and minimum sample preparation. The release and persistence of compounds in vivo when foods are eaten or drunk, cannot be evaluated by sensory evaluation (Newman *et al.*., 2014; Smyth & Cozzolino, 2013). The advantage of the electronic tongue over the other objective methods, including animal models, is that the results will be more representative of the human taste sensation (Anand *et al.*, 2007).

9.4 ELECTRONIC TONGUE AND BITTERNESS

An interesting application of the electronic tongue is in characterizing bitterness (Choi *et al.*, 2014). The development of bitterness can result in the loss of quality in the food product leading to economic losses. Therefore, it would of great value to be able to monitor its development throughout the food processing operation and production. A great advantage of using the electronic tongue for detecting bitterness is its ability to evaluate taste masking efficacy without requiring sensory test using human subjects (Choi *et al.*, 2014). From epidemiological evidence, in vivo and in vitro and clinical trial data suggest that bitter phytonutrients have disease prevention and health enhancement benefits (Shahidi, 2004). However, market studies showed that consumers select food on the basis of their taste and not for nutritional value or health benefits (Drewnowski & Gomez-Carneros, 2000). The production of food products that are tasty and healthy is a challenge unless you can get rid of unwanted bitter tastes or any off-flavors present. With the electronic tongue system you are able to quantify changes in bitter intensity in the presence of bitter masking compounds to obtain the best taste (Newman *et al.*, 2014; Takagi *et al.*, 2001). Consequently it is very promising method for detecting bitterness in food products.

Bitterness is generally considered an undesirable taste by most consumers so that extensive research has been undertaken for ways to reduce the compounds responsible in our foods. A wide range of physical, chemical and biochemical methods are available for measuring specific compounds eliciting bitterness in food products (Gil *et al.*, 2008). Analytical methods for evaluating bitter compounds include such chromatographic methods as HPLC (Jaskula, Goiris, De Rouck, Aerts & De Cooman, 2007) and gas chromatography (GC) that has been used for a long time (Verzele, Vanluche, & Vandyck, 1973). Multivariate data analysis can then be applied for evaluating bitterness. Uni-point measurement on a statistical basis is the most common technique used to describe bitter taste during descriptive analysis in which information related to temporal course of bitter taste is not prepared. Moreover, the evaluation is not directed at the time of changing bitterness in a dramatic way. Time-intensity (TI) method is a less frequently applied technique that rates bitterness intensity during the period of perception and evaluates the added bitter compounds in products. The multidimensional technique temporal dominance of sensations (TDS) is the most recently advanced approach that is used to evaluate bitterness in alcohol-reduced red wines (Martina *et al.*, 2007). In Taste dilution analysis (TDA), HPLC fractionates the taste extract and serial dilutions determine the taste threshold of the fractions (Frank, Ottinger & Hofmann, 2001).

The lower specificity to bitterness and the high cost of such techniques would hinder their use as promising tools for bitterness analysis.

9.5 EVALUATING BITTERNESS IN FOOD PRODUCTS USING ELECTRONIC TONGUES

Electronic tongues for bitterness evaluation have been successfully approved for various bitter drugs, such as H_1-antihistamines (Ito *et al.*, 2013), quinine hydrochloride (Uchida *et al.*, 2001) and different antibiotics (Sadrieh *et al.*, 2005; Uchida *et al.*, 2001). Besides the pharmaceutical industry, bitter taste evaluation using such systems has attracted considerable attention in the food industry over the past few years. Since a basic taste such as bitterness plays an essential role in determining the palatability and consumer acceptance of many food items research has been undertaken in this field on various foods and beverages; such as beer, wine, olive oil, tonic water and fruit juices (Baldwin, Bai, Plotto & Dea, 2011; Ciosek & Wroblewski, 2007; Escuder-Gilabert & Peris, 2010). Table 9.1 summarizes the main features of some of those studies.

One of the most important quality parameters in beer is bitterness, which is mainly caused by the formation of soluble iso-α-acids in hops (Verzele & De Keukeleire, 1992). Several studies have reported the application of electronic tongue systems to assess the bitter taste in beer. In a study conducted by Rudnitskaya *et al.* (2009), 50 Belgian and Dutch beers were evaluated for different sensory attributes including bitter intensity using a potentiometric electronic tongue. Principal component analysis (PCA) was applied to evaluate the capability of sensor's array with the data compared to the human sensory using a canonical correlation analysis (CCA). The results indicated that the electronic tongue system was an efficient tool for the fast screening and prediction of different taste attributes in beer including bitterness.

The capability of the same electronic tongue with 18 potentiometric sensors for the quantification of bitterness in the Belgian and Dutch beer samples was examined in the aqueous solutions of isomerizing hop extract. The electronic tongue sensors demonstrated an acceptable sensitivity to isomerized hop extract with good prediction ability for bitter taste in different beer samples. The electronic tongue system also proved its effectiveness in predicting other physicochemical parameters such as real extract, alcohol and polyphenol contents (Polshin *et al.*, 2010).

In the same context, an electronic tongue with an array of electroactive conducting polymers was developed by Arrieta *et al.* (2010) to generate mathematical models to predict the content of iso-α-acids and the alcoholic strength in commercial beers. Partial least square regression (PLS2) was used to construct the prediction models. To compare data high performance liquid chromatography (HPLC) analysis wasn conducted to quantify the iso-α-acids content. According to these results, building a mathematical model to predict the content of iso-α-acids and ethyl alcohol in commercial beers appeared feasible using the voltammetric electronic tongue.

A similar study on a different food product by Rudnitskaya *et al.* (2009) examined the relation between the intensity of bitter taste and the concentrations of phenolic compounds such as catechin, epicatechin, gallic and caffeic acids and quercetin in red wines. A set of 39 single cultivar Pinotage wines, including 13 samples with medium to high bitterness were analyzed using an electronic tongue with potentiometric chemical sensors. The data from electronic tongue system were analyzed using the partial least squares discriminate analysis (PLS-DA) regression model. The pH of wine samples was adjusted to 7 before measurement because the system's array of sensors showed sensitivity to the majority of the studied phenolic compounds at that pH level. This proved to be a rapid analytical tool for assessing

Table 9.1 Summary of selected studies with application of electronic tongue systems in bitterness analysis in food products.

Sample	**Type of sensors**	**Study Design**	**Reference**
Beer (Belgian and Dutch beers)	Potentiometric chemical sensors	PCA	(Rudnitskaya *et al.*, 2009)
Beer (Belgian and Dutch beers)	Potentiometric chemical sensors	CCA, PLS	(Polshin *et al.*, 2010)
Beer Commercial bears	Polypyrrole sensors	PCA, PLS	(Arrieta, Rodriguez-Mendez, de Saja, Blanco, & Nimubona, 2010)
Wine (Pinotage wines)	Potentiometric chemical sensors	one-way analysis of variance, PLS-DA	(Rudnitskaya *et al.*, 2010)
Wine (Italian wines: Barbera d'Asti and Gutturnio)	Potentiometric cross-sensitive chemical sensors	ANN, PLS	(Legin *et al.*, 2003)
Wine	Potentiometric chemical sensors	PLS	(Vlasov, Legin, Rudnitskaya, Di Natale, & D'Amico, 2005)
Wine	Potentiometric chemical sensors	PLS	(Schmidtke *et al.*, 2010)
Extra virgin olive oils	Voltammetric sensors	PLS	(Rodriguez-Mendez *et al.*, 2008)
Extra virgin olive oils	Enzyme-based biosensors (employing tyrosinase or peroxidase)	Simple linear regression	(Busch, Hrncirik, Bulukin, Boucon, & Mascini, 2006)
Tonic water	Piezoelectric quartz crystal sensor with molecularly imprinted polymer coating	Simple linear regression	(Sun, Mo, Choy, Zhu, & Fung, 2008)

the bitter taste of wines, as potentiometric electronic tongue systems were able to discriminate between bitter and control wines and also predict bitterness intensity with high accuracy.

A similar study was conducted by Legin *et al.*(2003) for recognition, quantitative analysis and flavor assessment of Italian red wines. Twenty samples of Barbera d'Asti and 36 samples of Gutturnio wine were analyzed using an electronic tongue with a sensor array of 23 potentiometric cross-sensitive chemical sensors. The electronic tongue system was efficient in determining the concentration of sulphur dioxide, glycerol and polyphenols in different wine samples, with an average prediction error less than 12%. It was also capable to predict human sensory scores with high precision (13% and 8% for Barbera d'Asti wines and Gutturnio wines, respectively).

Apart from beer and wine samples, the electronic tongue has been used for determining taste compounds in other foodstuffs such as olive oil. The latter is an example for a food in which bitterness may serve a quality feature in some grades. An electronic tongue with an array of chemically modified voltammetric electrodes was developed by Rodríguez-Méndez *et al.*(2008), to investigate the content of phenolic compounds, the main compounds responsible for the bitter taste in extra virgin olive oils. The mentioned electrodes contained electroactive materials such as phthalocyanines and conducting polymers. Partial least square discriminate analysis (PLS-DA) was conducted to interpret the relationships between voltammograms and the sensory or chemical analysis data. High correlations (PLS, $R^2 > 0.99$) were obtained and proved that the array of sensors presented could be used as a promising analytical device for predicting the polyphenolic content and the bitterness index of extra virgin olive oils.

The bitterness and pungency of virgin olive oil using electronic tongues was also reported by Busch *et al.*(2006) in which two amperometric enzyme-based biosensors (employing tyrosinase or peroxidase) were used. The study assessed the feasibility of using enzyme-based biosensors as rapid analytical tools for the prediction of sensory properties of virgin olive oil (a premium quality grade of olive oil). The feasibility of the biosensors was assessed by measuring a set of samples and comparing the results with data obtained by the reference HPLC method and panel sensory scores. Their results showed that enzyme-based biosensors were capable of successfully analyzing the total content of phenolics in virgin olive oils.

A combination of an electronic nose, electronic tongue and an electronic eye were applied all together to characterize the bitter taste intensity of 25 extra virgin olive oils from three different varieties. Analyzing the data by PCA indicated that combining the electronic tongue with other artificial sense technologies increased the discrimination efficiency significantly. Partial least squares-discriminate analysis showed good correlations (PLSDA, $R^2 > 0.9$) between the electronic tongue data and the polyphenolic content and the bitterness index which was scored by a human panel taste for various olive oil samples (Apetrei *et al.*, 2010).

The application of electronic tongues for determining taste causing compounds in beverages was reported by Sun *et al.* (2008). Using a piezoelectric quartz crystal sensor array based on molecularly imprinted polymer coating they determined the content of quinine and saccharine in bitter drinks such as tonic water. In comparison to the sensory panel scores, the electronic tongue sensor array showed a satisfactory repeatability and a high sensitivity for detecting the change in bitter tats of tonic water at practical concentrations.

Electronic tongues can also be applied to tonic water as well as fruit juices. Ding and co-workers (2010) used an Astree electronic tongue to determine the bitterness index of six sweet orange juices from *Citrus sinensis* Osbeck variety. Their results indicated a positive correlation between the limonin compounds of the sweet oranges with the relative bitterness value measured by the electronic tongue system.

9.6 CONCLUSION

Human perception of bitter taste varies widely due to such factors as genetic makeup, age, health status, and other environmental factors. It is important to be able to satisfy individual taste preferences and needs in an era of individualized medicine, nutrition and every product that directly affects human health. To reach that level of efficiency requires firstly developing sensors capable of detecting difference in bitterness perception in different products among individual. Secondly, the need to develop rapid, non-invasive technologies to analyze bitterness in foods with the electronic tongue. Artificial sense technologies such as "the electronic tongue systems" have been successfully developed and utilized as rapid analytical tools for qualitative and quantitative evaluation of bitter compounds and bitter tastes in various food products and beverages is promising. Depending on the type of the chemical sensors, the electronic tongue can be a used to classify of a wide range of food items especially those which are known for their bitterness preferences by consumers such as coffee, cocoa, tea and related products. The rapid developments in this field have been quite remarkable with further developments and improvements expected in the near future.

REFERENCES

Anand, V., Kataria, M., Kukkar, V., Saharan, V., & Choudhury, P.K. (2007). The latest trends in the taste assessment of pharmaceuticals. *Drug Discov. Today* **12**, 257–265.

Arrieta, A. A., Rodriguez-Mendez, M. L., de Saja, J. A., Blanco, C. A., & Nimubona, D. (2010). Prediction of bitterness and alcoholic strength in beer using an electronic tongue. *Food Chem.* **123**(3), 642–646. doi: DOI 10.1016/j.foodchem.2010.05.006

Baldwin, E. A., Bai, J. H., Plotto, A., & Dea, S. (2011). Electronic Noses and Tongues: Applications for the Food and Pharmaceutical Industries. *Sensors* **11**(5), 4744-4766. doi: DOI 10.3390/s110504744

Busch, J. L., Hrncirik, K., Bulukin, E., Boucon, C., & Mascini, M. (2006). Biosensor measurements of polar phenolics for the assessment of the bitterness and pungency of virgin olive oil. *J. Agric. Food. Chem.* **54**(12), 4371–4377. doi: 10.1021/jf060103m

Choi, D.H., Kim, N.A., Nam, T.S., Lee, S., & Jeong, S.H. (2014). Evaluation of taste-masking effects of pharmaceutical sweeteners with an electronic tongue system. *Drug Dev. Ind. Pharm.* **40**, 308–317.

Ciosek, P., Brudzewski, K., & Wróblewski, W. (2006). Milk classification by means of an electronic tongue and Support Vector Machine neural network. *Meas. Sci. Technol.* **17**, 1379

Ciosek, P., & Wroblewski, W. (2007). Sensor arrays for liquid sensing--electronic tongue systems. *Analyst* **132**(10), 963–978. doi: 10.1039/b705107g

Claire Turner, A.R. (2003). Monitoring batch fermentations with an electronic tongue. *J. Biotechnol.* **103**, 87–91.

Codinachs, L. M. I., Kloock, J. P., Schoning, M. J., Baldi, A., Ipatov, A., Bratov, A., & Jimenez-Jorquera, C. (2008). Electronic integrated multisensor tongue applied to grape juice and wine analysis. *Analyst* **133**(10), 1440–1448. doi: DOI 10.1039/b801228h

Cram, A., Breitkreutz, J., Desset-Brèthes, S., Nunn, T., Tuleu, C., and European Paediatric Formulation Initiative (EuPFI) (2009). Challenges of developing palatable oral paediatric formulations. *Int. J. Pharm.* **365**, 1–3.

Davies, E.H., & Tuleu, C. (2008). Medicines for children: a matter of taste. *J. Pediatr.* **153**, 599–604, 604.e1–e2.

Del Valle, M. (2012). Sensor arrays and electronic tongue systems. *Int. J. Electrochem.* **2012**, 1–11.

Di Natalea, C., Paolesseb, r., Macagnanoc, A., Mantinia, A., D'Amicoa, A., Legind, A., … Vlasovd, Y. (2000). Electronic nose and electronic tongue integration for improved classification of clinical and food samples. *Sens. Actuators, B*, **64**(1-3), 15–21.

Drewnowski, A., & Gomez-Carneros, C. (2000). Bitter taste, phytonutrients, and the consumer: a review. *Am. J. Clin. Nutr.* **72**, 1424–1435.

Escuder-Gilabert, L., & Peris, M. (2010). Review: Highlights in recent applications of electronic tongues in food analysis. *Anal. Chim. Acta* **665**(1), 15–25. doi: DOI 10.1016/j.aca.2010.03.017

Frank, O., Ottinger, H., & Hofmann, T. (2001). Characterization of an intense bitter-tasting 1H,4H-quinolizinium-7-olate by application of the taste dilution analysis, a novel bioassay for the screening and identification of taste-active compounds in foods. *J. Agric. Food Chem.* **49**(1), 231–238. doi: DOI 10.1021/jf0010073

Gallardo, J., Alegret, S., & Del Valle, M. (2005). Application of a potentiometric electronic tongue as a classification tool in food analysis. *Talanta* **66**(5), 1303–1309. doi: 10.1016/j.talanta.2005.01.049

Ghasemi-Varnamkhasti, M., Mohtasebi, S. S., & Siadat, M. (2010). Biomimetic-based odor and taste sensing systems to food quality and safety characterization: An overview on basic principles and recent achievements. *J. Food Eng.* **100**, 377–387.

Gil, L., Barat, J.M., Escriche, I., Garcia-Breijo, E., Martínez-Máñez, R., & Soto, J. (2008). An electronic tongue for fish freshness analysis using a thick-film array of electrodes. *Microchim. Acta.* **163**, 121–129.

Ito, M., Ikehama, K., Yoshida, K., Haraguchi, T., Yoshida, M., Wada, K., & Uchida, T. (2013). Bitterness prediction of H-1-antihistamines and prediction of masking effects of artificial sweeteners using an electronic tongue. *Int. J. Pharm.* **441**(1-2), 121–127. doi: DOI 10.1016/j.ijpharm.2012.11.047

Jain, H., Panchal, R., Pradhan, P., Patel, H., & Pasha, T. Y. (2010). Electronic tongue: A new taste sensor. *Int. J. Pharm. Sci. Rev. Res.* **5**, 91–96.

Jaskula, B., Goiris, K., De Rouck, G., Aerts, G., & De Cooman, L. (2007). Enhanced quantitative extraction and HPLC determination of hop and beer bitter acids. *J. Ins. Brew.* **113**(4), 381–390.

Kirsanov, D., Legin, E., Zagrebin, A., Ignatieva, N., Rybakin, V., & Legin, A. (2014). Mimicking Daphnia magna bioassay performance by an electronic tongue for urban water quality control. *Anal. Chim. Acta.* **824**, 64–70.

Kobayashi, Y., Habara, M., Ikezazki, H., Chen, R. G., Naito, Y., & Toko, K. (2010). Advanced Taste Sensors Based on Artificial Lipids with Global Selectivity to Basic Taste Qualities and High Correlation to Sensory Scores. *Sensors* **10**(4), 3411–3443. doi: DOI 10.3390/s100403411

Kumar, R., Bhondekar, A. P., Kaur, R., Vig, S., Sharma, A., & Kapur, P. (2012). A simple electronic tongue. *Sens. Actuators, B* **171**, 1046–1053. doi: DOI 10.1016/j.snb.2012.06.031

Legin, A., Rudnitskaya, A., & Vlasov, Y. (2002). Electronic Tongues: Sensors, Systems, Applications. *Sens. Update* **10**, 143–188.

Legin, A., Rudnitskaya, A., Lvova, L., Vlasov, Y., Di Natale, C., & D'Amico, A. (2003). Evaluation of Italian wine by the electronic tongue: recognition, quantitative analysis and correlation with human sensory perception. *Anal. Chim. Acta* **484**(1), 33–44. doi: Doi 10.1016/S0003-2670(03)00301-5

Martina, V., Ionescu, K., Pigani, L., Terzi, F., Ulrici, A., Zanardi, C., & Seeber, R. (2007). Development of an electronic tongue based on a PEDOT-modified voltammetric sensor. *Anal. Bioanal. Chem.* **387**(6), 2101–2110. doi: DOI 10.1007/s00216-006-1102-1.

Miyanaga, Y., Inoue, N., Ohnishi, A., Fujisawa, E., Yamaguchi, M., & Uchida, T. (2003). Quantitative prediction of the bitterness suppression of elemental diets by various flavors using a taste sensor. *Pharm. Res.* **20**, 1932–1938.

Newman, J., Egan, T., Harbourne, N., O'Riordan, D., Jacquier, J.C., & O'Sullivan, M. (2014). Correlation of sensory bitterness in dairy protein hydrolysates: Comparison of prediction models built using sensory, chromatographic and electronic tongue data. *Talanta* **126**, 46–53.

Palit, M.; Tudu, B.; Bhattacharyya, N.; Dutta, A.; Dutta, P. K.; Jana, A.; Bandyopadhyay, R.; Chatterjee, A., Comparison of multivariate preprocessing techniques as applied to electronic tongue based pattern classification for black tea. *Anal. Chim. Acta* 2010 **675** (1), 8–15.

Parra, V., Arrieta, Á.A., Fernández-Escudero, J.A., García, H., Apetrei, C., Rodríguez-Méndez, M.L., & Saja, J.A. de (2006). E-tongue based on a hybrid array of voltammetric sensors based on phthalocyanines, perylene derivatives and conducting polymers: Discrimination capability towards red wines elaborated with different varieties of grapes. *Sens. Actuators, B* **115**, 54–61.

Pein, M., Eckert, C., Preis, M., & Breitkreutz, J. (2013). New protocol for alphaAstree electronic tongue enabling full performance qualification according to ICH Q2. *J. Pharm. Biomed. Anal* **83**, 157–163.

Peris, M., & Escuder-Gilabert, L. (2013). On-line monitoring of food fermentation processes using electronic noses and electronic tongues: a review. *Anal. Chim. Acta* **804**, 29–36.

Polshin, E., Rudnitskaya, A., Kirsanov, D., Legin, A., Saison, D., Delvaux, F., … Lammertyn, J. (2010). Electronic tongue as a screening tool for rapid analysis of beer. *Talanta* **81**(1-2), 88–94. doi: DOI 10.1016/j.talanta.2009.11.041

Rodriguez-Mendez, M. L., Parra, V., Apetrei, C., Villanueva, S., Gay, M., Prieto, N., … de Saja, J. A. (2008). Electronic tongue based on voltammetric electrodes modified with materials showing complementary electroactive properties. Applications. *Microchim. Acta* **163**(1-2), 23–31. doi: DOI 10.1007/s00604-007-0907-8

Rudnitskaya, A., Polshin, E., Kirsanov, D., Lammertyn, J., Nicolai, B., Saison, D., … Legin, A. (2009). Instrumental measurement of beer taste attributes using an electronic tongue. *Anal. Chim. Acta* **646**(1-2), 111–118. doi: DOI 10.1016/j.aca.2009.05.008

Rudnitskaya, A., Rocha, S. M., Legin, A., Pereira, V., & Marques, J. C. (2010). Evaluation of the feasibility of the electronic tongue as a rapid analytical tool for wine age prediction and quantification of the organic acids and phenolic compounds. The case-study of Madeira wine. *Anal. Chim. Acta* **662**(1), 82–89. doi: DOI 10.1016/j.aca.2009.12.042

Sadrieh, N., Brower, J., Yu, L., Doub, W., Straughn, A., Machado, S., ... Buhse, L. (2005). Stability, dose uniformity, and palatability of three counterterrorism drugs - Human subject and electronic tongue studies. *Pharma. Res.* **22**(10), 1747–1756. doi: DOI 10.1007/s11095-005-6387-x

Schmidtke, L. M., Rudnitskaya, A., Saliba, A. J., Blackman, J. W., Scollary, G. R., Clark, A. C., ... Legin, A. (2010). Sensory, Chemical, and Electronic Tongue Assessment of Micro-oxygenated Wines and Oak Chip Maceration: Assessing the Commonality of Analytical Techniques. *J. Agric. Food Chem.* **58**(8), 5026–5033. doi: DOI 10.1021/jf904104f

Shahidi, F. (2004). Functional foods: their role in health promotion and disease prevention. *J. Food Sci.* R146–149.

Simões da Costa, A.M., Delgadillo, I., & Rudnitskaya, A. (2014). Detection of copper, lead, cadmium and iron in wine using electronic tongue sensor system. *Talanta* **129**, 63–71.

Smyth, H., & Cozzolino, D. (2013). Instrumental methods (spectroscopy, electronic nose, and tongue) as tools to predict taste and aroma in beverages: advantages and limitations. *Chem. Rev.* **113**, 1429–1440.

Sun, H., Mo, Z. H., Choy, J. T. S., Zhu, D. R., & Fung, Y. S. (2008). Piezoelectric quartz crystal sensor for sensing taste-causing compounds in food. *Sens. Actuators, B* **131**(1), 148–158. doi: DOI 10.1016/j.snb.2007.12.014

Tahara, Y., & Toko, K. (2013). Electronic Tongues-A Review. *IEEE Sens. J.* **13**(8), 3001–3011. doi: Doi 10.1109/Jsen.2013.2263125

Takagi, S., Toko, K., Wada, K., & Ohki, T. (2001). Quantification of suppression of bitterness using an electronic tongue. *J. Pharm. Sci.* **90**, 2042–2048.

Uchida, T., Kobayashi, Y., Miyanaga, Y., Toukubo, R., Ikezaki, H., Taniguchi, A., ... Matsuyama, K. (2001). A new method for evaluating the bitterness of medicines by semi-continuous measurement of adsorption using a taste sensor. *Chem. Pharm. Bull.* **49**(10), 1336–1339. doi: DOI 10.1248/cpb.49.1336

Verzele, M., & De Keukeleire, D. (1992). *Chemistry and analysis of hop and beer bitter acids.* Amsterdam, The Netherlands: Elsevier.

Verzele, M., Vanluche. E, & Vandyck, J. (1973). Gas-Chromatography of Beer Bitter Acids. *Anal. Chem.* **45**(8), 1549–1552. doi: DOI 10.1021/ac60330a059

Vlasov, Y., Legin, A., Rudnitskaya, A., Di Natale, C., & D'Amico, A. (2005). Nonspecific sensor arrays ("electronic tongue") for chemical analysis of liquids (IUPAC Technical Report). *Pure Appl. Chem.* **77**(11), 1965–1983. doi: DOI 10.1351/pac200577111965

Woertz, K., Tissen, C., Kleinebudde, P., & Breitkreutz, J. (2011). Taste sensing systems (electronic tongues) for pharmaceutical applications. *Int. J. Pharm.* **417**, 256–271.

SECTION IV

METHODS FOR REMOVING BITTERNESS IN FUNCTIONAL FOODS AND NUTRACEUTICALS

10 Methods for Removing Bitterness in Functional Foods and Nutraceuticals

Erin Goldberg, Jennifer Grant, Michel Aliani, and Michael N. A. Eskin

10.1 INTRODUCTION

Bitterness is the most complex and poorly understood of the five basic tastes (sweet, salty, sour, umami, bitter) (Drewnowski, 2001) due to the large number and diversity of T2R bitter receptors (Ley *et al.*, 2008). It is believed that bitterness was an evolutionary mechanism designed to warn humans against ingesting potentially harmful substances (Rodgers *et al.*, 2006). However, in certain products such as coffee and beer (see Chapter 5), a small amount of bitterness is acceptable and even considered desirable (Drewnowski, 2001; Rodgers *et al.*, 2006). Nevertheless, for most food products, bitterness is unacceptable, rendering these products unpalatable.

Numerous compounds are responsible for producing bitter tastes in foods. These include amino acids, peptides (in fermented products), esters, lactones, phenols, polyphenols (in plants such as spinach, kale, mustard greens, radicchio, cabbage and Brussels sprouts), flavonoids, terpenes, caffeine (in tea, coffee and cocoa), organic and inorganic salts and saccharin (Drewnowski, 2001; Maehashi & Huang, 2009). Thus, determining which compound or mixture of compounds is responsible for bitterness in specific foods is an extremely complex task. In addition, many of these bitter compounds are considered healthful due to their high antioxidant potential, making them beneficial ingredients for use in 'functional foods'. Functional foods may appear conventional, but when consumed go beyond their normal contribution to nutrition by also providing health benefits, such as disease prevention (Milner, 1999). Bitterness could decrease compliance and consumption of functional foods in clinical trials, therefore, it is crucial that it be removed or reduced to ensure these foods are similar, if not superior, to conventional foods in terms of palatability and acceptance. Controlling bitterness of foods could also potentially reduce overall food waste. In general, there are three approaches to suppressing bitterness; physiochemical interactions in a food or beverage matrix, oral peripheral physiological interactions with receptor cells (e.g., via receptor inhibitors),

Bitterness: Perception, Chemistry and Food Processing, First Edition. Edited by Michel Aliani & Michael N. A. Eskin.

and central cognitive mixture suppression (e.g., via taste to taste and taste to aroma interactions) (Keast, 2008).

This chapter focuses on recent advances in masking or reducing bitterness for the development of functional foods, including temperature treatment, alteration of physical or chemical characteristics, addition of bitter masking agents as well as the use of bitter blockers (Table 10.1). The mechanism behind bitterness perception or the physical methods of bitterness reduction, such as microencapsulation or spray drying, will not be discussed.

10.2 REDUCING AND REMOVING BITTER COMPONENTS

10.2.1 Physical methods

10.2.1.1 *Temperature treatment*

Subjecting foods to different temperature treatments has been found to reduce the presence of bitterness in some foods. For example, Bhandari and Kawabata (2005) subjected wild yam (*Dioscorea* spp.) to boiling, baking and pressure cooking in an attempt to reduce bitterness caused by the presence of furanoid norditerpenes (diosbulbins A and B). Boiling proved to be the most effective cooking method for reducing bitterness in wild yams, although the other methods still reduced bitterness to some degree.

Heat treatment was also found to reduce bitterness caused by phenolic compounds in olive oils. García and colleagues (2001) reported that heating olives at 30, 40, 45 and 50°C for 24 hours or at 40°C for 24, 48, and 72 hours, effectively reduced bitterness in olive oil samples compared to samples heated at 40°C for 2 hours. While the heat treatment did not affect the acidity or oxidative stability of the olive oil (García *et al.*, 2001), there was a reported decrease in phenolic content.

Alternatively, studies examining the effects of cold temperature on olive oil yielded positive results in reducing bitterness. Storage of olive oil at 5°C effectively reduced the bitterness of olive oil but was time dependent. The length of storage time varied from 2 to 8 weeks depending on the type of olive used in each sample of oil (Yousfi *et al.*, 2008). Bett-Garber and colleagues (2011) found that fresh-cut cantaloupes stored at 4°C had the lowest intensity for off-flavors including bitterness, followed by the samples at 10°C, and the samples held at 4°C for 24 h before a temperature increase to 10°C. The samples held at 4°C for 48 h before a change to 10°C had the highest intensity of off-flavors. They found that cold storage temperature changes seem to enhance processes that produce off-flavors (rancid, musty and bitter) that could be the result of amino acid breakdown and/or cell wall leakage.

Table 10.1 Reducing, Removing and Masking Bitter Compounds.

Reference	Study Subjects	Bitter compound(s)	Experimental Design	Study Outcome	Mechanisms
Bertoldi *et al.*, 2004	25 trained panelists	Amino acids and peptides	• Various levels of glucose and NaCl added to dark tuna samples. All samples were inoculated with *Lactobacillus Casei* and fermented for 30 days at 10°C. • pH, lactic acid, total bacteria counts were measured. • Sensory analysis with one identified control sample and 5 coded samples with different treatment. • Bitterness measured on a scale (1–9).	• Increased glucose in sample resulted in a ↑ production of lactic acid and ↑ pH. • ↑ NaCl in sample resulted in ↓ production of lactic acid. • Bitterness ↓ with ↑ lactic acid production.	Suggestions: 1) enzymatic action on amino acids and hydrophobic peptides during fermentation; or 2) acidity disguised bitterness of dark meat.
Bhandari *et al.*, 2005	N/A	Terpenes (furanoid norditerpenes)	• Bitter compounds were extracted from wild yam using HPLC in the raw state and after applying various cooking methods (boiling, baking and pressure cooking) to compare changes in the presence of bitter compounds after cooking.	• Cooking wild yams appeared to ↓ the number of bitter compounds present. • Boiling ↓ bitterness compounds the most.	

(continued)

Table 10.1 *(Continued)*

Reference	Study Subjects	Bitter compound(s)	Experimental Design	Study Outcome	Mechanisms
Cravotto *et al.*, 2005	12 trained panelists	Lactones and terpenes (Cynaropicrin and grosheimin)	• Isolation of cynaropicrin and grosheimin from artichoke. • Bitterness measured with and without microwave irradiation.	• Bitterness ↓ when polarity of molecule ↑ • Opening of lactone ring supressed bitter taste. • Role of exomethylene in bitterness.	None suggested.
Fontoin *et al.*, 2008	16 trained panelists	Phenols	• Taste panel rated bitterness of red wines that had altered pH, ethanol level or tartaric acid concentration on a scale (0–10).	• No effect of pH. • ↑ tartaric acid slight effect on bitterness. • ↑ ethanol had significant influence on bitterness.	• Suggested that non-significant effect of pH could be attributed to interaction with astringency perception.
García *et al.*, 2001	12 analytical panelists	Phenol	• Heat treatment of olive oils (30, 40, 45, 50°C for 24 hours) and (40°C for 24, 48°C for 72 hours). • Oil extracted after heating and pH and oxidative stability. • Bitterness measured on a scale (0–5).	• Bitterness ↓ with all heat treatments. • ≤40°C for 24 hours had no effect on acidity or oxidative stability of olive oil.	• Content of hydroxytyrosol related to bitterness in olive oil but the relationship is different for each variety of olive.

Keast, 2008	33 untrained volunteers	Caffeine	• Bitterness measured on a general labeled magnitude scale.	• Bitterness ↓ using both oral, peripheral and central cognitive strategies. • Zinc lactate best to ↓ bitterness; however, it can also ↓ sweetness.	• Zinc ions are modulating an extracellular GPCR's but further studies are needed.
Keast & Breslin, 2005	20 employees of Monell Chemical Senses Center	Tetralone (bitter agent in beer)	• Bitterness measured using a general labeled magnitude scale.	• Zinc sulfate significantly ↓ bitterness of quinine HCL, tetralone, and denatonium. • Zinc sulfate did not inhibit bitterness of sucrose octa-acetate, pseudoephedrine and dextromethorphan • A combination of Na-cyclamate and zinc sulfate was most effective at ↓ bitterness.	• Suggested that bitterness suppression is a result of the action of zinc cations on a peripheral component of bitter taste physiology (probably taste receptors)
Kola *et al.*, 2010	N/A	Terpene (triterpenoid-limonin)	• Limonin determined using high performance liquid chromatography (HPLC). • Titration with NaOH to determine total titratable acidity. • Digital pH meter to assess pH. • Brix with an Abbe refractometer to measure total soluble solids. • Visual titration to determine ascorbic acid content.	• Amberlite XAD-16HP (a debittering agent) was found to have no significant effects on pH, acidity, total soluble solids and ascorbic acid content of orange juice but did ↓ limonin (bitter) content. • Dowex Optipore L285 reduced the titratable acidity and so increased the pH and soluble solid content accordingly.	

(*continued*)

Table 10.1 *(Continued)*

Reference	Study Subjects	Bitter compound(s)	Experimental Design	Study Outcome	Mechanisms
Tamamoto *et al.*, 2010	11 untrained panelists	Not identified.	Test 1 • Ginseng added to water solution to mimic energy drink. • 5 treatments applied; 1) masking- flavor treatment (citrus), 2) blocking agent- addition of Revolver®, 3) ingredient interaction treatment-taurine,4) enzyme modification- Rapsidase, and 5) complexation treatment- γ-CD • Sensory panelists ranked the bitterness of the five treatments and one control in order from least to most bitter. Test 2 • Most bitter compounds identified as blocking agent and complexation treatment. • Sensory panelists ranked the bitterness on a scale (1–10) for 7 samples of varying amounts of γ-CD and again for 7 samples of varying amounts of Revolver®	• There was a significant ↓ in bitterness when concentration of γ-CD ↑. • No significant ↓ in bitterness was observed for when concentration of Revolver® ↑ • Researchers concluded that Revolver® not an effective bitterness reducer for ginseng beverages.	• Authors suggested that effectiveness of γ-CD on bitterness reduction may be related to the size of γ-CD and bitterness molecules.

Yousfi *et al.*, 2008	8 trained panelists	Phenols	• Samples of olive oil from several varieties of olives subjected to cold storage at 5°c for up to 8 weeks. • At several stages during storage sensory panel assessed sensory quality and bitterness intensity. • Oxidative stability measured using rancimat method. • Phenolic content analysed using HPLC.	• Level of bitterness decreased gradually during cold storage • Length of cold storage needed to obtain mild bitterness intensity varied between varieties of olive. • Oils that exhibited higher content of phenolic compounds presented the lowest values of stability.	• Authors suggested that stability is due to the dependency of oil stability to factors other than phenolic compound content, such as fatty acid composition or tocopherol content.
Aliani *et al.*, 2011	11 trained panelists		• Bitterness of snack bars and muffins with and without flax measured on a 15 cm unstructured line scale.	• Orange cranberry and apple spice muffin with flax had ↑ bitterness compared to non-flax muffin. • Orange cranberry and cappuccino chocolate chip snack bars with flax had significantly ↑ bitter taste than those without flax.	• Expectations of bitterness (i.e. ginger flavoring) played a role in perception.
Ares *et al.*, 2009	8 trained panelists		• Bitterness extracts of two South-American plants measured on a 10 cm unstructured line scale.	• Sucrose and sucralose caused a ↓ in the bitterness of the antioxidant extracts. • The effectiveness of polydextrose in ↓ the bitterness of the extracts was lower than that of sucrose or sucralose. • Milk fat (0 and 3.2%) ↓ the bitterness of extracts from A satureioides.	Effect of polydextrose: • ↑ viscosity and the possibility of making hydrogen bonds with polyphenols, which could diminish the interaction of polyphenols with taste receptors or salivary proteins.

(*continued*)

Table 10.1 *(Continued)*

Reference	Study Subjects	Bitter compound(s)	Experimental Design	Study Outcome	Mechanisms
					Effect of milk: • Migration of polyphenolic compounds to the fat phase. • Milk proteins may complex with polyphenolic compounds yielding them insoluble or at incapable of interacting with the taste receptors.
Kranz *et al.*, 2010	11 trained panelists		• Protocol threshold test, DIN 10959. • Protocol ranking tests, panel rank sums, Friedman's test. • Protocol scale test, DIN 10967.	• Although sucrose and sodium cyclamate did ↓ perception of bitterness, a strong remaining impression of bitterness occurred.	
Ishizaka *et al.*, 2008	8 participants		• 5-point rating scale.	• As the intensity of sweetness, sourness and saltiness ↑, bitterness ↓ in prednisone powder (PP) samples.	• Mechanism unclear (caused multiple pathways). • Carbonated beverages suppressing bitterness caused by interaction between the bitter substance and carbon dioxide.

Mukai *et al.*, 2004	12 participants		• 5-point rating scale.	• Strawberry, apple and vanilla aromas contributed to the bitterness inhibition of branched-chain amino acid solutions.
Bett-Garber *et al.*, 2011	Trained panelists		• Cantalope was processed, packaged and stored for 7 days and maintained at 4°C and 10°C or transferred from 4 to 10°C after 24 or 48 hours.	• Musty, rancid/painty flavors and bitterness ↑ due to temperatures changes. • Slight changes in storage temperature can result in sensory quality problems that were not observed at constant storage temperatures.
Miyashita & Etoh, 2013	N/A	Catechins	• Sub-critical water extraction (SWE) was used to remove bitterness from green tea extracts. • Taste sensor analyzer used to measure bitterness.	• SWE ↓ bitterness while retaining the original fragrance in the green tea extract. • Produced ↑ amounts of arginine and water-soluble pectin, which are thought to mask bitterness of green tea.

(continued)

Table 10.1 *(Continued)*

Reference	Study Subjects	Bitter compound(s)	Experimental Design	Study Outcome	Mechanisms
Homma *et al.*, 2012		Oily fraction of natural white mold cheese, Baraka	• Oily fraction of cheese was extracted and fractionated using silica gel column chromatography. • Fractions were characterized by thin-layer chromatography and nuclear magnetic resonance spectrometry. • Bitterness was quantitated using a method based on subjective equivalents.	• Fatty acids containing fractions had the highest bitterness masking activity against quinine hydrochloride. • Oleic acid masked bitterness.	• Oleic masked bitterness by forming a complex with bitter compounds.
Zhang *et al.*, 2014	N/A	Catechins	• Catechins were subjected to 200°C for 15 min. in a low-moisture Maillard models of glycine (gly) and a reducing sugar (D-glucose, D-xylose or D-galactose). • Analyzed with LC-MS isotopic labeling techniques.	• 8 reaction products identified. • One spiro product was reported to suppress the perceived bitterness intensity of a caffeine solution. • Spiro product was identified in cocoa and ↑ during bean roasting.	

Gaudette & Pickering, 2012a	15 trained panelists	Catechins	• 2 bitter blockers (β-cyclodextrin, homoeriodictyol sodium salt), 2 sweeteners (sucrose, rebaudioside A) and 2 odorants (vanillan and black tea) were assessed for their efficacy at modifying bitterness of catechiin in model aqueous solutions. • Bitterness assessed using a 15 cm visual analogue scale.	• β-cyclodextrin with either sweetener effectively ↓ bitterness (68% reduction). • Odorants were not effective at modifying bitterness of catechins.
Sharafi *et al.*, 2013	37 participants	Vegetables (Brussels sprouts, kale and asparagus)	• Vegetables were misted with bitter masking agents (1.33 M sodium acetate, 10 and 32 mM sodium chloride, 3.2 mM aspartame)	• Aspartame was the most effective at supressing bitterness. • Sodium acetate ↓ bitterness for all vegetables and ↑ perceived sweetness. • 32 mM ↓ bitterness the most for kale.
Koprivnjak *et al.*, 2009	16 trained panelists	Hydrophilic phenolic compounds in virgin olive oil	• Granular soy lecithin (source of phospholipids was added to virgin olive oil (2.5 to 30 g/kg). • Bitterness accessed with sensory difference tests and the bitterness index (K_{225}).	• The addition of phospholipids in a range of 5 to 10 g/kg ↓ bitterness.

(continued)

Table 10.1 *(Continued)*

Reference	Study Subjects	Bitter compound(s)	Experimental Design	Study Outcome	Mechanisms
Jensen *et al.*, 2001	Trained panelists	Lactates from ground chicken breast tissue	• 3% or 4% sodium lactate or potassium lactate, sucrose, dextrose, lactitol, sorbitol or lysine were individually added to ground chicken breast and heated in water bath to 72°C and evaluated by trained panel at 45°C.	• Samples with potassium lactate were more bitter and less sweet than those with sodium lactate. • Samples with 4% sodium lactate were more bitter than those with 3% sodium lactate. • Bitterness ↓ in samples containing sodium lactate when 1% sucrose or lactitol were added.	
Binello *et al.*, 2004	12 trained panelists		• Synthesized bitter blockers were covalently bonded to cyclodextrins with chitosans via succinyl or maleyl bridges to produce β-cyclodextrin and γ-cyclodextrin adducts. • The bitter masking capabilities of adducts were analyzed in caffeine and natural bitter extracts (artichoke leaves, aloe and gentian).	• β-cyclodextrin chitosan adduct exhibited the ↑ efficacy by ↓ bitterness.	

Gaudette & Pickering, 2012b	12 trained panelists		• High and low concentrations of 5 bitter blockers; β-cyclodextrin, homoeriodictyol sodium salt (HED), zinc sulphate monohydrate, magnesium sulphate and carboxymethycellulose sodium salt were added to high and low concentrations of +(−)catechins and caffeine. • Bitterness measure on a 15 cm line scale.	• β-cyclodextrin and HED were that most effective at ↓ bitterness of +(−)catechins. • HED was the more effective at ↓ bitterness of caffeine.
Kim *et al.*, 2015			• Five soybean derived umami peptides (Glu-Asp, Glu-Glu, Glu-Ser, Asp-Glu-Ser,Glu-Gly-Ser) were present with a bitter substance (salicin) on ionized calcium flux signaling assay using human hTAS2R16 expressing cells.	• All peptides reduced the response of the bitter taste receptor induced by salicin in a non-competitive manner. • Glu-Glu was the strongest inhibitor and was more effective than a known hTAS2R16 antagonist, probenecid.
Tárrega *et al.*, 2012	10 trained panelists		• UHT-treated, low lactose functional milk containing with or without American ginseng with and without vanilla flavor and sucralose were prepared and evaluated by panelists on a scale.	• Bitterness ↑ with the addition of ginsenosides and thermal treatment. • Bitterness attributed to ginseng-extract ↓ with the addition of vanilla flavor and sucralose.

Overall, temperature fluctuations should be avoided and a constant cool temperature should be maintained. The use of cold storage is likely to be favorable over heat treatment to minimize the degradation of beneficial compounds; however, further research is needed to determine optimal temperature treatment. Additionally, some beneficial compounds are known to increase with exposure to heat treatment, as in the case of canolol in the oil fractions of toasted canola seeds and meals (Mayengbam *et al.*, 2013). It is apparent that the effectiveness of any temperature treatment is product-specific, and general recommendations are unclear at this time without further research.

10.2.1.2 *Ion exchange and adsorbent resins*

Bitter phenolic compounds are routinely adsorbed to resins, trapped on polymers, precipitated, extracted with solvents, or converted to non-bitter compounds (Drewnowski & Gomez-Carneros, 2000).

A recent study by Kola and colleagues (2010), compared the effectiveness of two ion exchange and adsorbent resins (Amberlite XAD-16HP and Dowex Optipore L285) on removing bitterness in orange juice. They found that both agents effectively reduced the bitterness of the terpene limonin in orange juice; however, Dowex caused alterations to the acidity, thereby decreasing the soluble solids content and increasing the pH value. On the other hand, Amberlite did not alter the quality characteristics of orange juices and therefore may be the recommended adsorbent resin for the beverage.

10.2.1.3 *Extraction with sub-critical water*

Miyashita and Etoh (2013) showed that sub-critical water extraction (SWE) was very effective in removing catechins from green tea. While catechins are known for their health-promoting effects they are highly bitter and astringent. Hydrothermal and pressurized extraction techniques as well as SWE were all shown to effectively extract over 300 mg of catechins per 100 mL of green tea extracts. Unlike the other techniques, SWE is considered to be more environmentally friendly as it only uses standard water and requires low temperature and pressure. In addition to reducing bitterness and astringency, SWE was the only technique to retain the original fragrance of the green tea extract.

10.2.2 Chemical methods

10.2.2.1 *Fermentation*

To reduce bitterness, researchers have also attempted to alter the physical and chemical structure of the compounds responsible for the bitterness. For example, Bertoldi and colleagues (2004) reduced the bitterness of dark tuna meat using lactic acid fermentation with *Lactobacillus Casei* subsp. *Casei ATCC 393* in the presence of varying levels of glucose and sodium. They observed that fermentation increased the level of lactic acid, which effectively reduced bitterness in tuna. Higher glucose levels were shown to enhance fermentation while higher sodium levels suppressed fermentation. These researchers also reported an inverse relationship between the increase in pH of tuna and the decrease in bitterness, possibly due to the enzymatic action on amino acids and hydrophobic peptides during fermentation or because the acidity masked the bitterness. However, researchers examining the effect of physical alterations on the phenolic content of red wine concluded that changes in pH had no effect on bitterness. They did find that increasing tartaric acid concentration slightly decreased bitterness perception in red wines whereas increasing ethanol content increased bitterness perception (Fontoin *et al.*, 2008). The conflicting results from these studies (Bertoldi *et al.*, 2004; Fontoin *et al.*, 2008) demonstrated that the food matrix is critical in bitterness development, perception and masking.

10.2.2.2 *Aging and polymerization of phenols*

Aging of wine reduces bitterness because phenols continue to polymerize and eventually precipitate. Young red wines sold without being aged sometimes have high residual sugar concentrations (1–3%) to reduce bitterness. Sugar has also been added to wines to reduce bitter taste (Noble, 1998).

Cravotto and colleagues (2005) isolated bitter compounds from artichoke leaves (cynaropicrin and grosheimin) by subjecting them to microwave irradiation. This resulted in chemical alterations of these compounds including esterification and opening of the lactone ring. They found that the increased presence of oxygenated polar groups, opening of the lactone ring and probably the reduction of exomethylenes resulted in reduction of bitterness for these compounds.

10.2.2.3 *Alkalization*

Taste dilution analysis (TDA) of roasted cocoa nibs showed that in addition to alkaloids (theobromine and caffeine), a series of bitter tasting 2,5-diketopiperazines and monomeric and oligomeric flavan-3-ols, were key inducers of bitterness in consumed roasted beans (Stark *et al.*, 2005, 2006; Stark & Hofmann, 2005). Stark and Hofmann (2006) subsequently demonstrated, for the first time that alkalization of cocoa induced the non-enzymatic C-glycosylation of flavan-3-ols to form flavan-3-ol-C-glycosides. These newly formed C-glycosides modified the bitter taste profile by decreasing the bitter taste intensity of the alkalized cocoa powder. The ability of C-glycosides to suppress bitterness was recently confirmed by Zhang and colleagues (2014) who subjected catechins to 200°C for 15 minutes in low moisture Maillard models composed of glycine (gly) and a reducing sugar (D-glucose, D-xylose or D-galactose). Using isotopic labeling techniques, they identified eight reaction products including six flavan-3-ol-spiro-C-glycosides. Of these spiro products, one was found to significantly suppress the bitterness of a caffeine solution. The concern with many of these techniques is that they may also reduce the beneficial effects associated with such compounds as polyphenols.

Therefore, in order to not compromise the nutrition of the food of interest, masking of bitterness through the addition of ingredients may be a better alternative for developing functional foods.

10.2.3 Masking techniques

10.2.3.1 *Flavorings*

Taste "masking" occurs when the perceived sensory signal is overridden with a competing sensory signal (Thorngate, 1997). Ideally, non-bitter and agreeable tastes will dominate (Sun-Waterhouse & Wadhwa, 2013) when masking techniques are utilized. For example, without the citrus flavor in grapefruit juice, the bitterness of naringin and limonin would render it unpalatable (Szejtli & Szente, 2005). Similarly, the taste of caffeine in cola beverages is overwhelmed by high doses of sucrose, sweeteners and acid (Ley *et al.*, 2008).

Aliani and colleagues (2011) incorporated ground brown flaxseed, rich in antioxidants, omega-3 fatty acids and lignans into muffins (17%) and snack bars (32%) and used a trained sensory panel to determine aroma and flavor attributes. The control snack bars contained stabilized bran and toasted wheat germ while the control muffins were made with whole-wheat flour. Orange cranberry, gingerbread raisin and cappuccino chocolate flavorings were used in the snack bars, whereas orange cranberry and apple spice flavorings were used in the muffins. The bitter attribute was strongly related to grain/flax flavor, as depicted in bi-plots from principal component analysis (PCA). Both the orange cranberry and apple spice muffin formulations with flaxseed had significantly higher bitter taste compared to the non-flax muffin. The snack bars containing flax with the orange cranberry and cappuccino chocolate chip flavorings had significantly higher bitter taste while the bitter intensity for the flax and non-flax gingerbread raisin snack bar was not significant. The authors attributed this effect to the gingerbread raisin flavoring, which may have naturally imparted bitterness to the non-flax bar, as it was higher than for the non-flax bars with the other two flavorings. Because the bitterness rating was still quite low on the unstructured 15-centimeter line scale, Aliani and colleagues (2011) suggested that the use of flavorings where a small amount of bitterness is expected (in ginger, for example) may reduce the bitterness effect caused by the addition of flax. This data showed that none of the flavorings completely masked bitterness, although expectations of bitterness before consuming the sample influenced perception, and in turn, acceptance of the food product. While both ingredients in the cappuccino chocolate flavor were expected to impart bitter taste to some degree, they could not significantly reduce the bitterness caused by flax. In contrast, the addition of the gingerbread raisin flavoring did reduce the bitterness of flax. Thus, certain ingredients worked better than others but the exact reason for this is not fully understood.

Ares and colleagues (2009) used a trained sensory panel to determine bitterness in antioxidant extracts of two native South-American plants, *Achyrocline satureioides* and *Baccharis trimera* with added sucrose, sucralose, polydextrose or milk (0 or 3.2% fat). All four ingredients were effective in reducing the bitterness of these antioxidant extracts and the concentration of polyphenols in each antioxidant extract examined. The authors explained the reduction of bitterness by considering the mutual suppression of bitterness and sweetness due to a central cognitive effect (Calvino & Garrido, 1991; Keast & Breslin, 2003). One of the mechanisms suggested by Keast (2008) for the reduction of bitterness by milk was the migration of polyphenolic compounds to the lipid phase. However, Ares and colleagues (2009) challenged this theory as both 0% and 3.2% fat milk showed the same reduction of bitterness. Therefore, the effect could be attributed to milk proteins, which may complex with polyphenolic compounds yielding them insoluble or at least making the compounds incapable of interacting with the taste receptors (Keast, 2008). Further explanation for polydextrose as an effective bitterness suppressant was potentially due to an increase in viscosity and hydrogen bonds with polyphenols (Plug & Haring, 1993).

Kranz and colleagues (2010) fortified fruit smoothies with olive leaf extract (OLE), and added sodium cyclamate, sodium chloride or sucrose to mask bitterness. Using a trained panel to determine sensory differences, only sucrose significantly masked the bitterness of OLE in smoothies. While sodium cyclamate appeared to reduce bitterness, increasing its concentration did not lead to a strong reduction of bitterness, as observed for sucrose. The results of the scale test indicated that OLE polyphenol concentrations above the sensory detection threshold could be partly reduced by sodium cyclamate (39.9% bitterness reduction) and sucrose (24.9% bitterness reduction). Addition of sodium chloride could not mask bitterness, in fact, a slight increase of bitterness was observed when sodium chloride was added. Therefore, these results suggest that sodium cyclamate should be preferred for bitterness masking since it was the most effective, and does not contribute to additional calories like sucrose. Despite this finding, sucrose may be preferred due to potential health concerns related to cyclamate consumption. The U.S. Food and Drug Administration banned the use of cyclamate in 1970 following incidences of bladder cancer (Sharma & Lewis, 2010). Kranz and colleagues (2010) suggested that the increased sweetness impression of the smoothies may have interfered with bitterness perception because human signal transduction pathways for sweetness are opposed to bitterness pathways (Kinnamon & Margolskee, 1996). One methodological difference between this study and the studies by Aliani and colleagues (2011) and Ares and colleagues (2009) is the use of an unstructured line scale to determine sensory differences. Kranz and colleagues (2010) used ranking and scale tests with the panelists selected based on their ability to perceive bitterness. Because panelists were not randomly chosen, the findings of this study may not be applicable to all populations, since bitterness sensitivity and perception among individuals vary widely.

10.2.3.2 *Amino acids*

Mukai and colleagues (2004) tested the enteral solution Aminoleban EN improved formulations with increased particle size branched-chain amino acids (BCAA) L-isoleucine, L-leucine and L-valine with added flavors. Fruit flavor was deemed the most effective followed by pineapple and apple flavors. Coffee and green tea did not change perception, probably because of their already bitter nature. The combination of sweetness and sourness decreased bitterness, potentially due to the citric acid content.

10.2.3.3 *Aroma additions*

It is possible to manipulate bitterness perception through use of aroma additions because much of what is perceived as taste is influenced by the aroma of food (i.e., flavor). Mukai and colleagues (2007) used a 5-point rating scale for panelists to define sensory attributes and found that strawberry, apple and vanilla aromas evoked an image of sweetness, which contributed to the bitterness inhibition of BCAA solutions. Earlier research from Mukai and colleagues (2004) found similar results where green tea and coffee aromas did not have this effect. Additionally, the strawberry aroma, which evoked an image of both sweetness and sourness was particularly successful in inhibiting bitterness, more so than either vanilla (which also evoked both sweetness and sourness although the sourness intensity was lower) or apple (which did not evoke sourness at all). Based on results from both studies the evocation of *both* sweetness and sourness by an aroma and/or flavor is necessary for effective bitterness suppression. Of course, this is in a simple beverage matrix and results may not be transferable to other food products.

In another example of a simple matrix, Ishizaka and colleagues (2008) attempted to suppress the bitterness of prednisolone powder (PP), a drug used in the treatment of respiratory and renal failure and nephritis, by inclusion of PP into various commercial beverages. The authors found that as the intensity of sweetness, sourness and saltiness increased, bitterness decreased. Exclusively sour or salty agents were unable to improve palatability while agents that are very sweet and also contain some sourness and a little saltiness improved palatability the most. They concluded that the addition of sweetness to acidic beverages was the most effective way of suppressing the bitterness of PP. These results agreed with the findings of Mukai and colleagues (2007) where they determined that a small amount of sourness added to a predominantly sweet product could effectively suppress bitterness. While Mukai and colleagues (2007) demonstrated how aroma can affect perception of bitter flavor, Ishizaka and colleagues (2008) only examined taste. Interestingly, they reported a negative effect of milk, finding that it reduced palatability and increased bitterness which is contrary to that reported by Ares and colleagues (2009). Although, both studies used different measurement scales to determine bitterness, a different result may have been found if Ishizaka *et al.* (2008) had used a 10-centimeter unstructured line scale. Despite this, Ishizaka and colleagues (2008) also found that coffee milk and condensed milk significantly reduced bitterness. These conflicting results demonstrate that bitterness perception can be largely influenced by food matrix, as shown in previous studies. Contrary to findings by Ares and colleagues (2009) that 0 and 3.2% milk fat had no difference in bitterness, Madsen and Ardü (2001) found that bitterness masking in cheese was greater with a higher fat content. Once again, matrix and product differences are vast and must be explored further.

Bechoff (2014) tested Hibiscus (*Hibiscus sabdariffa L.*) beverages and found that the concentration of polyphenols was significantly correlated to bitter taste. This observation was in agreement with that reported by Jaegar and colleagues (2009) and Lawless and colleagues (2012). The bitter taste was significantly more pronounced in infusions compared to syrups, and bitter taste was negatively associated with total sugars, which suggest that addition of sugar would be masking bitter taste. In another study, Jaeger and colleagues (2009) found that panelists tasting beverages containing polyphenols from berry fruit and/or cocoa responded well to increasing levels of sucrose; however, there was no perceived bitterness difference between beverages with 7% and 10.5% added sucrose. The point at which the reduction of bitterness has reached a plateau is critical for food developers to not overuse calorically dense ingredients like sucrose.

Often the presence of limonin and nomilin caused by Huanglongbing (HLB) infected fruit can contribute to bitterness in orange juice. Therefore, Dea and colleagues (2013) tested whether sucrose or citric acid could reduce bitterness. They found that only sucrose could mask it, which is commonly used in industry but poses other problems as a less healthful product. A potentially good alternative to using sucrose is a combination approach using an artificial sweetener and an aroma attribute. Tárrega and colleagues (2012) found that sucralose and vanilla aroma decreased bitterness caused by triterpenoid peptides found in ginseng-enriched milk, but not enough to be significant. As suggested by Tamamoto and colleagues (2010), and in agreement with Aliani and colleagues (2011), more congruent or related flavors such as chocolate, citrus and coffee would be more effective. The reason for this is because of their inherent bitterness which garners a certain level of acceptance. Even cyclodextrins, which form complexes with bitter molecules, would be equally effective. In fact, cyclodextrin dissolves flavonoids and subsequently masks the bitter taste of citrus juice, while maintaining the bioactivity of the flavonoids (Drewnowski & Gomez-Carneros, 2000). Gaudette and Pickering (2012a) found that sweeteners (sucrose and Rebaudioside A) used in combination with ß-cyclodextrin effectively reduced the bitterness of (+)-catechin. Rebaudioside A was equally as effective as sucrose. Central cognitive mechanism, from sweetness, and a physiological effect from β-cyclodextrin encapsulating (+)-catechin, was consistent with the reduction in bitterness observed with zinc sulfate and sodium-cyclamate (Keast & Breslin, 2005).

10.2.3.4 *Gluconate and acetate*

Gluconate and acetate are examples of sodium salts low in perceived salty taste found to be successful at masking bitterness (Keast *et al.*, 2001, 2004). Sodium gluconate added to a quinine-hydrogen chloride solution effectively reduced bitterness, but not in caffeine or naringin solutions, possibly due to a depression in gustatory nerve activity (Narukawa *et al.*, 2012). Gluconate, although effective in quinine type bitterness, may not be feasible for a range of compounds due to its specificity. A benefit of using gluconate is its promotion of *Lactobacillus bifidus* growth which may improve gut health. Sodium salts in conjunction with L-arginine have also been used to reduce peptide bitterness (Ogawa *et al.*, 2004). In a different food product, Sharafi and colleagues (2013) found that aspartame misted onto vegetables (kale, Brussels sprouts and asparagus) was more effective in reducing bitterness compared to sodium acetate or sodium chloride. This demonstrated that a sweetener could be used to sufficiently reduce bitterness, while not contributing to a sweet taste.

10.2.3.5 *Lipids: phospholipids and fatty acids*

Koprivnjak and colleagues (2009) examined the potential of phospholipids to attenuate the pronounced bitterness associated with the high content of hydrophilic phenolic compounds in virgin olive oil (VOO). The addition of granular soy lecithin in the range of 5-10 g/kg significantly increased sweetness and decreased bitterness with only slight changes in olive fruit and green odor notes. The treated oil; however, could not be sold as natural VOO but as a functional food. Nevertheless, the authors pointed out the need to study the impact of added phospholipids on the shelf life of the product.

The potential use of fatty acids as masking agents was recently reported by Homma and colleagues (2012). They identified these compounds in a natural white mold cheese called Baraka, and were found to be capable of masking bitterness. Using sensory tests, they reported a close relationship between masking activity and the high free fatty acid content of Baraka cheese. This was the first study reporting the bitter masking properties of free fatty acids. Based on their results it appeared that oleic acid effectively masked bitterness by complexation with the bitter compounds.

10.2.3.6 *Zinc, lactate, and acetate*

Keast (2008) reported that the addition of zinc lactate to caffeine products such as coffee and chocolate was effective in reducing the bitterness. An earlier study by Keast and Breslin (2005) examined the effect of zinc sulfate on various bitter compounds including tetralone (responsible for bitterness in beer) and several pharmaceutical compounds. They discovered that zinc sulfate reduced the bitterness in tetralone, quinine-hydrogen chloride (anti-malarial drug), and denatonium but had no effect on pseudoephedrine (stimulant), sucrose octa-actetate (acetylated derivative of sucrose) and dextromethorphan (antitussive drug).

Jensen and colleagues (2001) tested various masking agents to effectively reduce bitterness imparted by adding 3% or 4% sodium lactate (SL) or potassium lactate (PL) to ground chicken breast. They determined that PL imparted more bitterness compared to SL, and the higher concentration was higher in bitterness. Of all the masking agents tested (sucrose, dextrose, lactitol, sorbitol or lysine), they found that bitterness was decreased when 1% sucrose or lactitol was used in SL samples. It should also be noted that sweetness did not increase along with a subsequent bitterness decrease, suggesting that sweeteners like sucrose can be used without affecting the overall sensory profile.

The studies described have demonstrated that masking of bitterness through ingredient addition is important in functional food development and by far the most practical method because it does not compromise the bioactive components in the food product. A concern with masking is that although addition of substances such as sucrose is quite effective at masking bitterness, its use in larger amounts can decrease the total nutritional value of the food as well as increasing its glycemic index. Therefore, masking techniques using safe, non-caloric ingredients are essential. Also, there is a clear potential for the use of aroma compounds to reduce bitterness perception, which may reduce the need for undesirable ingredients such as sucrose. Aroma addition in combination with smaller amounts of sweeteners and sour ingredients may also prove to be effective.

It is also important that masking agents be used in appropriate applications so as to not change the overall flavor profile of the particular food. As an example, using high amounts of sucrose to mask bitterness in a savory dish is just as unacceptable as using sodium chloride in a sweet dessert.

10.2.4 Bitter blockers

Recent studies have highlighted the potential of bitter blockers as a healthier alternative to adding sugar and salt and are capable of reducing bitterness of phenols and caffeine. Overcoming the unpleasant bitter flavors associated with many medications could increase compliance as well as the acceptance of bitter-tasting healthy and functional foods. Five bitter blockers with different efficacies have so far been published (Slack *et al.*, 2010; Brockoff *et al.*, 2011; Grene *et al.*, 2011; Roland *et al.*, 2014).

Binello and colleagues (2004) synthesized several bitter-taste blockers by covalently bonding cyclodextrins (CDs) with chitosans via succinyl or maleyl bridges. The chitosan-cyclodextrin adducts (β and γ-CDs chitosans) were then investigated for their ability to reduce the bitterness of caffeine and natural bitter extracts (artichoke leaves, aloe and gentian). Twelve trained panelists reported that β-CD-chitosan exhibited the greatest efficacy by significantly decreasing bitterness.

Gaudette and Pickering (2012b) later examined the ability of five bitter inhibiting compounds, β-cyclodextrin (β-CYCLO), homoeriodictyol sodium salt (HED), zinc sulphate monohydrate ($ZnSO_4$), magnesium sulphate ($MgSO_4$) and carboxymethylcellulose sodium salt (CMC) on the bitterness of (+)-catechin and caffeine. Using 12 trained panelists, β-CYCLO and HED were the most effective in reducing bitterness of (+)-catechin compared to the other bitter blockers. However, only HED proved effective in reducing the bitterness of caffeine. Nevertheless, irrespective of the different formulations considerable bitterness still remained.

Further work by Gaudette and Pickering (2012a) examined the ability of two bitter blockers, β-CYCLO and HED, two sweeteners, sucrose and Rebaudioside A, and two odorants, vanillin and black tea aroma, to reduce the bitterness and astringency of (+)-catechin. Compared with the other treatments, a combination of sweetener and β-CYCLO was most effective in decreasing bitterness and astringency and together could be used in formulations of functional foods. The sweetener, Rebaudioside A was as effective as sucrose in inhibiting bitterness and could replace sucrose in low and non-sugar food and beverage products suitable for those with diabetes. The addition of odorants; however, had no effect on either the bitterness or astringency of (+)-catechin.

Certain peptides elicit a bitter taste, as discussed in Chapter 6, which is reflected by their amino acid composition and their ability to activate the bitter taste receptors. Maehashi and colleagues (2009) showed that bitter peptides, as well as other bitter compounds, activate the human bitter receptors hTAS2Rs. Not all peptides are bitter; however, as umami peptides have been reported to attenuate bitterness. As discussed earlier in this chapter, both sugar and salt can suppress bitterness. The ability of umami to interact with other tastes can also result in the suppression of bitterness (Arai *et al.*, 1973; Noguchi *et al.*, 1975; Kemp & Beauchamp, 1994; Tokita & Boughter, 2012). In addition to monosodium glutamate (MSG), umami-active acidic oligopeptides, glutamic acid (Glu)-aspartic acid (Asp), Glu-Glu, Glu-Serine (Ser), and Glu-Glu-Glu, were shown over 40 years ago, long before the establishment of umami, to suppress the bitterness of protein hydrolysate (Arai *et al.*, 1972, 1973). A recent study by Kim and colleagues (2015) determined whether such umami peptides suppressed bitterness by inhibiting the binding of the bitter ligand to the human taste receptor, hTAS2R16. Five soybean derived umami peptides, Glu-Asp, Glu-Glu, Glu-Ser, Asp-Glu-Ser, and Glu-Gly-Ser, were examined for their ability to suppress the bitterness of salicin in an ionized calcium flux signaling assay using human hTAS2R16-expressing cells. All five peptides markedly reduced the response to the bitter taste receptor (hTAS2R16) by salicin. Of all the peptides examined, Glu-Glu proved to be the strongest inhibitor and was even more effective than probenecid, a known hTAS2R16 antagonist. This is the first study that clearly established the interaction between bitter and umami taste at the taste receptor level.

Pydi and colleagues (2014) elucidated the ligand binding pocket of the human bitter taste receptor T2R4. They found that the binding of T2R4 to its agonist quinine involved seven amino acids located in the extracellular side of transmembrane 3 (TM3), TM4, extracellular loop 2 (ECL2) and ECL3. As a result, they screened a number of amino acid derivatives including, L-ornithine β-alanine (OA) and γ-amino butyric acid (GABA), for their ability to bind T2R4. Since Ley (2008) suggested OA could mask the bitter taste of potassium salt and GABA the bitter taste of quinine, caffeine, coca and chocolate, Pydi and colleagues (2014) hypothesized that these amino acid derivatives may act at the receptor level as T2R blockers. Two novel bitter blockers GABA and Na, Na-*Bis*(carboxymethyl)-L-lysine (*B*MCL) were identified with respective IC_{50} values of 3.2 ± 0.3 and 59 ± 18 nM. Pharmacological characterization using T2R-CAMs showed GABA acted as an antagonist while BMCL acted as an inverse agonist on T2R4. Both shared the same orthostatic site in T2R4 as the agonist quinine and involved signature residues Ala90 and Lys270. The authors point out the tremendous potential of these bitter blockers for eliminating the bitter taste of healthful foods.

During the screening of extracts from selected plants native to Ohio, Li and colleagues (2014) obtained a chloroform-soluble extract from Canada Goldenrod (*Solidago canadensis*) with hTAS2R31 antagonistic activity. Further characterization of the extract resulted in identification of a new labdane diterpenoid, solidagol, together with six known terpenoids and a triterpenoid. Of these compounds, 3β-acetoxycopalic acid, the first member of the labdane diterpene class, was found to exhibit inhibitory activity against hTAS2R31 activation. Thus *S. canadensis* could be a good source for the development of bitterness-masking agents.

10.3 CONCLUSION

Eliminating bitterness is of great interest and value to the food industry. As demonstrated in this chapter, bitterness is complex. The perception of bitterness varies enormously between individuals while the compounds responsible for bitterness in food are numerous and varied. Researchers have successfully identified methods for masking and reducing bitterness in products but have yet to establish a universal or gold standard method for eliminating bitter compounds. Additionally, because many healthful compounds are bitter, and small amounts of bitterness can be considered desirable in many cases. Some methods have only been tested on a few compounds and the effects of these methods on other bitter compounds have not yet been established. More positively, some methods have been useful for masking or reducing some bitter compounds while having little or no effect. The most commonly used techniques for modifying bitterness in functional food systems were reviewed by Gaudette and Pickering (2013). It is evident that further research is needed to identify new and effective ways of masking or reducing bitterness in complex food matrices.

Additionally, the method of bitterness suppression must be scrutinized to ensure the compound(s) present in the functional foods will not decrease its perception of healthfulness by consumers (i.e., artificial sweeteners, sugar and salt is of concern). It is likely that a combination of techniques to decrease bitterness would ensure the best possible outcome by addressing various aspects of bitterness perception. Ideally, the most promising formulation should be simple, economical, involve the fewest steps possible, have no effect on nutrient bioactivity and/or bioavailability and use techniques in which safety and efficacy have been clearly demonstrated.

REFERENCES

Aliani, M., Ryland, D. & Pierce, G.N. (2011). Effect of flax addition on the flavor profile of muffins and snack bars. *Food Res. Inter.* **44**(8), 2489–2496.

Arai, S., Yamashita, M. & Fujimaki, M. (1972). Glutamyl oligopeptides as factors responsible for tastes of a proteinae-modified soybean protein. *Agric. Biol. Chem.* **36**, 1253–1256.

Arai, S., Yamashita, M., Noguchi, M. & Fujimaki, M. (1973). Tastes of L-glutamyl oligopeptides in relaxation to their chromatographic properties. *Agric. Biol. Chem.* **37**, 151–156.

Ares, G., Barreiro, C., Deliza, R. & Gámbaro, A. (2009). Alternatives to reduce the bitterness, astringency and characteristic flavour of antioxidant extracts. *Food Res. Inter.* **42**(7), 871–878.

Bechoff, A., Cissé, M., Fliedel, G., Declemy, A., Ayessou, N., Akissoe, N., Touré, C., Bennett, B., Pintado, M. & Pallet, D. (2014). Relationships between anthocyanins and other compounds and sensory acceptability of Hibiscus drinks. *Food Chem.* **148**, 112–119.

Bertoldi, F.C., Sant'Anna, E.S. & Beirão, L.H. (2004). Reducing the bitterness of tuna (Euthynnus pelamis) dark meat with Lactobacillus casei subsp. casei ATCC 393. *Food Technol. Biotechnol.* **42**(1), 41–45.

Bett-Garber, K., Greene, J., Lamikanra, O., Ingram, D. & Watson, M. (2011). Effect of storage temperature variations on sensory quality of fresh-cut cantaloupe melon. *J. Food Qual.* **34**(1), 19–29.

Bhandari, M.R. & Kawabata, J. (2005). Bitterness and toxicity in wild yam (Dioscorea spp.) tubers of Nepal. *Plant Foods. Hum. Nutr.* **60**(3), 129–135.

Binello, A., Gravotto, G., Nano, G.M. & Spagliardi, P. (2004). Synthesis of chitosan-cyclodextrin adducts and evaluation of their bitter-masking properties. *Flavor Frag. J.* **19**(5): 394–400.

Brockhoff, A., Behrens, M., Roudnitsky, N., Appendino, G. Avonto, C. & Meyerhof, W. (2011). Receptor agonism and antagonism of dietary bitter compounds. *J. Neurosci.* **31**, 14775–14782.

Calvino, A. & Garrido, D. (1991). Spatial and temporal suppressive behavior of bitter-sweet mixtures. *Percept. Motor Skills.* **73**(3f), 1216–1216.

Cravotto, G., Nano, G.M., Binello, A., Spagliardi, P. & Seu, G. (2005). Chemical and biological modification of cynaropicrin and grosheimin: A structure-bitterness relationship study. *J. Sci. Food Agric.* **85**(10), 1757–1764.

Dea, S., Plotto, A., Manthey, J.A., Raithore, S., Irey, M. & Baldwin, E. (2013). Interactions and thresholds of limonin and nomilin in bitterness perception in orange juice and other matrices. *J. Sens. Stud.* **28**(4), 311–323.

Drewnowski, A. (2001). The science and complexity of bitter taste. *Nutr. Rev.* **59**(6), 163–169.

Drewnowski, A. & Gomez-Carneros, C. (2000). Bitter taste, phytonutrients, and the consumer: A review. *The American J. Clin. Nutr.* **72**(6), 1424–1435.

Fontoin, H., Saucier, C., Teissedre, P. & Glories, Y. (2008). Effect of pH, ethanol and acidity on astringency and bitterness of grape seed tannin oligomers in model wine solution. *Food Qual. Pref.* **19**(3), 286–291.

García, J.M., Yousfi, K., Mateos, R., Olmo, M. & Cert, A. (2001). Reduction of oil bitterness by heating of olive (Olea europaea) fruits. *J. Agric. Food Chem.* **49**(9), 4231–4235.

Gaudette, N.J. & Pickering, G.J. (2012a). Optimizing the orosensory properties of model functional beverages: The influence of novel sweeteners, odorants, bitter blockers, and their mixtures on (+)-catechin. *J. Food Sci.* **77**(6), S226–S232.

Gaudette, N.J. and Pickering, G.J. (2012b). The efficacy of bitter blackers on health-relevant bitterants. *J. Funct. Foods.* **4**, 177–184.

Gaudette, N.J. & Pickering, G.J. (2013). Modifying bitterness in functional food systems. *Crit. Rev. Food Sci. Nutr.* **53**(5), 464–481.

Grene, T.A., Aarcon, S., Thomas, A., Berdougo, E., Doranz, B.J., Breslin, P.A. & Rucker, J.B. (2011). Probenecid inhibits the human bitter taste receptor TAS2R16 and suppresses bitter reception of salicin. *PLoSONE*, **6**, e20123.

Homma, R., Yamashita, H., Funaki, J., Ueda, R., Sakurai, T., Ishimanu, Y., Abe, K. and Asakura, T. (2012). Identification of bitterness-masking compounds from cheese. *J. Agric. Food Chem.* **60**(18), 4492–4499.

Ishizaka, T., Okada, S., Tokuyama, E., Mukai, J. & Uchida, T. (2008). Suppression of bitterness and improvement of palatability of commercial prednisolone powder. *Chem. Pharmaceut. Bull.* **56**(10), 1395–1399.

Jaeger, S.R., Axten, L.G., Wohlers, M.W. & Sun-Waterhouse, D. (2009). Polyphenol-rich beverages: Insights from sensory and consumer science. *J. Sci. Food Agric.* **89**(14), 2356–2363.

Jensen, J.M., Prestat, C. & Brewer, M.S. (2001). A research not flavor-masking agent effects on sensory characteristics of lactate-containing chicken model systems. *J. Muscle Foods.* **12**(4), 317–324.

Keast, R.S. & Breslin, P.A. (2003). An overview of binary taste–taste interactions. *Food Qual. Pref.* **14**(2), 111–124.

Keast, R.S., Breslin, P.A. & Beauchamp, G.K. (2001). Suppression of bitterness using sodium salts. *CHIMIA Int. J. Chem.* **55**(5), 441–447.

Keast, R.S., Canty, T.M. & Breslin, P.A. (2004). The influence of sodium salts on binary mixtures of bitter-tasting compounds. *Chem. Sens.* **29**(5), 431–439.

Keast, R.S.J. (2008). Modification of the bitterness of caffeine. *Food Qual. Pref.* **19**(5), 465–472.

Keast, R.S.J. & Breslin, P.A.S. (2005). Bitterness suppression with zinc sulfate and Na-cyclamate: A model of combined peripheral and central neural approaches to flavor modification. *Pharmaceut. Res.* **22**(11), 1970–1977.

Kemp, S.E and Beauchamp, G.K. (1994). Flavor modification by sodium chloride and monosadium glutamate. *J. Food Sci.* **59**, 682–686.

Kim, M.J., Son, J.J., Kim, V., Misaka, T. and Rhyu, M-R. (2015). Umami-bitter interactions: The suppression of bitterness by umami peptides via human bitter receptor. *Biochem. Biophys. Res. Commun.* **456**, 586–590.

Kinnamon, S.C. & Margolskee, R.F. (1996). Mechanisms of taste transduction. *Curr. Opin. Neurobiol.* **6**(4), 506–513.

Kola, O., Kaya, C., Duran, H. & Altan, A. (2010). Removal of limonin bitterness by treatment of ion exchange and adsorbent resins. *Food Sci. Biotechnol.* **19**(2), 411–416.

Koprivnjak, G., Skevin, D., Petricevic, S., Bubola, K.B. and Mokrovcak, Z. (2009). Bitterness, odor properties and volatile compounds of virgin olive oil with phospholipids addition. *LWT-Food Sci.Technol.* **42**, 50–55.

Kranz, P., Braun, N., Schulze, N. & Kunz, B. (2010). Sensory quality of functional beverages: Bitterness perception and bitter masking of olive leaf extract fortified fruit smoothies. *J. Food Sci.* **75**(6), S308–S311.

Lawless, L.J., Threlfall, R.T., Howard, L.R. & Meullenet, J. (2012). Sensory, compositional, and color properties of nutraceutical-rich juice blends. *American J. Enol. Viticult.* **63**(4), 529–537.

Ley, J.P. (2008). Masking taste by molecules. *Chemosens. Percept.* **1**, 58–77.

Ley, J.P., Paetz, S., Blings, M., Hoffmann-Lücke, P., Bertram, H. & Krammer, G.E. (2008). Structural analogues of homoeriodictyol as flavor modifiers. Part III: Short chain gingerdione derivatives. *J. Agric. Food Chem.* **56**(15), 6656–6664.

Li, J., Pan, L., Fletcher, J.N., Lv, W., Deng, Y., Vincent, M.A., Slack, J.P., McCluskey, T.S., Jia, Z., Cushman, M. & Kinghorn, A.D. (2014). In vitro evaluation of potential bitterness-masking terpenoids from Canada Golodenrod (*Solidago canadensis*). *J. Nat. Prod.* **77**, 1739–1743.

Madsen, J.S. & Ardü, Y. (2001). Exploratory study of proteolysis, rheology and sensory properties of Danbo cheese with different fat contents. *Int. J. Dairy.* **11**, 423–431.

Maehashi, K., Matano, M., Wang, H., Vo, L. A., Yamamoto, Y. and Huang, L. (2009). Bitter peptides activate hTAS2Rs, the human bitter receptors. *Biochem. Biophys. Res. Commun.* **365**, 851–855.

Maehashi, K. & Huang, L. (2009). Bitter peptides and bitter taste receptors. *Cell. Mol. Life Sci.* **66**(10), 1661–1671.

Mayengbam, S., Khattab, R. & Thiyam-Hollander, U. (2013). Effect of conventional and microwave toasting on sinapic acid derivatives and canolol content of canola. *Curr. Nutr. Food Sci.* **9**(4), 321–327.

Milner, J.A. (1999). Functional foods and health promotion. *The J. Nutr.* **129**(7), 1395S–1397S.

Miyashita, T. & Etoh, H. (2013). Improvement of the bitterness and astringency of green tea by sub-critical water extraction. *Food Sci. Technol. Res.* **19**, 471–478.

Mukai, J., Miyanaga, Y., Ishizaka, T., Asaka, K., Nakai, Y., Tsuji, E. & Uchida, T. (2004). Quantitative taste evaluation of total enteral nutrients. *Chem. Pharmaceut. Bull.* **52**(12), 1416–1421.

Mukai, J., Tokuyama, E., Ishizaka, T., Okada, S. & Uchida, T. (2007). Inhibitory effect of aroma on the bitterness of branched-chain amino acid solutions. *Chem. Pharmaceut. Bull.* **55**(11), 1581–1584.

Narukawa, M., Tsujitani, T., Ueno, Y., Nakano-Ooka, K., Miyamoto-Kokubo, N., Sawano, S. & Hayashiy, Y. (2012). Evaluation of the suppressive effect on bitter taste of gluconate. *Biosci. Biotechnol. Biochem.* **76**(12), 2282–2288.

Noble, A.C. (1998). Why do wines taste bitter and feel astringent? In *Chemistry of Wine Flavor*. (Eds. Waterhouse, A.L. & Ebeler, S.) 156–164. American Chemical Society, Washington, D.C.

Noguchi, M., Yamashita, M., Arai, S. and Fujimaki, M. (1975). On the bitter-masking activity of glutamic acid-rich oligopeptide fraction. *J. Food Sci.* **40**, 367–369.

Ogawa, T., Nakamura, T., Tsuji, E., Miyanaga, Y., Nakagawa, H., Hirabayashi, H. & Uchida, T. (2004). The combination effect of L-arginine and NaCl on bitterness suppression of amino acid solutions. *Chem. Pharmaceut. Bull.* **52**(2), 172–177.

Plug, H. & Haring, P. (1993). The role of ingredient-flavour interactions in the development of fat-free foods. *Trends Food Sci. Technol.* **4**(5), 150–152.

Pydi, S.P., Sobotkiewicz, T., Bhullar, B.P., Loewen, M.C. & Chelikani, P. (2014). Amino acids as bitter taste receptor (T2R) blockers. *J. Biol. Chem.* **289**(4), 25054–25066.

Rodgers, S., Glen, R.C. & Bender, A. (2006). Characterizing bitterness: Identification of key structural features and development of a classification model. *J. Chem. Inform. Model.* **46**(2), 569–576.

Roland, W.S., Gouka, R.J., Gruppen, H., Driesse, M., van Buren, L., Smit, G. & Vincken, J.P. (2014). Methoxyflavonones as bitter taste receptor blockers for hTAS2E39. *PloSOne*, **9**, e94451.

Sharafi, M., Hayes, J.E. & Duffy, V.B. (2013). Masking vegetable bitterness to improve palatability depends on vegetable type and taste phenotype. *Chemosen. Percep.* **6**(1), 8–19.

Sharma, S. & Lewis, S. (2010). Taste masking technologies: A review. *International J. Pharm. and Pharmaceut. Sci.* **2**(2), 6–13.

Slack, J.P., Brockhoff, A., Batram, C., Menzel, S., Sonnabend, C., Born, S., Galindo, M.M., Kohl, S.,Thalmann, S., Ostopovici-Halip, L., Simons, C.T., Ungureanu, I., Dioneveld, K, Bologa, C.G., Behrens, M. Furrer, S.,Oprea, T.I. & Meyerhof, W. (2010). Modulation of bitter taste perception by a small molecule hTAS2R antagonist. *Curr. Biol.* **20**, 1104–1109.

Stark, T., Bareuther, S. & Hofmann, T. (2005). Sensory-guided decomposition of roasted cocoa nibs (*Theobroma cacoa*) and structure determination of taste-active polyphenols. *Journal of Agriculture and Food Chem.* **53**, 5407–5418.

Stark, T. & Hofmann, T. (2005). Isolation, structure determinationm, synthesis, and sensory activity of N-phenylpropenyl-L-amino acids fron cocoa (*Theobroma cacoa*). *J. Agric. Food Sci.* **53**, 5419–5428.

Stark, T. & Hofmann, T. (2006). Application of a molecular sensory science approach to alkalized cocoa (Theobroma cacao): Structure determination and sensory activity of nonenzyymatically C-glycosylated flavan-3-ols. *J. Agric. Food Chem.* **54**, 9610–9621.

Stark, T., Justus, H. & Hofmann, T (2006). Quantitative analysis of *N*-phenylpropenyl-L-amino acids in roasted coffe and cocoa powder by means of a stable isotope dilution assay. *J. Agric. Food Chem.* **54**, 2859–2867.

Sun-Waterhouse, D. & Wadhwa, S.S. (2013). Industry-relevant approaches for minimising the bitterness of bioactive compounds in functional foods: A review. *Food Bioprocess Technol.* **6**(3), 607–627.

Szejtli, J. & Szente, L. (2005). Elimination of bitter, disgusting tastes of drugs and foods by cyclodextrins. *European J. Pharm. Biopharm.* **61**(3), 115–125.

Tamamoto, L.C., Schmidt, S.J. & Lee, S. (2010). Sensory properties of ginseng solutions modified by masking agents. *J. Food Sci.* **75**(7), S341–S347.

Tárrega, A., Salvador, A., Meyer, M., Feuillère, N., Ibarra, A., Roller, M., Terroba, D., Madera, C., Iglesias, J. & Echevarría, J. (2012). Active compounds and distinctive sensory features provided by American ginseng (Panax quinquefolius L.) extract in a new functional milk beverage. *J. Dairy Sci.* **95**(8), 4246–4255.

Thorngate, J.H. (1997). The physiology of human sensory response to wine: A review. *Am. J. Enol. Viticul.* **48**(3), 271–279.

Tokita, K. and Boughter Jr., J.D. (2012). Sweet=bitter and umami-bitter taste interactions in single parabrachial neurons in C57BL/6J mice. *J. Neurophysiol.* **108**, 2179–2190.

Yousfi, K., Cayuela, J.A. & García, J.M. (2008). Reduction of virgin olive oil bitterness by fruit cold storage. *Journal of Agricultural and Food Chem.* **56**(21), 10085–10091.

Zhang, L., Xia, Y. & Peterson, D.G. (2014). Identification of bitter-modulating Maillard-catechin reaction products. *Journal of Agriculture and Food Chem.* **62**, 8470–8477.

Index

Note: Page numbers in bold refer to tables and figures.

Bitterness: Perception, Chemistry and Food Processing, First Edition. Edited by Michel Aliani & Michael N. A. Eskin.

14/11/17